Polymeric Carriers for Biomedical and Nanomedicine Application

Polymeric Carriers for Biomedical and Nanomedicine Application

Editors

Sofia Lima
Salette Reis

MDPI • Basel • Beijing • Wuhan • Barcelona • Belgrade • Manchester • Tokyo • Cluj • Tianjin

Editors
Sofia Lima
Department of Chemistry,
Applied Chemistry Laboratory
Faculty of Pharmacy, University
of Porto
Porto
Portugal

Salette Reis
Department of Chemistry,
Applied Chemistry Laboratory
Faculty of Pharmacy, University
of Porto
Porto
Portugal

Editorial Office
MDPI
St. Alban-Anlage 66
4052 Basel, Switzerland

This is a reprint of articles from the Special Issue published online in the open access journal *Polymers* (ISSN 2073-4360) (available at: www.mdpi.com/journal/polymers/special_issues/carrier_nanomedicine).

For citation purposes, cite each article independently as indicated on the article page online and as indicated below:

LastName, A.A.; LastName, B.B.; LastName, C.C. Article Title. *Journal Name* **Year**, *Volume Number*, Page Range.

ISBN 978-3-0365-1173-3 (Hbk)
ISBN 978-3-0365-1172-6 (PDF)

© 2021 by the authors. Articles in this book are Open Access and distributed under the Creative Commons Attribution (CC BY) license, which allows users to download, copy and build upon published articles, as long as the author and publisher are properly credited, which ensures maximum dissemination and a wider impact of our publications.

The book as a whole is distributed by MDPI under the terms and conditions of the Creative Commons license CC BY-NC-ND.

Contents

About the Editors vii

Sofia A. Costa Lima and Salette Reis
Polymeric Carriers for Biomedical and Nanomedicine Application
Reprinted from: *Polymers* **2021**, *13*, 1261, doi:10.3390/polym13081261 1

Raquel G. D. Andrade, Bruno Reis, Benjamin Costas, Sofia A. Costa Lima and Salette Reis
Modulation of Macrophages M1/M2 Polarization Using Carbohydrate-Functionalized Polymeric Nanoparticles
Reprinted from: *Polymers* **2020**, *13*, 88, doi:10.3390/polym13010088 5

Antonella Rozaria Nefeli Pontillo, Evangelia Konstanteli, Maria M. Bairaktari and Anastasia Detsi
Encapsulation of the Natural Product Tyrosol in Carbohydrate Nanosystems and Study of Their Binding with ctDNA
Reprinted from: *Polymers* **2020**, *13*, 87, doi:10.3390/polym13010087 23

Hazem Abdul Kader Sabbagh, Samer Hasan Hussein-Al-Ali, Mohd Zobir Hussein, Zead Abudayeh, Rami Ayoub and Suha Mujahed Abudoleh
A Statistical Study on the Development of Metronidazole-Chitosan-Alginate Nanocomposite Formulation Using the Full Factorial Design
Reprinted from: *Polymers* **2020**, *12*, 772, doi:10.3390/polym12040772 49

Jason Thomas Duskey, Cecilia Baraldi, Maria Cristina Gamberini, Ilaria Ottonelli, Federica Da Ros, Giovanni Tosi, Flavio Forni, Maria Angela Vandelli and Barbara Ruozi
Investigating Novel Syntheses of a Series of Unique Hybrid PLGA-Chitosan Polymers for Potential Therapeutic Delivery Applications
Reprinted from: *Polymers* **2020**, *12*, 823, doi:10.3390/polym12040823 73

Hyo Jung Shin, Hyewon Park, Nara Shin, Hyeok Hee Kwon, Yuhua Yin, Jeong-Ah Hwang, Song I Kim, Sang Ryong Kim, Sooil Kim, Yongbum Joo, Youngmo Kim, Jinhyun Kim, Jaewon Beom and Dong Woon Kim
p47phox siRNA-Loaded PLGA Nanoparticles Suppress ROS/Oxidative Stress-Induced Chondrocyte Damage in Osteoarthritis
Reprinted from: *Polymers* **2020**, *12*, 443, doi:10.3390/polym12020443 87

Alexandre Ferreira Lima, Isabel R. Amado and Liliana R. Pires
Poly(D,L-lactide-*co*-glycolide) (PLGA) Nanoparticles Loaded with Proteolipid Protein (PLP) —Exploring a New Administration Route
Reprinted from: *Polymers* **2020**, *12*, 3063, doi:10.3390/polym12123063 101

Yanling Zhang, Majella E. Lane and David J. Moore
An Investigation of the Influence of PEG 400 and PEG-6-Caprylic/Capric Glycerides on Dermal Delivery of Niacinamide
Reprinted from: *Polymers* **2020**, *12*, 2907, doi:10.3390/polym12122907 111

Aditee Ghose, Bushra Nabi, Saleha Rehman, Shadab Md, Nabil A. Alhakamy, Osama A. A. Ahmad, Sanjula Baboota and Javed Ali
Development and Evaluation of Polymeric Nanosponge Hydrogel for Terbinafine Hydrochloride: Statistical Optimization, In Vitro and In Vivo Studies
Reprinted from: *Polymers* **2020**, *12*, 2903, doi:10.3390/polym12122903 123

Louise Van Gheluwe, Eric Buchy, Igor Chourpa and Emilie Munnier
Three-Step Synthesis of a Redox-Responsive Blend of PEG–*block*–PLA and PLA and Application to the Nanoencapsulation of Retinol
Reprinted from: *Polymers* **2020**, *12*, 2350, doi:10.3390/polym12102350 . . . . **143**

Maria Victoria Cano-Cortes, Jose Antonio Laz-Ruiz, Juan Jose Diaz-Mochon and Rosario Maria Sanchez-Martin
Characterization and Therapeutic Effect of a pH Stimuli Responsive Polymeric Nanoformulation for Controlled Drug Release
Reprinted from: *Polymers* **2020**, *12*, 1265, doi:10.3390/polym12061265 . . . . **161**

About the Editors

Sofia Lima

Sofia Lima is an assistant researcher at REQUIMTE, University of Porto, with a great interest in polymers as drug delivery systems. She completed her PhD in chemistry at the University of Porto, and her research is focused on the study of cell–nanomaterial interactions using biochemical and biophysical techniques to unravel related mechanisms of action. Improving the knowledge on nanomaterial safety is crucial to ensure their emerging introduction in consumer goods (food and health). During the last 10 years, she has particularly been dedicated to the development of drug delivery systems able to overcome the barriers (skin and gastrointestinal) leading to the exploitation of natural compounds as tools to enhance permeation and to thus modulate biological processes, including the immune system.

Salette Reis

Salette Reis is the coordinator of the NanoPlatforms Research Group of LAQV, REQUIMTE (https://www.requimte.pt/laqv/); is the leader of the Molecular Biophysics and Biotechnology Unit of REQUIMTE at the University of Porto; and is a Cathedratic Professor at the Faculty of Pharmacy of the University of Porto. Her research activities are focused on the study of the complex interplay between drugs and lipid membranes. In this field, she is interested in the study of membrane biophysical changes related to the mechanism of action and toxicity of drugs and in the study of the effect of drugs on the activity of membrane enzymes involved in inflammation. Salette Reis is also involved in the study of the role of the membrane biophysical properties on membrane peroxidation and on the mechanism of action of antioxidants. Recently, her research activities have also been based on the development of drug nanocarrier systems to overcome the disadvantages of classical therapies.

Editorial

Polymeric Carriers for Biomedical and Nanomedicine Application

Sofia A. Costa Lima * and Salette Reis *

LAQV, REQUIMTE, Departamento de Ciências Químicas, Faculdade de Farmácia, Universidade do Porto, Rua de Jorge Viterbo Ferreira, 228, 4050-313 Porto, Portugal
* Correspondence: slima@ff.up.pt (S.A.C.L.); shreis@ff.up.pt (S.R.)

Citation: Lima, S.A.C.; Reis, S. Polymeric Carriers for Biomedical and Nanomedicine Application. *Polymers* **2021**, *13*, 1261. https://doi.org/10.3390/polym13081261

Received: 25 March 2021
Accepted: 7 April 2021
Published: 13 April 2021

Publisher's Note: MDPI stays neutral with regard to jurisdictional claims in published maps and institutional affiliations.

Copyright: © 2021 by the authors. Licensee MDPI, Basel, Switzerland. This article is an open access article distributed under the terms and conditions of the Creative Commons Attribution (CC BY) license (https://creativecommons.org/licenses/by/4.0/).

Polymeric carriers play a key role in modern biomedical and nanomedicine applications. Polymers can be obtained from natural or synthetic sources and have been exploited given their chemistry to achieve interaction with living tissues and cells. Different types of carriers can be produced for drug delivery, namely, micelles, nanoparticles, dendrimers, sponges, hydrogels, and microneedles. With different coatings, appropriate adhesion and targeting features can be designed. Polymeric carriers allow the incorporation or conjugation of both hydrophilic and hydrophobic molecules and have tunable chemical and physical features that allow effective drug protection from degradation or denaturation. Other features are noteworthy in polymeric carriers, like their generally good biocompatibility and the ability to exhibit a slow and controlled dug release, allowing for the use in biomedical applications.

This Special Issue provides an encompassing view on the state of the art of polymeric carriers, showing how current research is dealing with new stimuli-responsive systems for cancer therapies and biomedical challenges, namely, overcoming the skin barrier. The published papers cover topics ranging from novel production methods and insights on hybrid polymers to applications as diverse as nanoparticles, hydrogels, and microneedles to antifungal skin therapy, peptide and siRNA delivery, enhanced skin absorption of bioactive molecules, and anticancer therapy. This Special Issue contains one review paper on modulation of macrophage polarization mediated by carbohydrate-functionalized polymeric nanoparticles [1]. A couple of polymeric carriers targeting macrophages have been reviewed in terms of production methods and conjugation approaches. The role of mannose receptor in the polarization of macrophages is highlighted as strategies for infectious diseases and cancer therapies as well as prevention actions.

Taking advantage of polysaccharides' physicochemical features, Pontillo et al. designed new biocompatible and cost-effective carriers for tyrosol, a bioactive natural product present in olive oil and white wine [2]. A chitosan based nanosystem was obtained using the ionic gelation method, while for β-cyclodextrin (βCD), the kneading method was employed. Additionally, coating of the tyrosol–βCD inclusion complex with chitosan led to a sustained release of tyrosol and slowed down the initial burst effect observed from the inclusion complex. The nanosystems were extensively characterized after optimized production based on a two- or three-factor, three-level Box–Behnken experimental design. Moreover, the interaction of tyrosol and the corresponding nanosystems with ctDNA was evaluated. Data suggest that tyrosol is a ctDNA groove binder, which was confirmed by molecular modeling studies. The same mode of binding was found only for the tyrosol/βCD and tyrosol/βCD/chitosan nanosystems. Nanocomposites of chitosan and alginate were exploited by Sabbagh et al. to deliver metronidazole [3]. Optimization of the formulation was obtained using a full factorial design to study the effect of chitosan and alginate polymer concentrations and calcium chloride concentration on drug loading efficiency, particle size, and zeta potential. These dependent variables were affected by the chitosan, alginate, and calcium chloride concentrations, while zeta potential depended only on the alginate and calcium chloride concentrations. The applied mathematical models revealed that the developed response surface methodology models were statistically significant and adequate for all conditions. High correlation values were determined between the experimental data and predicted ones. The optimized nanocomposites were physiochemically characterized by X-ray diffraction, Fourier-transform infrared spectroscopy, thermal gravimetric analysis, scanning electron microscopy, and in vitro drug release studies. Overall, the optimized nanocomposites could be effective in sustaining the metronidazole release for a prolonged period. Hybrid nanosystems have been studied by Duskey et al. to increase the applicability of poly(lactic-co-glycolic acid) (PLGA) in drug delivery [4]. A series of unique PLGA–chitosan hybrid polymers with tailored and tunable physicochemical characteristics were obtained with two different synthetic methods: solid-phase synthesis on a film or in solution chemical reaction with polycaprolactone as intermediate. The hybrid polymers were physiochemically characterized using nuclear magnetic resonance, Fourier-transform infrared spectroscopy, and dynamic scanning calorimetry. A sodium dodecyl sulfate (SDS) salting-out reaction led to a chitosan SDS intermediate that is soluble in organic solvents, and consequently, a new series of PLGA–chitosan copolymers with different molar ratios were produced. The unique series of PLGA–chitosan hybrids with various molar rapports and solubilities represent the expansion of the PLGA delivery system for the protection and delivery of a wide range of previously noncompatible drugs either as nanoparticles formed through chitosan self-assembly techniques (for those still soluble in acidic solutions) or for the encapsulation in stable and nontoxic films for long-term controlled release (for those insoluble in biological solutions).

Shin et al. developed PLGA nanoparticles as siRNA carriers to overcome ROS/oxidative stress-induced chondrocyte damage in osteoarthritis [5]. A double emulsion technique allowed the successful incorporation of siRNA p47phox within PLGA nanoparticles. The nanosystem was physicochemically characterized and evaluated in chondrocytes and in an osteoarthritis in vivo model. The formulated PLGA nanoparticles provided a sustained release of siRNA, which could reduce dosing frequency to a weekly regimen. Inhibition of p47phox by nanoparticles delivered siRNA-attenuated pain behavior, cartilage damage, and ROS production in knee joints with induced osteoarthritis. The developed polymeric nanosystem may represent a promising novel therapeutic avenue for the treatment of osteoarthritis. PLGA nanoparticles were explored for peptide delivery by Lima et al. [6]. A peptide from the myeloid proteolipid protein (PLP) was encapsulated in PLGA nanoparticles and further incorporated within polymeric microneedle patches for an effective skin delivery. Trehalose was included to preserve the nanoparticles during the freeze-drying process. Polydimethylsiloxane molds were used to obtain poly(vinyl alcohol)–poly(vinyl pyrrolidone) microneedles to carry the freeze-dried PLP-loaded PLGA nanoparticles. Microneedle patches with 550 µm height and 180 µm diameter allowed the peptide release in physiological media. The achieved outcomes motivate the exploitation of this strategy as a new antigen-specific therapy, providing minimally invasive administration of PLP-loaded nanoparticles into the skin. Struggling with skin drug delivery, Zhang et al. studied the effect of poly(ethylene glycol) (PEG) 400 and PEG-6-caprylic/capric glycerides on the dermal absorption of niacinamide [7]. Binary and ternary systems composed of PEGs or PEG derivatives combined with other solvents were studied for skin delivery of niacinamide. Porcine skin permeation assays over 24 hours revealed improved performance of all designed vehicles in relation to PEG 400. High skin retention was observed for these vehicles when compared with the neat solvents investigated. Hence, these results indicate PEG 400 as a useful tool to deliver the bioactive agents to the skin, instead of through the skin. According to the bioactive agent, skin retention may be more interesting than skin permeation. Skin retention of terbinafine was investigated by Ghose et al. through the design of polymeric nanosponge hydrogel. The antifungal agent was incorporated in Box–Behnken-design-optimized nanosponge formulations [8]. In vitro drug release from the nanosponge incorporated into the hydrogel was higher than the drug suspension or the marked formulation. Antifungal activity, nonirritancy, and no erythema or edema confirmed the promising application of the developed nanosponge hydrogel for efficient topical delivery of terbinafine hydrochloride.

Stimuli-responsive nanosystems have been designed to control the release of active molecules into the intended site of action. Van Gheluwe et al. applied a three-step synthesis of a redox-responsive blend of poly(ethylene glycol)–*block*–poly(lactide) (PEG–*block*–PLA) and poly(lactide) (PLA) to deliver retinol in the skin [9]. The selection of short-length polymers to incorporate the lipophilic active molecule allowed for achieving a high loading and rapid release of retinol. Stimuli responsiveness of the nanosystem was confirmed in vitro in the presence of L-glutathione. Good biocompatibility of black nanocarriers was observed in human keratinocytes, and low toxicity was detected in the presence of retinol. The redox-responsive blend of PEG–*block*–PLA and PLA were assembled by nanoprecipitation in smart nanocarriers able deliver other retinoid molecules for the treatment of skin diseases, like acne, photoaging, psoriasis vulgaris, melisma, and skin cancers. Cano-Cortes et al. investigated the drug covalent conjugation to the polymeric nanosystem based on PEGylated polystyrene pH-responsive polymer [10]. Doxorubicin was selected to evaluate this pH-responsive approach to cancer therapy. An efficient loading was achieved upon covalent conjugation of doxorubicin to cross-linked polystyrene nanoparticles, allowing selective drug release under acidic pH values. Breast and lung cancer cell lines were studied to determine the efficiency of cellular uptake, therapeutic activity, and genotoxicity effect. The pH-responsive polymeric nanosystems exhibited better antitumor activity in relation to free doxorubicin. The implemented chemical strategy could be further applied to other molecules and types of cancer.

Data Availability Statement: No new data were created or analyzed in this study. Data sharing is not applicable to this article.

Acknowledgments: The guest editors would like to thank all contributors of this Special Issue in the Polymers journal (MDPI). Special thanks to all reviewers who help us to ensure the quality of each published article in this Special Issue; special thanks to the editor in chief and assistant editorial team of Polymers for helping us to complete this work. The guest editors are thankful for the support from FEDER funds through the COMPETE 2020 Operational Programme for Competitiveness and Internationalisation (POCI), Portugal 2020, and national funds through FCT/MCTES in the framework of the project POCI 01 0145-FEDER-030834, and Base Funding UIDB/50006/2020. Sofia Lima thanks the Portuguese Foundation for Science and Technology (FCT) for the financial support for her work contract through the Scientific Employment Stimulus-Individual Call (CEECIND/01620/2017).

Conflicts of Interest: The authors declare no conflict of interest.

References

1. Andrade, R.G.D.; Reis, B.; Costas, B.; Lima, S.A.C.; Reis, S. Modulation of Macrophages M1/M2 Polarization Using Carbohydrate-Functionalized Polymeric Nanoparticles. *Polymers* **2021**, *13*, 88.
2. Pontillo, A.R.N.; Konstanteli, E.; Bairaktari, M.M.; Detsi, A. Encapsulation of the Natural Product Tyrosol in Carbohydrate Nanosystems and Study of Their Binding with ctDNA. *Polymers* **2021**, *13*, 87.
3. Sabbagh, H.A.K.; Hussein-Al-Ali, S.H.; Hussein, M.Z.; Abudayeh, Z.; Ayoub, R.; Abudoleh, S.M. A Statistical Study on the Development of Metronidazole-Chitosan-Alginate Nanocomposite Formulation Using the Full Factorial Design. *Polymers* **2020**, *12*, 772. [CrossRef] [PubMed]
4. Duskey, J.T.; Baraldi, C.; Gamberini, M.C.; Ottonelli, I.; Da Ros, F.; Tosi, G.; Forni, F.; Vandelli, M.A.; Ruozi, B. Investigating Novel Syntheses of a Series of Unique Hybrid PLGA-Chitosan Polymers for Potential Therapeutic Delivery Applications. *Polymers* **2020**, *12*, 823. [CrossRef] [PubMed]
5. Shin, H.J.; Park, H.; Shin, N.; Kwon, H.H.; Yin, Y.; Hwang, J.-A.; Kim, S.I.; Kim, S.R.; Kim, S.; Joo, Y.; et al. p47phox siRNA-Loaded PLGA Nanoparticles Suppress ROS/Oxidative Stress-Induced Chondrocyte Damage in Osteoarthritis. *Polymers* **2020**, *12*, 443. [CrossRef] [PubMed]
6. Lima, A.F.; Amado, I.R.; Pires, L.R. Poly(d,l-lactide-co-glycolide) (PLGA) Nanoparticles Loaded with Proteolipid Protein (PLP)—Exploring a New Administration Route. *Polymers* **2020**, *12*, 3063. [CrossRef] [PubMed]
7. Zhang, Y.; Lane, M.E.; Moore, D.J. An Investigation of the Influence of PEG 400 and PEG-6-Caprylic/Capric Glycerides on Dermal Delivery of Niacinamide. *Polymers* **2020**, *12*, 2907. [CrossRef] [PubMed]

8. Ghose, A.; Nabi, B.; Rehman, S.; Md, S.; Alhakamy, N.A.; Ahmad, O.A.A.; Baboota, S.; Ali, J. Development and Evaluation of Polymeric Nanosponge Hydrogel for Terbinafine Hydrochloride: Statistical Optimization, In Vitro and In Vivo Studies. *Polymers* **2020**, *12*, 2903. [CrossRef] [PubMed]
9. Van Gheluwe, L.; Buchy, E.; Chourpa, I.; Munnier, E. Three-Step Synthesis of a Redox-Responsive Blend of PEG–block–PLA and PLA and Application to the Nanoencapsulation of Retinol. *Polymers* **2020**, *12*, 2350. [CrossRef] [PubMed]
10. Cano-Cortes, M.V.; Laz-Ruiz, J.A.; Diaz-Mochon, J.J.; Sanchez-Martin, R.M. Characterization and Therapeutic Effect of a pH Stimuli Responsive Polymeric Nanoformulation for Controlled Drug Release. *Polymers* **2020**, *12*, 1265. [CrossRef] [PubMed]

Modulation of Macrophages M1/M2 Polarization Using Carbohydrate-Functionalized Polymeric Nanoparticles

Raquel G. D. Andrade [1], Bruno Reis [2,3], Benjamin Costas [1,*], Sofia A. Costa Lima and Salette Reis [2]

1. LAQV, REQUIMTE, Departamento de Ciências Químicas, Faculdade de Farmácia, Universidade do Porto, Rua de Jorge Viterbo Ferreira, 228, 4050-313 Porto, Portugal; up201305657@fc.up.pt
2. Centro Interdisciplinar de Investigação Marinha e Ambiental (CIIMAR), Universidade do Porto, Avenida General Norton de Matos, S/N, 4450-208 Matosinhos, Portugal; breis@ciimar.up.pt (B.R.); bcostas@ciimar.up.pt (B.C.); shreis@ff.up.pt (S.R.)
3. Instituto de Ciências Biomédicas Abel Salazar (ICBAS-UP), Universidade do Porto, Rua de Jorge Viterbo Ferreira n° 228, 4050-313 Porto, Portugal
* Correspondence: slima@ff.up.pt

Abstract: Exploiting surface endocytosis receptors using carbohydrate-conjugated nanocarriers brings outstanding approaches to an efficient delivery towards a specific target. Macrophages are cells of innate immunity found throughout the body. Plasticity of macrophages is evidenced by alterations in phenotypic polarization in response to stimuli, and is associated with changes in effector molecules, receptor expression, and cytokine profile. M1-polarized macrophages are involved in pro-inflammatory responses while M2 macrophages are capable of anti-inflammatory response and tissue repair. Modulation of macrophages' activation state is an effective approach for several disease therapies, mediated by carbohydrate-coated nanocarriers. In this review, polymeric nanocarriers targeting macrophages are described in terms of production methods and conjugation strategies, highlighting the role of mannose receptor in the polarization of macrophages, and targeting approaches for infectious diseases, cancer immunotherapy, and prevention. Translation of this nanomedicine approach still requires further elucidation of the interaction mechanism between nanocarriers and macrophages towards clinical applications.

Keywords: glyconanoparticles; immunotherapy; infectious diseases; mannose receptors; nutraceuticals

1. Introduction

Nanomedicine aims to improve health and life welfare with nanosized materials. Nanoparticles can be designed for drug delivery by modulating surface properties and composition to improve therapeutic effect and targeting specificity. Active targeting can be obtained with surface functionalization of the nanoparticles using specific ligands to reach the target of interest [1]. Taking advantage of this receptor-mediated specificity will reduce toxicity and side-effects to healthy tissues.

Macrophages are innate immune cells widely present in the body acting to maintain homeostasis and to resist pathogen invasion [2]. Macrophages are distributed according to their functions, surface-expressed markers, and secreted cytokines in M1/M2-polarized phenotypes. However, the simplicity of M1/M2 dichotomy of macrophage activation is too broad to explain all the actual states of the macrophages as a response to several stimuli. To have a proper description of the macrophage activation, it is generally accepted to include information on macrophage source, type of activators, and markers [3]. An imbalance in the M1/M2 ratio weakens the immune response and leads to inflammation. Hence, macrophages constitute an important player in the therapeutic strategies against infections, inflammatory conditions, and cancer. Receptors frequently expressed on the surface of macrophages constitute a potential target for nanomedicine-based approaches. Macrophage scavenger receptor, Toll-like receptors, glucan receptor, folate receptor, and mannose receptor are among the most used surface receptors of macrophages [4].

Mannose receptor (MR) is composed by several domains that allows recognition to various molecules of the carbohydrate family and contributes to receptor-mediated endocytosis. Upon internalization, nanocarriers can elicit macrophage polarization in vivo. Different types of polymeric based carriers (e.g., nanoparticles, micelles, dendrimers) are emerging as macrophage-targeted delivery systems [5]. Nanocarriers can also reset the macrophage activation state, as it is the case of the conversion of M2 phenotype to M1 in tumor-associated macrophages [6]. Understanding the interaction mechanisms between nanoparticles and macrophages is essential to a successful and effective nanocarrier's design towards a therapeutic or prevention strategy.

2. Polymeric Nanoparticles as Biomedical Delivery Devices

Over the past few decades, the development of new strategies that surpasses the problems associated with conventional diagnosis and therapies have gained great importance on the scope of nanomedicine. One of the main goals in this field is to design nanoparticles capable of a targeted delivery and controlled release of bioactive compounds to a specific site, increasing its therapeutic effect while minimizing its side effects [7,8]. Several types of nanoparticles can be prepared from different building blocks like lipids, proteins, metals, and polymers [7–9]. Polymeric nanoparticles have gained great importance as biocompatible drug delivery systems given their simplicity and low-cost production [10]. The use of polymeric nanoparticles in drug delivery has many advantages over the use of other types of nanocarriers: a growing choice of biodegradable and biocompatible polymers, higher encapsulation efficiencies, higher stability in physiological conditions, improved drug bioavailability, and simpler preparation (for more detailed information on the synthesis methods see ref. [11]).

The design of drug delivery systems needs to consider several characteristics, namely, hydrophobicity, size, surface charge, biological interactions/toxicity, and biodegradability. A wide variety of natural or synthetic polymers are available for the preparation of the nanoparticles [12,13]. To produce nanoparticles, the most commonly used natural polymers include chitosan, a linear polysaccharide extracted from the exoskeletons of marine crustaceans [14], alginate that is isolated from brown algae [15], and gelatin obtained by hydrolyzed collagen [16]. Natural polymers have the advantage of combining biological properties like mimicking the extracellular matrix, allowing to sustain cell growth in tissue engineering applications, and tunable mechanical properties like stimuli-responsiveness, degradation, swelling, and crosslinking capabilities [13–15]. However, the application of natural polymers is often hampered by contaminants and batch-to-batch variability. Other constraints involve low hydrophobicity that compromises lipophilic drugs encapsulation, and a rapid drug release from the matrix [17,18].

Limitations of natural polymers can be overcome with the use of synthetic polymers, which are more reproducible in manufacture and more stable. Polymeric nanoparticles obtained with synthetic polymers allow drug-controlled release for a period of days up to several weeks [18]. Drawbacks associated with these type of nanoparticles involve their limited aqueous solubility and the need of surfactants to form stable suspensions [19]. The outcome of the nanoparticle as a drug delivery system can be modulated in the composition not only in the nature of the polymer, but also in the molecular weight, copolymer composition, and selected surfactant. To produce a targeted drug release within the body, the nanoparticles can be considered with additional properties to respond to external or internal stimuli such as redox state or pH [20]. Polylactic acid (PLA), poly(glycolic acid) (PGA), and their copolymers (PLGA) represent the most extensively used and studied synthetic polymers for drug delivery [21–23]. The presence of ester linkages in their backbones make these polyesters biodegradable. In fact, in a living organism, these polymers suffer a hydrolysis, and the resulting products are easily metabolized in the Krebs cycle and eliminated as carbon dioxide and water [17]. Also widely applied in the production of nanoparticles is poly-ε-caprolactone (PCL) that allows a slow degradation rate in comparison with PLA and PLGA, and thus is more adequate for long-term drug delivery.

Poly(alkylcyanoacrylate) (PACA) is an interesting polymer whose properties can be controlled by the side of the introduced chains, being that the longer the side-chains the longer the half-life of the nanoparticles [17].

Depending on the preparation method, used polymers and desired application, different polymeric nanocarriers can be obtained such as polymer-drug conjugates, polymeric micelles, polymeric nanogels, and dendrimers [24]. Two types of polymer nanoparticles can be obtained for drug delivery: nanocapsules, composed of a liquid or semisolid core covered by a polymer membrane; or nanospheres that consist in a solid polymer matrix [11,12,23–25]. The drugs can be either entrapped in nanoparticles or adsorbed at the surface. In nanocapsules, the drug can be encapsulated in the inner core, while in nanospheres it is uniformly dispersed in the polymer matrix (Figure 1). These represent versatile tools for surface modification, as well as shape, size, and even optical characteristics. In nanomedicine, core–shell polymeric nanoparticles are also interesting, as the polymeric platform allows a second shell, usually a solid, which may confer smart properties to the nanoparticle (e.g., pH sensitive, thermo- and enzyme-responsive) [26–28].

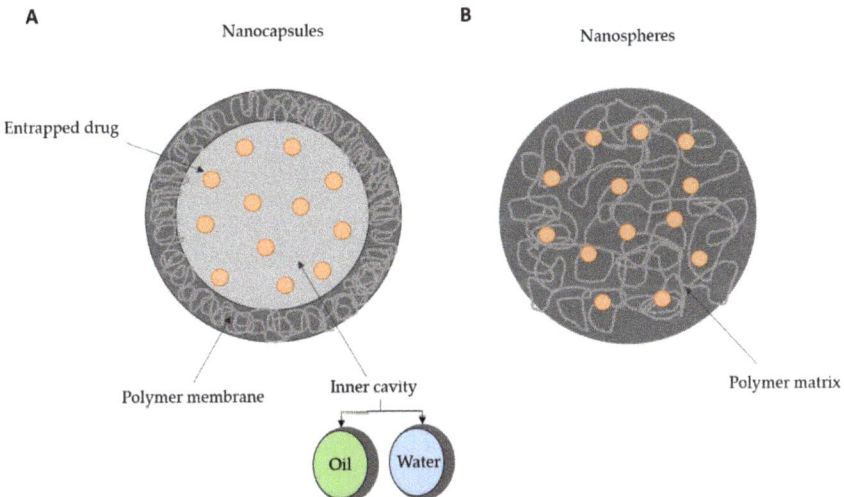

Figure 1. Schematic representation of the two types of polymeric nanoparticles: nanocapsules (**A**) and nanospheres (**B**). Nanocapsules comprise an inner cavity, composed of water or a semi solid (oil), and covered with a polymer membrane, while in nanospheres the entire mass is a polymer matrix. Drug molecules can be entrapped in both types of nanoparticles.

3. Production Methods for Polymeric Nanoparticles and Surface Properties Modifications

Currently, there are several methods developed and well-implemented for the preparation of polymeric nanoparticles. At first, one needs to ponder on (i) the physicochemical properties of the bioactive compound to be delivered, (ii) the nature and type of polymer, (iii) the target and biological environment, and (iv) the administration route. Based on this information it is possible to select the most adequate production method among emulsification-solvent evaporation, nanoprecipitation, emulsification reverse salting-out, and emulsification solvent diffusion. These polymerization processes allow production of nanoparticles with control of physicochemical and biological properties of the nanoparticles that are formed (Figure 2). At least two steps are involved in these conventional production methods: (i) polymer dissolution in an organic solvent followed by emulsification in an aqueous phase, and (ii) solvent evaporation to obtain the nanoparticles [13,29]. Polymeric nanoparticles can also be produced using monomers in an emulsion or as a micellar suspension by interfacial poly-condensation [13,17].

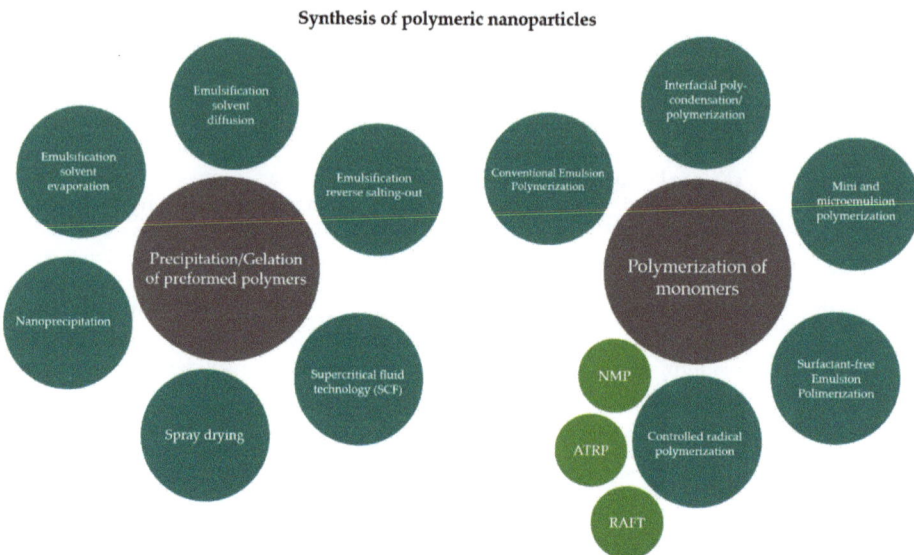

Figure 2. Diagram representing the options of production methods to obtain polymeric nanoparticles. Abbreviations: NMP (nitroxide-mediated polymerization); ATRP (atom transfer radical polymerization); RAFT (reversible addition and fragmentation transfer chain polymerization).

Hydrophilicity of the drug delivery systems represents an important feature to be considered for biological application. In fact, upon intravenous administration, hydrophobic nanoparticles are taken as foreign and the organism removes them from circulation to the excretion organs (liver, spleen, and lymph nodes) using the mononuclear phagocytic system [30]. If the intended treatment targets one of these organs, hydrophobic nanoparticles are the best solution. When aiming different targets, systemic circulation needs to occur, so the delivery systems reaches the diseased site. In this case, the surface of the nanoparticles must be modified with hydrophilic polymers to prevent the action of the mononuclear phagocytic system and phagocytosis. Hydrophilic nanoparticles will have long circulation times and reduced nonspecific distribution [31,32]. The list of hydrophilic polymers is long and include polyethylene glycol (PEG), poly-vinyl pyrrolidone (PVP), pluronics (poly-ethylene oxides), poloxamers, vitamin E TPGS, polysorbate 20, polysorbate 80, and polysaccharides (e.g., dextran) [33]. A protective layer can be obtained at the surface of the nanoparticles with these hydrophilic compounds, by adsorption or grafting shield groups. In some cases, PEG can be incorporated as copolymer [30,34,35]. The most used hydrophilic polymer for nanoparticles' surface modification is PEG. The nature (flexible chains) and the physicochemical (hydrophilicity) features of this polymer as well as the presence of functional groups able to prevent plasma proteins binding are the reasons for this success [36]. A significant decrease in the opsonization and macrophage internalization of nanoparticles was observed with PEG coating, which lead to an enhanced long-term blood circulation. PEGylated nanoparticles promote a higher drug uptake by target tissues when compared to non-PEGylated ones [37–39].

In sum, a crucial feature of polymeric nanoparticles is their surface modification in order to improve drug delivery. On the one hand, the addition of a stealth layer (PEG, PVA, polysorbate) at the surface of nanoparticles allows an increased blood circulation time, avoiding the binding of opsonins and the rapid clearance from the mononuclear phagocytic system, and on the other hand, the functionalization at the surface with targeting ligands (proteins, peptides, antibodies [40–42]) improves the specificity of the treatment [43,44].

4. Carbohydrate-Functionalized Polymeric Nanoparticles

As stated before, polymeric nanoparticles have excellent features that make them promising delivery systems for therapeutic applications. A higher specificity of drug delivery to a certain site of action is achieved when targeting ligands are incorporated in these nanocarriers. The functionalization of nanoparticles with carbohydrates, also known as glyconanoparticles, plays a key role in receptor-mediated delivery, as it allows to establish specific interactions with carbohydrate-binding proteins (lectins) [45,46]. Besides molecular recognition, sugars can act as colloidal stabilizers [47], reduce toxicity [48] and immunogenicity [49] and unlike PEG, increase circulation time in the bloodstream without compromising cellular uptake [50].

Glycopolymers can be prepared either by post-polymerization modification, which consists in the functionalization of a preformed polymeric backbone, or in the polymerization of glycosylated monomers [51,52] that can be performed by several synthetic routes that provide controllable architectures, stereochemistry, and molecular weights, such as free radical polymerization (ring-opening polymerization (ROP)), ionic polymerization, controlled radical polymerization (nitroxide-mediated polymerization (NMP), atom transfer radical polymerization (ATRP), reversible addition fragmentation chain transfer (RAFT), and enzyme-mediated polymerization [51,53–55]. Here, we will focus on the post-functionalization of polymeric nanoparticles with carbohydrates, as it allows the attachment of pendant carbohydrate moieties (Figure 3), making it ideal for targeted delivery.

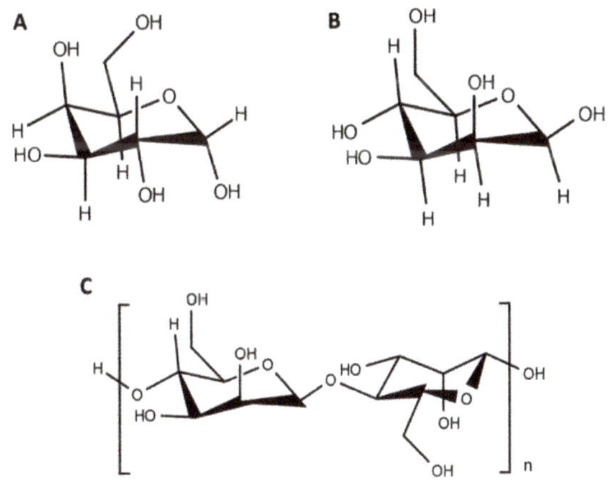

Figure 3. Chemical structure of some carbohydrates commonly used to produce glyconanoparticles. (**A**) Galactose; (**B**) mannose; (**C**) mannan.

The coupling of a ligand to a nanoparticle can be achieved either by electrostatic interactions or by covalent conjugation strategies [56,57]. The last requires the presence of reactive functional groups (amine, carboxyl, sulfhydryl, hydroxyl, azide-reactive groups) at the surface of nanoparticle that enable conjugation with ligands [58]. A very popular method used for chemical conjugation is the carbodiimide method, which consists of the activation of carboxylate functional groups that react with primary amines to form amide bonds [58,59]. In this case, a direct conjugation is performed, but sometimes linkers are used. For instance, Kim and collaborators used N,N'-dicyclohexyl carbodiimide (DCC) for a two-step coupling reaction of a galactose moiety to polymeric nanoparticles composed of cholic acid and diamine-terminated poly(ethylene glycol) as a linker [60]. Palmioli and co-workers also described the functionalization of PLGA with sugar entities bearing a 2-(2-aminoethoxy)ethanol linker through amide bond using N,N'-diisopropylcarbodiimide and NHS [61].

Crucho and colleagues produced a polymeric conjugate composed of PLGA modified with sucrose and cholic acid moieties [62]. The functionalization of the PLGA backbone was made through esterification using DCC/NHS reactions, and then sucrose and cholic acid-functionalized PLGA nanoparticles were obtained by nanoprecipitation. Sucrose addition provided colloidal stability to the nanoparticles, demonstrated by the decrease of the negative surface charge. Rieger and collaborators reported a simple method for the preparation of mannose-functionalized PLA NPs [63]. The synthesis approach consisted in the co-nanoprecipitation evaporation of a mannosylated poly(ethylene oxide)-b-poly(ε-caprolactone) (PEO-b-PCL) diblock copolymer with PLA. The amphiphilic copolymers bearing the mannose moieties worked as surface modifiers and were able to specifically bind to MR.

Freichels and co-workers prepared crosslinked hydroxyethyl starch (HES) nanocapsules, which is a hydroxyethylated glucose polymer, functionalized with (oligo)mannose [64]. The preparation of the nanocapsules consisted in interfacial addition of HES with 2,4-toluene diisocyanate (TDI) in inverse miniemulsion. This procedure leaves an amount of non-reacted amine groups that were used to perform the functionalization with three types of mannose molecules: a-D-mannopyranosylphenyl isothiocyanate, 3-O-(a-D-mannopyranosyl)-D-mannose (di-mannose), and $\alpha 3,\alpha 6$-mannotriose (tri-mannose). The amine groups on the surface of nanocapsules were used to react directly with mannose isothiocyanate while di- and tri-mannose were coupled through reductive amination. The obtained delivery systems exhibit a specific binding to agglutinin and the presence of a PEG linker showed to increase the interaction to the receptor, due to a higher accessibility of the sugar molecule.

Kim et al. developed a siRNA delivery system composed of PEI, PEG, and mannose [65]. PEI molecules were used to form the polymer/siRNA polyplex, PEG was used as a stabilizer, and mannose as a targeting ligand for macrophages. Here, two different functionalization methods were performed: one in which PEG and mannose molecules were directly linked to PEI backbone (mannose-PEI-PEG), and another in which mannose chains were conjugated to PEI using a PEG spacer, i.e., mannose was linked to PEG before reaction of mannose-PEG chains to PEI backbone. In these reactions, like the ones described before, α-D-mannopyranosylphenyl isothiocyanate was used for mannosylation and PEG was conjugated to PEI via glutaraldehyde linkage. The researchers also found that the location in which mannose ligands are conjugated affect the cytotoxicity of nanocarriers. Table 1 resumes examples of glycoproteins produced with electrostatic interactions and covalent conjugation strategies identifying the ligand and the target defined for the nanocarriers.

Table 1. List of developed carbohydrate-functionalized polymeric nanoparticles.

Polymeric Nanocarrier Composition	Carbohydrate Ligand	Functionalization Strategy	Target Tissue/Cells	Ref
Cholic acid and PEG	Galactose	N,N'-dicyclohexyl carbodiimide reaction	Liver-specific delivery	[65]
PLGA NPs	Galactose Glucose Mannose	N,N'-diisopropylcarbodiimide/NHS reaction	-	[61]
PLGA NPs	Sucrose Cholic acid	DCC/NHS reactions	-	[62]
PLA and PEO-b-PCL diblock copolymer NPs	Mannose	Nanoprecipitation-evaporation approach	Mannose receptors	[63]

Table 1. Cont.

Polymeric Nanocarrier Composition	Carbohydrate Ligand	Functionalization Strategy	Target Tissue/Cells	Ref
Hydroxyethyl starch nanocapsules	Mannose Dimannose Trimannose	Mannose: -Amine to isothiocyanate group reaction Dimannose and trimannose: -Reductive amination	Agglutinin (mannose receptor)	[64]
PEI-PEG NPs	Mannose	Binding of mannose to PEI-PEG NPs Binding of mannose to PEI NPs via a PEG spacer	Macrophage cells	[65]
PLGA NPs	Mannose Mannan Mannoseamine	DCC/NHS/EDA reaction	Macrophages Leishmania-infected mice	[66]
6-Amino-6-deoxy-curdlan	Mannose	Amine to isothiocyanate group reaction	Mouse peritoneal macrophages	[67]

5. Macrophages

5.1. Functions and Polarization State

The mononuclear phagocytic system, also designated as the reticuloendothelial system, is composed of monocytes in the blood and macrophages in the tissues and is part of the innate immune system. During the hematopoiesis process, mature monocytes circulate for about 8 h, grow, and end up in specific tissues, as macrophages [68].

Macrophages are present throughout the body resident in tissues and also motile, known as free or wandering macrophages. They can originate from circulating monocytes, but also from embryonic hematopoietic stem cells or yolk sac [69]. Macrophages play relevant roles in the immune response, as they act in tissue development, inflammation related to pathogens, cancer, and organ transplantation. During phagocytosis, macrophages engulf pathogens, mediated by receptors on macrophage surface that bind to the fragment crystallizable (Fc) region of molecule from the pathogen. This process leads to the formation of a phagosome that merges with the lysosome where the target is digested. Macrophages act as antigen presenting cells, when displaying foreign material or parts of antigens on its surface in association with class II major histocompatibility complex (MHC) molecules. This triggers T-cells, and consequently, the adaptive immunity. Likewise, macrophages can secrete several cytokines involved in the immune response, homeostasis, and inflammation, which modulate their function and surface marker expression [70].

Macrophages are polarized to respond to alterations in their environment, being classified as M1 macrophages and M2 macrophages [71]. Contact with pathogen-associated molecular patterns (PAMPs), such as bacterial lipopolysaccharide (LPS) from *Escherichia coli* (Gram-negative) or peptidoglycan (PGN) from *Staphylococcus aureus* (Gram-positive) drives macrophages polarization towards M1 phenotype, with the ability to elicit proinflammatory response and production of interleukin (IL) 6 (IL-6), IL-12, and tumor necrosis factor-alpha (TNF-α), all pro-inflammatory cytokines. Alternatively, activated macrophages are produced in the presence of the Th2 cytokines IL-4 and/or IL-13, which can lead macrophage polarization to M2, characterized by anti-inflammatory responses and tissue repair abilities [72].

Regulation of macrophage polarization phenotype is reversible and modulates their immune function. An important feature in this mechanism is the expression of the cell surface markers. M1 macrophages overexpress CD80, CD86, and CD16/32, while M2 exhibits more arginase-1 and mannose receptor (CD206).

5.2. Macrophage Polarization Mediated by Nanocarriers

To date, several nanocarriers were able to induce inflammatory and immune responses in vitro and in vivo [73–75]. Nanocarriers can be internalized by macrophages inducing changes at the cell surface as well as secretion of cytokines and chemokines [76]. Understanding the mechanism of interaction between nanocarriers and macrophages will contribute to an effective design of nanocarriers for a specific therapeutic strategy. Macrophage-mediated therapies are emerging as a promising and effective approach towards the treatment of several diseases. In particular, uptake of nanocarriers by macrophages implies interaction between nanocarriers' surface and macrophage cell membrane. Therefore, the formed membrane-bound vesicle will have a size, composition, and internal environment according to the internalization, resulting in endosomes, phagosomes, or macropinosomes. In fact, the uptake mechanisms can be described as phagocytosis, micropinocytosis, endocytosis mediated by clathrin or by caveolin, or independent from both [77]. Passive and active targeting approaches can be designed to achieve the intended effect. Size and surface of the nanocarrier govern passive targeting, while for an active targeting the surface of the nanocarrier requires functionalization with a specific ligand towards a particular surface cell receptor. Carbohydrate-coated nanocarriers have been exploited to target mannose receptors expressed in macrophages and dendritic cells (antigen presenting cells, APCs) [51].

Conjugation of ligands at the surface of nanocarriers may modulate the immune system. The use of targeted nanocarriers elicit the maturation of APCs, with alterations at the surface expression of co-stimulatory molecules and in the secretion of cytokines that activate T-cell responses [78–80]. Active targeting of nanocarriers towards endocytic receptors present on macrophage surface can be achieved using C-type lectin receptors (CLR) or the mannose receptor CD206.

5.3. Mannose Receptor

CD206 or mannose receptor (MR) has the ability to recognize mannosylated or fucosylated glycoproteins and engulf them [81]. This 175 kDa endocytic receptor was first identified in rabbit alveolar macrophages and is a type I transmembrane receptor composed by an extracellular region containing a cysteine-rich (CR) domain that acts as second lectin domain, and a fibronectin type II (FNII) domain that is involved in collagen binding, and multiple C-type lectin-like domains (CTLDs) within a single polypeptide backbone where the binding of sugars terminated in D-mannose, L-fucose, or N-acetyl glucosamine occurs [82]. Based on their structure, CLR are grouped as transmembrane CLRs and soluble CLRs (collectins). Type I transmembrane CLRs include MR and ENDO180 (mannose receptor C type 2), while type II transmembrane CLRs include dendritic cell-specific intracellular adhesin molecule 3 grabbing non-integrin (DC-SIGN), langerin, and macrophage galactose type lectin (MGL) receptors [83].

MR expression is not restricted to resident macrophages and dendritic cells. It was also found on immature monocyte-derived dendritic cells [84], hepatic endothelial cells [85], tracheal smooth muscle cells [86], and kidney mesangial cells, among others [84]. Expression of this receptor is modulated by cytokines, immunoglobulin receptors, and pathogens [87]. MR synthesis is more rapid in the presence of immunoglobulins IgG2a and IgG2b [88]. Cytokines regulate MR expression as IL-4 [89], IL-13 [90], and IL-10 [91] enhance macrophages receptors expression, while interferon-γ (IFN-γ) [92] down-regulate MR expression and increase macrophage's activation.

Macrophages cell surface express about 10–30% of MR at steady state and the remaining 70–90% have an intracellular location. Early endosomes contain MR internalized and are able to send these receptors back to the cell surface through the interaction with the clathrin-mediated endocytic machinery [80]. This mechanism is mediated by small intracellular vesicles (below 0.2 μm) and drive the MR to be recycled to the macrophage membrane or delivered into late endosomes, filled with lysosomal hydrolases. Here, under acidic pH and hydrolase-rich environment, the final degradation of the internalized cargo happens. The ability of nanoparticles to modulate the macrophage state through MR was described for several authors (Table 1). For example, chitin and mannose-coated beads improved the production of tumor necrosis factor-alfa (TNF-α), IFN-γ, and IL-12 by murine spleen cells in relation to non-coated beads [93].

MR is also expressed in DCs and actively contributes to antigen recognition and processing. Evidences confer MR an important part in the antigen-internalization mechanism in DCs. For example, bovine serum albumin coated with mannose enhanced the uptake and presentation of this antigen to T cells [94,95].

Macrophages are responsible for the internalization and degradation of pathogens, acting as pattern recognition receptors, given the highly conserved C-type lectin receptors, in a calcium dependent manner. Thus, this first line of defense binds to carbohydrate molecules (e.g., mannose, fucose, and N-acetyl glucosamine) present on the surface of a wide variety of pathogens, including *Candida albicans* [81], *Leishmania donovani* [96], and *Mycobacterium tuberculosis* [97].

5.4. Mannose Receptor-Targeted Nanocarriers Interactions with Macrophages

Targeting MR in macrophages using polysaccharides or glycoproteins containing mannose or fucose residues has been exploited to develop nanocarrier-based macrophage-mediated therapies [98]. Mannose-based glycopolymers exhibited an increased internalization by macrophages in comparison to galactose-containing glycoprolymers [99]. Given the variability of ligand–target interaction according to the activation and differentiation state of macrophages, studies should consider various types of carbohydrate moieties. The design of the nanocarriers should also pay attention to their surface charge, as it affects macrophage binding affinity. Anionic sialic acid is present on macrophages surface and enhances phagocytosis of positively charged nanocarriers [100]. Nanocarriers coated with albumin, folic acid, or cholesterol are easily internalized by caveolin-mediated endocytosis which prevents lysosomal degradation. However, mechanisms of uptake are interchangeable and blocking a path may "open" another endocytic path, which poses a challenge in the design of a nanocarrier (Figure 4) [5,101].

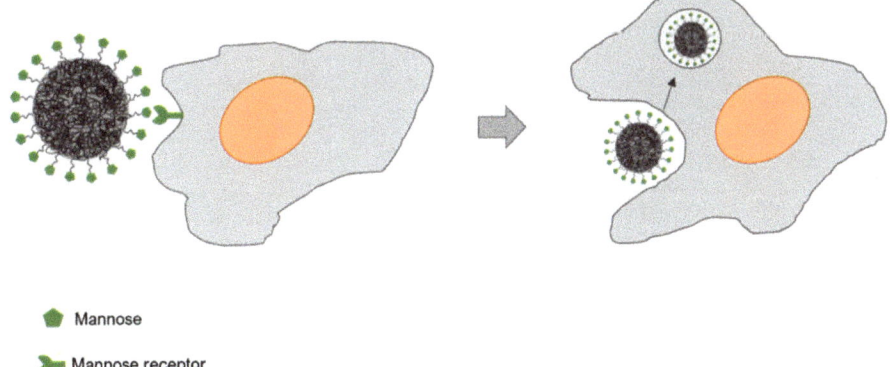

Figure 4. Glyconanoparticle interaction with macrophages through receptor mediated endocytosis mechanism.

5.4.1. Mannose Receptor-Targeting Nanocarriers towards Infection Resolution

Macrophages are host cells of many intra-cellular pathogens (bacteria, parasites, and virus) causing infectious diseases that could be managed with nanocarriers targeting MR. Recent examples of carbohydrate-based polymeric nanocarriers towards macrophages are described and shown in Table 2.

Tuberculosis is the bacterial infection responsible for more deaths worldwide. The treatment regimen involves oral administration of rifampicin, isoniazid, pyrazinamide, and ethambutol for long periods, usually over six months. The completion rate is highly dependent of patient compliance, but interruptions may occur due to adverse side-effects. Hence, new therapeutic approaches which are more efficient, with less side-effects and shorter duration of treatment are envisaged [102]. Aminoglycoside antibiotics are used against mycobacterial infections, but usually are not highly membrane permeable eliciting adverse side effects. Chitosan nanoparticles loaded with aminoglycoside were produced with dextran sulphate as counter ion to shield the positive charge of the antibiotic. In vivo results showed effective killing of intracellular *M. tuberculosis* upon oral administration of antibiotic-loaded nanocarriers [103]. Isoniazid, an anti-tuberculostatic agent, was incorporated in mannosylated gelatin nanoparticles. Macrophages were effectively targeted by these nanoparticles, as assessed by flow cytometry [104]. For rifampicin, several examples of nanocarriers have been described. Rifampicin was loaded in dendrimers able to enhance alveolar macrophage uptake and drug release at pH 5 [105], and also in flower-like polymeric micelles which surface was modified with hydrolyzed galactomannan [106]. The latter combined mannose and galactose were both recognized by CLRs. A complex nanocarrier based on poly(epsilon-caprolactone)-*b*-poly(ethylene-glycol)-*b*-poly(epsilon-caprolactone) flower-like polymeric micelles (PMs) coated with chitosan or GalM-h/chitosan was produced allowing higher intracellular levels of rifampicin in murine macrophages, relative to its free and chitosan-loaded forms.

The protozoa Leishmania is the causative agent of several infectious diseases upon invading macrophages in the liver and spleen (visceral leishmaniasis) or in the skin (cutaneous leishmaniasis). Leishmaniasis remains endemic in developing countries and without proper treatment leads to death. Pentavalent antimonials were the first anti-leishmanial agents used, but given their toxicity, treatment evolved to amphotericin B, miltefosine, pentamidine, primaquine, paromomycin, and even natural compounds (e.g., amarogentin and andrographolide) [107]. Treatment is hampered by the intracellular localization of the protozoa inside the phagolysosome. The US FDA-approved poly(d,l-lactide-coglycolide) (PLGA) polymer was functionalized with carbohydrate moieties (mannose, mannan, and mannosamine) to identity the most effective in targeting macrophages infected with *Leishmania*. In vitro data obtained with murine primary macrophages evidenced the immunemodulatory properties of the nanocarriers, with activation of macrophages and production of pro-inflammatory cytokines, upon clathrin-mediated endocytosis. Amphotericin B-loaded on mannan-functionalized PLGA nanocarriers confirmed in vivo efficacy in relation to Fungizone© alone, in a visceral leishmaniasis model [66]. MR was targeted by coating polyanhydride nanoparticles with carbohydrates (galactose and di-mannose) by Chavez-Santoscoy and co-workers [108]. The designed nanocarriers increased surface expression of markers in alveolar macrophages, enhanced the expression of MR, and promoted production of pro-inflammatory cytokines (IL-1b, IL-6, and TNF-a). Curcuminloaded mannosylated chitosan nanoparticles improved the drug mean residence time within infected macrophages [109]. Effective endocytosis mediated by MR lead to better pharmacokinetic parameters.

Table 2. Mannose receptor-targeting nanocarriers towards infection resolution.

Composition	Carbohydrate	Cargo	Advantages	Ref
Chitosan, dextran sulphate	-	Aminoglycoside	Oral administration allowed effective killing of intracellular *M. tuberculosis*	[103]
Gelatin	Mannose	Isoniazid	Effective targeting of macrophages	[104]
Poly(epsilon-caprolactone)-*b*-poly(ethylene-glycol)-*b*-poly(epsilon-caprolactone) and chitosan	Galactomannan	Rifampicin	Improved cellular internalization in murine macrophages	[106]

Table 2. Cont.

Composition	Carbohydrate	Cargo	Advantages	Ref
PLGA	Mannose, mannan and mannosamine	Amphotericin B	Improved in vivo efficacy against visceral leishmaniasis	[66]
Polyanhydride	Galactose and di-mannose	-	Increase production of pro-inflammatory cytokines	[108]
Chitosan	Mannose	Curcumin	Enhanced the drug residence time within infected macrophages	[109]
Gelatin	Mannose	Didanosine	Improved uptake by alveolar macrophages, and in vivo distribution mainly in the lungs, spleen and lymph nodes	[110]
Stearate-g-chitosan	Oligosaccharide	Lamiduvine	High cellular uptake with low toxicity in viral infected cells	[111]
Sialic acid and poly(propyleneimine)	Mannose	Zidovudine	Low cell toxicity and in vivo biodistribution on the lymph nodes	[112]

Targeted mannose-coated gelatin nanoparticles were produced to enhance therapeutic efficacy of didanosine towards human immunodeficiency virus [110]. Higher uptake by alveolar macrophages was observed with the mannose coating, and in vivo biodistribution studies revealed the presence of the nanocarriers in the spleen, lymph nodes, and lungs. Lamiduvine delivery towards HIV was improved with the incorporation in stearate-g-chitosan oligosaccharide polymeric micelles. The nanocarrier led to high internalization and low cytotoxicity in viral transfected cells [111]. Another antiretroviral drug, zidovudine, was incorporated within sialic acid and mannose dual-coated poly(propyleneimine) dendrimer [112]. This nanocarrier produced less cell toxicity and hemolysis, most probably related to the zidovudine-sustained release and enhanced internalization by macrophages. In vivo biodistribution revealed targeting to sialo-adhesin and carbohydrate receptors in the lymph nodes.

5.4.2. Mannose Receptor-Targeting Nanocarriers towards Tumor-Associated Macrophages

Macrophages accumulate in the tumor microenvironment, being designated as tumor-associated macrophages (TAM). These represent the major contribution of tumor immune escape, angiogenesis, growth, and metastasis [113]. Mannosylated nanocarriers can modulate macrophage polarization from M2 phenotype to the M1 phenotype enhancing antitumor immunity. Delivery of Toll-like receptor (TLR) agonists reset TAM polarization towards an antitumor M1 phenotype. Rodell and co-workers produced b-cyclodextrin nanoparticles containing a TLR7/8 agonist that reprogrammed TAM and, as a consequence, efficiently controlled tumor growth [114].

MR targeting can also contribute to improve gene delivery efficiency, by improving transfection and tissue specificity. Reeducation of TAM can be accomplished with delivery of siRNA, miRNA, or mRNA using mannosylated nanoparticles [115,116]. Likewise, chitosan nanoparticles allowed to deliver therapeutic DNA by MR-mediated endocytosis [117]. Experimental data highlights less cytotoxicity, improved gene transfection, and induction of IFN-γ production upon IL-12 gene delivery, in comparison to plain chitosan nanocarriers. IL-12-based gene delivery can be applied for cancer immunotherapy, as it elicits a Th1-type immunity and also cell-mediated immunity.

Instead of only modulating TAM polarization to control cancer progression, it is also possible to completely neutralize or kill them, with the delivery of cytotoxic compounds using TAM-targeted nanoparticles [118].

Nanoparticle-based immunotherapies represent a promising approach to target tumor environment, in particular TAM, instead of aiming for the tumor cells, preventing immune-mediated adverse-effects. Another application could be cancer vaccination by targeting immune cells in the lymph node.

5.4.3. Mannose Receptor-Targeting Nanocarriers towards Prevention Approaches

Oral delivery is the preferred route for drug/bioactive compounds administration, due to effects both at a local and systemic level, minimal invasiveness, and cost-effectiveness [119,120]. However, a question of bioavailability and efficacy emerges when these immunomodulatory compounds are orally administered in its free form. This can be attributed to compound degradation due to pH variation and enzymatic activity in the gastrointestinal (GI) tract or poor permeability across intestinal biological membranes [119,121,122]. Delivery systems such as carbohydrate-functionalized polymeric nanoparticles are able to provide protection from degradation in the GI tract, increase absorption by the intestinal epithelium due to its mucoadhesive properties (e.g., PLGA, chitosan, and alginate) and cell or tissue-targeted delivery and sustained release [121,123–126]. Gentamicin (GM) is an antibiotic that can only be administered in parenteral form or in topical formulations, and it cannot be orally administered due to enzymatic degradation and poor bioavailability. However, when GM was encapsulated in chitosan-functionalized PLGA nanoparticles and orally given to healthy rabbits, it not only reached the GI tract, as it was able to cross the membrane entering the blood stream [127]. Based upon these findings the authors concluded that biodegradable chitosan-functionalized PLGA nanoparticles are potential candidates for GM oral delivery. Furthermore, these polysaccharide polymers-based nanoparticles show unique physicochemical properties, namely, biocompatibility, biodegradability, non-toxicity, and low cost [128,129].

Immunomodulators targeting myeloid cells, particularly macrophages, are a proven strategy to improve the host immunological status and immune response. Several studies show that oral immunostimulation with bioactive compounds can be an effective prophylactic strategy to prevent infectious disease or curtail its effects [130–132]. As already mentioned, macrophages perform critical roles in innate immune response, including inflammation and tissue repair, pathogen elimination, and coordination of the adaptive immune response. Cell surface receptors that recognize polysaccharide residues such as mannose, galactose, or N-acetylglucosamine residues are paramount for macrophage activation and response. Carriers comprising a matrix of polysaccharide moieties, or surface ligands composed of carbohydrates, are suitable candidates for macrophage targeting or stimulation. A chitosan nanoparticle functionalized with a high molecular weight ulvan polysaccharide, activated Senegalese sole (*Solea senegalensis*) macrophages and triggered a stronger immune response than the ulvan extract free form. Ulvan is a complex polysaccharide composed of glucuronic acid and sulphated rhamnose, known to activate and induce a potent stimulating effect on macrophage oxidative burst [133]. It was hypothesized that ulvan stimulating properties improved in the chitosan/ulvan nanoparticles possibly due to particle endocytic uptake by macrophages [134]. Particle size is an important feature for cell uptake: when comparing microparticles to nanoparticles, the latter is generally having higher cell internalization rates, and thus can be utilized to target cellular and intracellular receptors due to their smaller size and mobility [135]. Furthermore, several studies explored mannose-functionalized nanoparticles recognition by the macrophage MR as a way to stimulate macrophages [67,136].

The potential to use orally delivered carbohydrate-functionalized polymeric nanoparticles to target macrophages is recognized, mostly because of the unique structural features of polysaccharides referred above. As research progresses in the field of nutraceuticals, these glyconanoparticles seem to be a highly suitable delivery system for biologically active compounds targeting macrophages.

6. Future Perspectives

Further application of carbohydrate-functionalized polymeric nanoparticles depends on more efficient production methods and improved selectivity towards macrophages or other defined targets (Table 3). The design should consider drug release rate to assure rapid release of the cargo at the target site. The amount of loaded cargo is also crucial, since a balance needs to be achieved between high capacity and safety of the total administered dose. Altogether, the product should be scalable and cost-effective to attract investors and industries. However, not all these requirements are currently met. In fact, the production methods are hardly reproducible, as the molecular weight, functional groups, and purity of polymers depends on the source and batch. More knowledge on the mechanism of interaction between glyconanoparticles and targeted macrophages will certainly allow to optimize these parameters and obtain a product for further translation. In fact, the potential of the carbohydrate-functionalized nanoparticles is highlighted by the increasing number of patents found on the World Intellectual Property Organization and recently discussed by Patil and Deshpande [73].

Table 3. A resume of the advantages and limitations of mannose receptor-targeting polymeric nanocarriers.

Advantages	Limitations
Surface chemistry can be controlled to reduce impact in the nanoparticles toxicity, immunogenicity, and biodistribution	Production of heterogeneous populations
Improved pharmacokinetics/pharmacodynamics profile	Nanoparticles stability during storage, in contact with blood and tissues
Effective internalization in targeted cells	Scale up and time of production, particularly for functionalized nanoparticles
Site-specific delivery with reduced side-effects	
High binding affinity for targeted cells	

7. Conclusions

Carbohydrates play a fundamental role in many aspects of receptor-mediated delivery and therapies. The insertion of carbohydrates in biodegradable polymeric nanoparticles enhances their biocompatibility and favors their use for biomedical applications. In this review, we focused on the preparation methods and use of carbohydrate-functionalized polymeric nanoparticles for macrophage targeting. The sugar moieties present in these nanocarriers are able of specifically interacting with receptors at the surface of macrophage cells and trigger immune responses. The study of this interaction makes the development of new macrophage-mediated therapies possible, with the mannose receptor binding being the most exploited, due to its abundant expression in dendritic cells and increased internalization. Mannose-targeting nanocarriers have shown to be effective in increasing the production of pro-inflammatory cytokines, in infection resolution, modulate tumor-associated macrophages' polarization, and improving nutraceuticals oral administration.

Author Contributions: Conceptualization, S.A.C.L.; resources, S.A.C.L., B.C., and S.R.; writing—original draft preparation, R.G.D.A., B.R., and S.A.C.L.; writing—review and editing, S.A.C.L., B.C., and S.R.; supervision, S.A.C.L. and B.C.; project administration, S.R. funding acquisition, S.A.C.L. All authors have read and agreed to the published version of the manuscript.

Funding: This research was partially supported by PT national funds provided by FCT–Foundation for Science and Technology through COMPETE POCI-01-0145-FEDER-030834 and National Funds (FCT) through project PTDC/QUI-COL/30834/2017.

Institutional Review Board Statement: Not applicable.

Informed Consent Statement: Not applicable.

Acknowledgments: The authors acknowledge the support obtained within the scope of UIDB/04423/20 and UIDP/04423/2020. SCL, BR, and BC are grateful for the funding from FCT/MEC (CEECIND/01620/ PD/BDE/129262/2017, and IF/00197/2015, respectively) financed by national funds. To all financing sources, the authors are greatly indebted.

Conflicts of Interest: The authors declare no conflict of interest.

References

1. Steichen, D.S.; Caldorera-Moore, M.; Peppas, N.A. A review of current nanoparticle and targeting moieties for the delivery of cancer therapeutics. *Eur. J. Pharm. Sci.* **2013**, *48*, 416–427. [CrossRef] [PubMed]
2. Davies, L.C.; Jenkins, S.J.; Allen, J.E.; Taylor, P.R. Tissue-resident macrophages. *Nat. Immunol.* **2013**, *14*, 986–995. [CrossRef] [PubMed]
3. Murray, P.J.; Allen, J.E.; Biswas, S.K.; Fisher, E.A.; Gilroy, D.W.; Goerdt, S.; Gordon, S.; Hamilton, J.A.; Ivashkiv, L.B.; Lawrence, T.; et al. Macrophage activation and polarization: Nomenclature and experimental guidelines. *Immunity* **2014**, *41*, 14–20. [CrossRef] [PubMed]
4. Mukhtar, M.; Ali, H.; Ahmed, N.; Munir, R.; Talib, S.; Khan, A.S.; Ambrus, R. Drug delivery to macrophages: A review of nano-therapeutics targeted approach for inflammatory disorders and cancer. *Expert Opin. Drug Deliv.* **2020**, *17*, 1239–1257. [CrossRef] [PubMed]
5. Elsabahy, M.; Wooley, K.L. Design of polymeric nanoparticles for biomedical delivery applications. *Chem. Soc. Rev.* **2012**, *41*, 2545–2561. [CrossRef]
6. Zanganeh, S.; Hutter, G.; Spitler, R.; Lenkov, O.; Mahmoudi, M.; Shaw, A.; Pajarinen, J.S.; Nejadnik, H.; Goodman, S.; Moseley, M.; et al. Iron oxide nanoparticles inhibit tumour growth by inducing pro-inflammatory macrophage polarization in tumour tissues. *Nat. Nanotechnol.* **2016**, *11*, 986–994. [CrossRef]
7. Chen, G.; Roy, I.; Yang, C.; Prasad, P. Nanochemistry and nanomedicine for nanoparticle-based diagnostics and therapy. *Chem. Rev.* **2016**, *116*, 2826–2885. [CrossRef]
8. Duncan, R.; Gaspar, R. Nanomedicine(s) under the microscope. *Mol. Pharm.* **2011**, *8*, 2101–2141. [CrossRef]
9. Khan, I.; Saeed, K.; Khan, I. Nanoparticles: Properties, applications and toxicities. *Arab. J. Chem.* **2019**, *12*, 908–931. [CrossRef]
10. Soppimath, K.S.; Aminabhavi, T.M.; Kulkarni, A.R.; Rudzinski, W.E. Biodegradable polymeric nanoparticles as drug delivery devices. *J. Control. Release* **2001**, *70*, 1–20. [CrossRef]
11. Kulkarni, A.; Rao, P. Synthesis of polymeric nanomaterials for biomedical applications. *Nanomater. Tissue Eng.* **2013**, 27–63.
12. Nagavarma, B.V.N.; Yadav, H.; Ayaz, A.; Vasudha, L.S.; Shivakumar, H.G. Different techniques for preparation of polymeric nanoparticles—A review. *Asian J. Pharm. Clin. Res.* **2012**, *5*, 16–23.
13. Vauthier, C.; Bouchemal, K. Methods for the preparation and manufacture of polymeric nanoparticles. *Pharm. Res.* **2009**, *26*, 1025–1058. [CrossRef] [PubMed]
14. Samrot, A.; Burman, U.; Philip, S.N.S.; Chandrasekaran, K. Synthesis of curcumin loaded polymeric nanoparticles from crab shell derived chitosan for drug delivery. *Inform. Med. Unlocked* **2018**, *10*, 159–182. [CrossRef]
15. Gheorghita Puscaselu, R.; Lobiuc, A.; Dimian, M.; Covasa, M. Alginate: From food industry to biomedical applications and management of metabolic disorders. *Polymers* **2020**, *12*, 2417. [CrossRef]
16. Zhao, X.; Lang, Q.; Yildirimer, L.; Lin, Z.; Cui, W.; Annabi, N.; Ng, K.; Dokmeci, M.; Ghaemmaghami, A.; Khademhosseini, A. Photocrosslinkable gelatin hydrogel for epidermal tissue engineering. *Adv. Healthc. Mater.* **2015**, *5*, 108–118. [CrossRef]
17. Lamprecht, A. *Nanotherapeutics: Drug Delivery Concepts in Nanoscience*; Pan Stanford Publishing: Singapore, 2009; p. 279, chapter xii.
18. Panyam, J.; Labhasetwar, V. Biodegradable nanoparticles for drug and gene delivery to cells and tissue. *Adv. Drug Deliv. Rev.* **2003**, *55*, 329–347. [CrossRef]
19. Heinz, H.; Pramanik, C.; Heinz, O.; Ding, Y.; Mishra, R.K.; Marchon, D.; Flatt, R.J.; Estrela-Lopis, I.; Llop, J.; Moya, S.; et al. Nanoparticle decoration with surfactants: Molecular interactions, assembly, and applications. *Surf. Sci. Rep.* **2017**, *72*, 1–58. [CrossRef]
20. Qiu, L.Y.; Bae, Y.H. Polymer architecture and drug delivery. *Pharm. Res.* **2006**, *23*, 1–30. [CrossRef]
21. Edlund, U.; Albertsson, A.C. Polyesters based on diacid monomers. *Adv. Drug Deliv. Rev.* **2003**, *55*, 585–609. [CrossRef]
22. Han, F.Y.; Thurecht, K.J.; Whittaker, A.K.; Smith, M.T. Biodegradable PLGA-based microparticles for producing sustained-release drug formulations and strategies for improving drug loading. *Front. Pharmacol.* **2016**, *7*, 185. [CrossRef] [PubMed]
23. Tyler, B.; Gullotti, D.; Mangraviti, A.; Utsuki, T.; Brem, H. Polylactic acid (PLA) controlled delivery carriers for biomedical applications. *Adv. Drug Deliver. Rev.* **2016**, *107*, 163–175. [CrossRef] [PubMed]
24. Vasile, C. *Polymeric Nanomaterials in Nanotherapeutic*, 1st ed.; Elsevier: San Diego, CA, USA, 2018.
25. Kumari, A.; Yadav, S.K.; Yadav, S.C. Biodegradable polymeric nanoparticles-based drug delivery systems. *Colloids Surf. B Biointerfaces* **2010**, *75*, 1–18. [CrossRef] [PubMed]
26. Chen, G.; Wang, Y.; Xie, R.; Gong, S. A review on core-shell structured unimolecular nanoparticles for biomedical applications. *Adv. Drug Deliv. Rev.* **2018**, *130*, 58–72. [CrossRef] [PubMed]
27. Wang, K.; Wen, H.-F.; Yu, D.-G.; Yang, Y.; Zhang, D.-F. Electrosprayed hydrophilic nanocomposites coated with shellac for colon-specific delayed drug delivery. *Mater. Des.* **2018**, *143*, 248–255. [CrossRef]

28. Huang, W.D.; Xu, X.; Wang, H.L.; Huang, J.X.; Zuo, X.H.; Lu, X.J.; Liu, X.L.; Yu, D.G. Electrosprayed ultra-thin coating of ethyl cellulose on drug nanoparticles for improved sustained release. *Nanomaterials* **2020**, *10*, 1758. [CrossRef]
29. Khan, I.; Gothwal, A.; Sharma, A.K.; Kesharwani, P.; Gupta, L.; Iyer, A.K.; Gupta, U. PLGA nanoparticles and their versatile role in anticancer drug delivery. *Crit. Rev. Ther. Drug Carrier Syst.* **2016**, *33*, 159–193. [CrossRef]
30. Storm, G.; Belliot, S.O.; Daemen, T.; Lasic, D.D. Surface modification of nanoparticles to oppose uptake by the mononuclear phagocyte system. *Adv. Drug Deliv. Rev.* **1995**, *17*, 31–48. [CrossRef]
31. Hans, M.L.; Lowman, A.L. Biodegradable nanoparticles for drug delivery and targeting. *Curr. Opin. Solid State Mater. Sci.* **2002**, *6*, 319–327. [CrossRef]
32. Alexis, F.; Pridgen, E.; Molnar, L.K.; Farokhzad, O.C. Factors affecting the clearance and biodistribution of polymeric nanoparticles. *Mol. Pharm.* **2008**, *5*, 505–515. [CrossRef]
33. Torchilin, V.P.; Trubetskoy, V.S. Which polymers can make nanoparticulate drug carriers long-circulating? *Adv. Drug Deliv. Rev.* **1995**, *16*, 141–155. [CrossRef]
34. Hoang Thi, T.T.; Pilkington, E.H.; Nguyen, D.H.; Lee, J.S.; Park, K.D.; Truong, N.P. The importance of poly(ethylene glycol) alternatives for overcoming PEG immunogenicity in drug delivery and bioconjugation. *Polymers* **2020**, *12*, 298. [CrossRef] [PubMed]
35. Owens, D.E.; Peppas, N.A. Opsonization, biodistribution, and pharmacokinetics of polymeric nanoparticles. *Int. J. Pharm.* **2006**, *307*, 93–102. [CrossRef] [PubMed]
36. Gref, R.; Domb, A.; Quellec, P.; Blunk, T.; Muller, R.H.; Verbavatz, J.M.; Langer, R. The controlled intravenous delivery of drugs using PEG-coated sterically stabilized nanospheres. *Adv. Drug Deliv. Rev.* **1995**, *16*, 215–233. [CrossRef]
37. Tobio, M.; Sanchez, A.; Vila, A.; Soriano, I.I.; Evora, C.; Vila-Jato, J.L.; Alonso, M.J. The role of PEG on the stability in digestive fluids and in vivo fate of PEG-PLA nanoparticles following oral administration. *Colloids Surf. B Biointerfaces* **2000**, *18*, 315–323. [CrossRef]
38. Calvo, P.; Gouritin, B.; Brigger, I.; Lasmezas, C.; Deslys, J.P.; Williams, A.; Andreux, J.P.; Dormont, D.; Couvreur, P. PEGylated polycyanoacrylate nanoparticles as vector for drug delivery in prion diseases. *J. Neurosci. Methods* **2001**, *111*, 151–155. [CrossRef]
39. Avgoustakis, K.; Beletsi, A.; Panagi, Z.; Klepetsanis, P.; Karydas, A.G.; Ithakissios, D.S. PLGA-mPEG nanoparticles of cisplatin: In vitro nanoparticle degradation, in vitro drug release and in vivo drug residence in blood properties. *J. Control. Release* **2002**, *79*, 123–135. [CrossRef]
40. Srinivasan, M.; Rajabi, M.; Mousa, S.A. Multifunctional nanomaterials and their applications in drug delivery and cancer therapy. *Nanomaterials* **2015**, *5*, 1690–1703. [CrossRef]
41. Ang, C.; Tan, S.; Zhao, Y. Recent advances in biocompatible nanocarriers for delivery of chemotherapeutic cargoes towards cancer therapy. *Org. Biomol. Chem.* **2014**, *12*, 4776–4806. [CrossRef]
42. Karra, N.; Nassar, T.; Ripin, A.; Schwob, O.; Borlak, J.; Benita, S. Antibody conjugated PLGA nanoparticles for targeted delivery of paclitaxel palmitate: Efficacy and biofate in a lung cancer mouse model. *Small* **2013**, *9*, 4221–4236. [CrossRef]
43. Coester, C.; Kreuter, J.; von Briesen, H.; Langer, K. Preparation of avidin-labelled gelatin nanoparticles as carriers for biotinylated peptide nucleic acid (PNA). *Int. J. Pharm.* **2000**, *196*, 147–149. [CrossRef]
44. Hua, S. Advances in oral drug delivery for regional targeting in the gastrointestinal tract—Influence of physiological, pathophysiological and pharmaceutical factors. *Front Pharmacol.* **2020**, *1*, 524. [CrossRef] [PubMed]
45. Yilmaz, G.; Becer, C. Glyconanoparticles and their interactions with lectins. *Polym. Chem.* **2015**, *6*, 5503–5514. [CrossRef]
46. Eddie Ip, W.K.; Takahashi, K.; Alan Ezekowitz, R.; Stuart, L.M. Mannose-binding lectin and innate immunity. *Immunol. Rev.* **2009**, *230*, 9–21. [CrossRef] [PubMed]
47. Cade, D.; Ramus, E.; Rinaudo, M.; Auzély-Velty, R.; Delair, T.; Hamaide, T. Tailoring of bioresorbable polymers for elaboration of sugar-functionalized nanoparticles. *Biomacromolecules* **2004**, *5*, 922–927. [CrossRef] [PubMed]
48. Vela-Ramirez, J.; Goodman, J.; Boggiatto, P.; Roychoudhury, R.; Pohl, N.; Hostetter, J.; Wannemuehler, M.; Narasimhan, B. Safety and biocompatibility of carbohydrate-functionalized polyanhydride nanoparticles. *AAPS J.* **2014**, *17*, 256–267. [CrossRef]
49. Lemarchand, C.; Gref, R.; Passirani, C.; Garcion, E.; Petri, B.; Müller, R.; Costantini, D.; Couvreur, P. Influence of polysaccharide coating on the interactions of nanoparticles with biological systems. *Biomaterials* **2006**, *2*, 108–118. [CrossRef]
50. Morille, M.; Passirani, C.; Letrou-Bonneval, E.; Benoit, J.; Pitard, B. Galactosylated DNA lipid nanocapsules for efficient hepatocyte targeting. *Int. J. Pharm.* **2009**, *379*, 293–300. [CrossRef]
51. Pramudya, I.; Chung, H. Recent progress of glycopolymer synthesis for biomedical applications. *Biomater. Sci.* **2019**, *7*, 4848–4872. [CrossRef]
52. Babiuch, K.; Stenzel, M.H. *Synthesis and Application of Glycopolymers*; John Wiley & Sons, Inc.: Hoboken, NJ, USA, 2014.
53. Zhang, Y.; Chan, J.; Moretti, A.; Uhrich, K. Designing polymers with sugar-based advantages for bioactive delivery applications. *J. Control. Release* **2015**, *219*, 355–368. [CrossRef]
54. Ladmiral, V.; Melia, E.; Haddleton, D. Synthetic glycopolymers: An overview. *Eur. Polym. J.* **2004**, *40*, 431–449. [CrossRef]
55. Miura, Y. Design and synthesis of well-defined glycopolymers for the control of biological functionalities. *Polym. J.* **2012**, *44*, 679–689. [CrossRef]
56. Boehnke, N.; Dolph, K.; Juarez, V.; Lanoha, J.; Hammond, P. Electrostatic conjugation of nanoparticle surfaces with functional peptide motifs. *Bioconjugate Chem.* **2020**, *31*, 2211–2219. [CrossRef]

57. Sidorov, I.; Prabakaran, P.; Dimitrov, D. Non-covalent conjugation of nanoparticles to antibodies via electrostatic interactions—A computational model. *J. Comp. Theor. Nanosci.* **2007**, *46*, 1103–1107. [CrossRef]
58. Hermanson, G. *Bioconjugate Techniques*, 3rd ed.; Amsterdam University Press: Amsterdam, The Netherlands, 2013.
59. Ulbrich, K.; Holá, K.; Šubr, V.; Bakandritsos, A.; Tuček, J.; Zbořil, R. Targeted drug delivery with polymers and magnetic nanoparticles: Covalent and noncovalent approaches, release control, and clinical studies. *Chem. Rev.* **2016**, *116*, 5338–5431. [CrossRef]
60. Kim, I.-S.; Kim, S.-H.; Cho, C.-S. Preparation of polymeric nanoparticles composed of cholic acid and poly(ethylene glycol) end-capped with a sugar moiety. *Macromol. Rapid Commun.* **2000**, *21*, 1272–1275. [CrossRef]
61. Palmioli, A.; La Ferla, B. Glycofunctionalization of poly(lactic-co-glycolic acid) polymers: Building blocks for the generation of defined sugar-coated nanoparticles. *Org. Lett.* **2018**, *20*, 3509–3512. [CrossRef]
62. Crucho, C.I.C.; Barros, M.T. Formulation of functionalized PLGA polymeric nanoparticles for targeted drug delivery. *Polym. Chem.* **2015**, *68*, 41–46. [CrossRef]
63. Rieger, J.; Freichels, H.; Imberty, A.; Putaux, J.-L.; Delair, T.; Jérôme, C.; Auzély-Velty, R. Polyester nanoparticles presenting mannose residues: Toward the development of new vaccine delivery systems combining biodegradability and targeting properties. *Biomacromolecules* **2009**, *10*, 651–657. [CrossRef] [PubMed]
64. Freichels, H.; Wagner, M.; Okwieka, P.; Meyer, R.G.; Mailänder, V.; Landfester, K.; Musyanovych, A. (Oligo)mannose functionalized hydroxyethyl starch nanocapsules: One route to drug delivery systems with targeting properties. *J. Mater. Chem. B* **2013**, *1*, 4338–4348. [CrossRef]
65. Kim, N.; Jiang, D.; Jacobi, A.; Lennox, K.; Rose, S.; Behlke, M.; Salem, A. Synthesis and characterization of mannosylated pegylated polyethylenimine as a carrier for siRNA. *Int. J. Pharm.* **2012**, *427*, 123–133. [CrossRef] [PubMed]
66. Barros, D.; Costa Lima, S.A.; Cordeiro-da-Silva, A. Surface functionalization of polymeric nanospheres modulates macrophage activation. *Nanomedicine* **2015**, *10*, 387–403. [CrossRef] [PubMed]
67. Ganbold, T.; Baigude, H. Design of mannose-functionalized curdlan nanoparticles for macrophage-targeted siRNA delivery. *ACS App. Mater. Interfaces* **2018**, *10*, 14463–14474. [CrossRef] [PubMed]
68. Haniffa, M.; Bigley, V.; Collin, M. Human mononuclear phagocyte system reunited. *Semin. Cell Dev. Biol.* **2015**, *41*, 59–69. [CrossRef]
69. Epelman, S.; Lavine, K.J.; Randolph, G.J. Origin and functions of tissue macrophages. *Immunity* **2014**, *41*, 21–35. [CrossRef]
70. Arango Duque, G.; Descoteaux, A. Macrophage cytokines: Involvement in immunity and infectious diseases. *Front. Immunol.* **2014**, *5*, 491. [CrossRef]
71. Martinez, F.O.; Gordon, S. The M1 and M2 paradigm of macrophage activation: Time for reassessment. *F1000 Prime Rep.* **2014**, *6*, 13. [CrossRef]
72. Gordon, S. Alternative activation of macrophages. *Nat. Rev. Immunol.* **2003**, *3*, 23–35. [CrossRef]
73. Patil, T.S.; Dehpande, A.S. Mannosylated nanocarriers mediated site-specific drug delivery for the treatment of cancer and other infectious diseases: A state of the art review. *J. Control. Release* **2020**, *320*, 239–252. [CrossRef]
74. Mosaiab, T.; Farr, D.C.; Kiefel, M.J.; Houston, T.A. Carbohydrate-based nanocarriers and their application to target macrophages and deliver antimicrobial agents. *Adv. Drug Deliv. Rev.* **2019**, *152*, 94–129. [CrossRef]
75. Sun, B.; Wang, X.; Ji, Z.; Li, R.; Xia, T. NLRP3 inflammasome activation induced by engineered nanomaterials. *Small* **2013**, *9*, 1595–1607. [CrossRef] [PubMed]
76. Hu, G.; Guo, M.; Xu, J.; Wu, F.; Fan, J.; Huang, Q.; Yang, G.; Lv, Z.; Wang, X.; Jin, Y. Nanoparticles targeting macrophages as potential clinical therapeutic agents against cancer and inflammation. *Front. Immunol.* **2019**, *10*, 1998. [CrossRef] [PubMed]
77. Sahay, G.; Alakhova, D.Y.; Kabanov, A.V. Endocytosis of nanomedicines. *J. Control. Release* **2010**, *145*, 182–195. [CrossRef]
78. Carrillo-Conde, B.; Song, E.H.; Chavez-Santoscoy, A.; Phanse, Y.; Ramer-Tait, A.E.; Pohl, N.L.; Wannemuehler, M.J.; Bellaire, B.H.; Narasimhan, B. Mannose-functionalized "pathogen-like" polyanhydride nanoparticles target C-type lectin receptors on dendritic cells. *Mol. Pharm.* **2011**, *8*, 1877–1886. [CrossRef]
79. Hamdy, S.; Haddadi, A.; Shayeganpour, A.; Samuel, J.; Lavasanifar, A. Activation of antigen-specific T cell-responses by mannan-decorated PLGA nanoparticles. *Pharm. Res.* **2011**, *28*, 2288–2301. [CrossRef]
80. Shrivastavaa, R.; Shukla, N. Attributes of alternatively activated (M2) macrophages. *Life Sci.* **2019**, *224*, 222–231. [CrossRef]
81. Ezekowitz, R.A.B.; Sastry, K.; Bailly, P.; Warner, A. Molecular characterization of the Human macrophage mannose receptor—Demonstration of multiple carbohydrate recognition-like domains and phagocytosis of teasts in cos-1 cells. *J. Exp. Med.* **1990**, *172*, 1785–1794. [CrossRef]
82. East, L.; Isacke, C.M. The mannose receptor family. *BBA Gen. Subj.* **2002**, *1572*, 364–386. [CrossRef]
83. Hu, J.; Wei, P.; Seeberger, P.; Yin, Y. Mannose-functionalized nanoscaffolds for targeted delivery in biomedical applications. *Chem. Asian J.* **2018**, *13*, 3448–3459. [CrossRef]
84. McKenzie, E.J.; Taylor, P.R.; Stillion, R.J.; Lucas, A.D.; Harris, J.; Gordon, S.; Martinez-Pomares, L. Mannose receptor expression and function define a new population of murine dendritic cells. *J. Immunol.* **2007**, *178*, 4975–4983. [CrossRef]
85. Linehan, S.A.; Martinez-Pomares, L.; Stahl, P.D.; Gordon, S. Mannose receptor and its putative ligands in normal murine lymphoid and nonlymphoid organs: In situ expression of mannose receptor by selected macrophages, endothelial cells, perivascular microglia, and mesangial cells, but not dendritic cells. *J. Exp. Med.* **1999**, *189*, 1961–1972. [CrossRef] [PubMed]

86. Lew, D.B.; Songumize, E.; Pontow, S.E.; Stahl, P.D.; Rattazzi, M.C. A mannose receptor mediates mannosyl-rich glycoprotein-induced mitogenesis in bovine airway smooth-muscle cells. *J. Clin. Invest.* **1994**, *94*, 1855–1863. [CrossRef] [PubMed]
87. Stahl, P.D.; Ezekowitz, R.A. The mannose receptor is a pattern recognition receptor involved in host defense. *Curr. Opin. Immunol.* **1998**, *10*, 50–55. [CrossRef]
88. Schreiber, S.; Blum, J.S.; Stenson, W.F.; MacDermott, R.P.; Stahl, P.D.; Teitelbaum, S.L.; Perkins, S.L. Monomeric IgG2a promotes maturation of bone-marrow macrophages and expression of the mannose receptor. *Proc. Natl. Acad. Sci. USA* **1991**, *88*, 1616–1620. [CrossRef]
89. Stein, M.; Keshav, S.; Harris, N.; Gordon, S. Interleukin 4 potently enhances murine macrophage mannose receptor activity: A marker of alternative immunologic macrophage activation. *J. Exp. Med.* **1992**, *176*, 287–292. [CrossRef]
90. Doyle, A.G.; Herbein, G.; Montaner, L.J.; Minty, A.J.; Caput, D.; Ferrara, P.; Gordon, S. Interleukin-13 alters the activation state of murine macrophages in vitro—Comparison with interleukin-4 and interferon-gamma. *Eur. J. Immunol.* **1994**, *24*, 1441–1445. [CrossRef]
91. Martinez-Pomares, L.; Reid, D.M.; Brown, G.D.; Taylor, P.R.; Stillion, R.J.; Linehan, S.A.; Zamze, S.; Gordon, S.; Wong, S.Y. Analysis of mannose receptor regulation by IL-4, IL-10, and proteolytic processing using novel monoclonal antibodies. *J. Leukoc. Biol.* **2003**, *73*, 604–613. [CrossRef]
92. Harris, N.; Super, M.; Rits, M.; Chang, G.; Ezekowitz, R.A. Characterization of the murine macrophage mannose receptor: Demonstration that the downregulation of receptor expression mediated by interferon-gamma occurs at the level of transcription. *Blood* **1992**, *80*, 2363–2373. [CrossRef]
93. Shibata, Y.; Metzger, W.J.; Myrvik, Q.N. Chitin particle-induced cell-mediated immunity is inhibited by soluble mannan: Mannose receptor-mediated phagocytosis initiates IL-12 production. *J. Immunol.* **1997**, *159*, 2462–2467.
94. Engering, A.J.; Cella, M.; Fluitsma, D.; Brockhaus, M.; Hoefsmit, E.C.; Lanzavecchia, A.; Pieters, J. The mannose receptor functions as a high capacity and broad specificity antigen receptor in human dendritic cells. *Eur. J. Immunol.* **1997**, *27*, 2417–2425. [CrossRef]
95. Tan, M.C.; Mommaas, A.M.; Drijfhout, J.W.; Jordens, R.; Onderwater, J.J.; Verwoerd, D.; Mulder, A.A.; van der Heiden, A.N.; Scheidegger, D.; Oomen, L.C.; et al. Mannose receptor-mediated uptake of antigens strongly enhances HLA class II-restricted antigen presentation by cultured dendritic cells. *Eur. J. Immunol.* **1997**, *27*, 2426–2435. [CrossRef] [PubMed]
96. Chakraborty, P.; Das, P.K. Role of mannose N-acetylglucosamine receptors in blood clearance and cellular attachment of *Leishmania donovani*. *Mol. Biochem. Parasit.* **1988**, *28*, 55–62. [CrossRef]
97. Kang, P.B.; Azad, A.K.; Schlesinger, L.S. The human macrophage mannose receptor directs *Mycobacterium tuberculosis* lipoarabinomannan-mediated phagosome biogenesis. *J. Exp. Med.* **2005**, *202*, 987–999. [CrossRef] [PubMed]
98. Irache, J.M.; Salman, H.H.; Gamazo, C.; Espuelas, S. Mannose-targeted systems for the delivery of therapeutics. *Expert Opin Drug Deliv.* **2008**, *5*, 703–724. [CrossRef]
99. Song, E.-H.; Manganiello, M.J.; Chow, J.-H.; Ghosn, B.; Convertine, A.J.; Stayton, P.S.; Schnapp, L.M.; Ratner, D.M. In vivo targeting of alveolar macrophages via RAFT-based glycopolymers. *Biomaterials* **2012**, *33*, 6889–6897. [CrossRef]
100. Nycholat, C.M.; Rademacher, C.; Kawasaki, N.; Paulson, J.C. In silico-aided design of a glycan ligand of sialo-adhesin for in vivo targeting of macrophages. *J. Am. Chem. Soc.* **2012**, *134*, 15696–15699. [CrossRef]
101. Adler, A.F.; Leong, K.W. Emerging links between surface nanotechnology and endocytosis: Impact on nonviral gene delivery. *Nano Today* **2010**, *5*, 553–569. [CrossRef]
102. Kaur, I.P.; Singh, H. Nanostructured drug delivery for better management of tuberculosis. *J. Control. Release* **2014**, *184*, 36–50. [CrossRef]
103. Lu, E.; Franzblau, S.; Onyuksel, H.; Popescu, C. Preparation of aminoglycoside-loaded chitosan nanoparticles using dextran sulphate as a counterion. *J. Microencapsul.* **2009**, *26*, 346–354. [CrossRef]
104. Nimje, N.; Agarwal, A.; Saraogi, G.K.; Lariya, N.; Rai, G.; Agrawal, H.; Agrawal, G.P. Mannosylated nanoparticulate carriers of rifabutin for alveolar targeting. *J. Drug Target.* **2009**, *17*, 777–787. [CrossRef]
105. Kumar, P.V.; Asthana, A.; Dutta, T.; Jain, N.K. Intracellular macrophage uptake of rifampicin loaded mannosylated dendrimers. *J. Drug Target.* **2006**, *14*, 546–556. [CrossRef] [PubMed]
106. Moretton, M.A.; Chiappetta, D.A.; Andrade, F.; das Neves, J.; Ferreira, D.; Sarmento, B.; Sosnik, A. Hydrolyzed galactomannan-modified nanoparticles and flower-like polymeric micelles for the active targeting of rifampicin to macrophages. *J. Biomed. Nanotechnol.* **2013**, *9*, 1076–1087. [CrossRef] [PubMed]
107. Wagner, V.; Minguez-Menendez, A.; Pena, J.; Fernández-Prada, C. Innovative solutions for the control of leishmaniases: Nanoscale drug delivery systems. *Curr. Pharm. Des.* **2019**, *25*, 1582–1592. [CrossRef] [PubMed]
108. Chavez-Santoscoy, A.V.; Roychoudhury, R.; Pohl, N.L.; Wannemuehler, M.J.; Narasimhan, B.; Ramer-Tait, A.E. Tailoring the immune response by targeting C-type lectin receptors on alveolar macrophages using "pathogen-like" amphiphilic polyanhydride nanoparticles. *Biomaterials* **2012**, *33*, 4762–4772. [CrossRef]
109. Chaubey, P.; Patel, R.R.; Mishra, B. Development and optimization of curcumin-loaded mannosylated chitosan nanoparticles using response surface methodology in the treatment of visceral leishmaniasis. *Expert Opin. Drug Deliv.* **2014**, 1–19. [CrossRef] [PubMed]
110. Jain, S.K.; Gupta, Y.; Jain, A.; Saxena, A.R.; Khare, P.; Jain, A. Mannosylated gelatin nanoparticles bearing an anti-HIV drug didanosine for site-specific delivery. *Nanomedicine* **2008**, 41–48. [CrossRef]

111. Li, Q.; Du, Y.-Z.; Yuan, H.; Zhang, X.-G.; Miao, J.; Cui, F.-D.; Hu, F.-Q. Synthesis of Lamivudine stearate and antiviral activity of stearic acid-g-chitosan oligosaccharide polymeric micelles delivery system. *Eur. J. Pharm. Sci.* **2010**, *41*, 498–507. [CrossRef] [PubMed]
112. Gajbhiye, V.; Ganesh, N.; Barve, J.; Jain, N.K. Synthesis, characterization and targeting potential of zidovudine loaded sialic acid conjugated-mannosylated poly(propyleneimine) dendrimers. *Eur. J. Pharm. Sci.* **2013**, *48*, 668–679. [CrossRef]
113. Mantovani, A.; Allavena, P. The interaction of anticancer therapies with tumor-associated macrophages. *J. Exp. Med.* **2015**, *212*, 435–445. [CrossRef]
114. Rodell, C.B.; Arlauckas, S.P.; Cuccarese, M.F.; Garris, C.S.; Li, R.; Ahmed, M.S.; Kohler, R.H.; Pittet, M.J.; Weissleder, R. TLR7/8-agonist-loaded nanoparticles promote the polarization of tumour-associated macrophages to enhance cancer immunotherapy. *Nat. Biomed. Eng.* **2018**, *2*, 578–588. [CrossRef]
115. Ryan, R.A.; Ortega, A.; Whitney, J.B.; Kumar, B.; Tikhomirov, O.; McFadden, I.D.; Yull, F.E.; Giorgio, T.D. Biocompatible mannosylated endosomal-escape nanoparticles enhance selective delivery of short nucleotide sequences to tumor associated macrophages. *Nanoscale* **2014**, *7*, 500–510.
116. Zhang, M.; Yan, L.; Kim, J.A. Modulating mammary tumor growth, metastasis and immunosuppression by siRNA-induced MIF reduction in tumor microenvironment. *Cancer Gene Ther.* **2015**, *22*, 463–474. [CrossRef] [PubMed]
117. Kim, T.H.; Nah, J.W.; Cho, M.H.; Park, T.G.; Cho, C.S. Receptor-mediated gene delivery into antigen presenting cells using mannosylated chitosan/DNA nanoparticles. *J. Nanosci. Nanotechnol.* **2006**, *6*, 2796–2803. [CrossRef]
118. Xia, W.; Hilgenbrink, A.R.; Matteson, E.L.; Lockwood, M.B.; Cheng, J.X.; Low, P.S. A functional folate receptor is induced during macrophage activation and can be used to target drugs to activated macrophages. *Blood* **2009**, *113*, 438–446. [CrossRef]
119. Hu, Q.; Bae, M.; Fleming, E.; Lee, J.-Y.; Luo, Y. Biocompatible polymeric nanoparticles with exceptional gastrointestinal stability as oral delivery vehicles for lipophilic bioactives. *Food Hydrocoll.* **2019**, *89*, 386–395. [CrossRef]
120. Thakur, A.; Foged, C. Nanoparticles for mucosal vaccine delivery. *Nanoeng. Biomater. Adv. Drug Deliv.* **2020**, *603*, 603–646.
121. Li, J.; Cai, C.; Li, J.; Li, J.; Li, J.; Sun, T.; Wang, I.; Wu, H.; Yu, G. Chitosan-based nanomaterials for drug delivery. *Molecules* **2018**, *23*, 2661. [CrossRef]
122. Tapia-Hernández, J.A.; Rodríguez-Felix, F.; Juárez-Onofre, J.E.; Ruiz-Cruz, S.; Robles-García, M.A.; Borboa-Flores, J.; Wong-Corral, F.J.; Cinco-Moroyoqui, F.J.; Castro-Enríquez, D.D.; Del-Toro-Sánchez, C.L. Zein-polysaccharide nanoparticles as matrices for antioxidant compounds: A strategy for prevention of chronic degenerative diseases. *Food Res. Int.* **2018**, *111*, 451–471. [CrossRef]
123. Jiménez-Fernández, E.; Ruyra, A.; Roher, N.; Zuasti, E.; Infante, C.; Fernández-Díaz, C. Nanoparticles as a novel delivery system for vitamin C administration in aquaculture. *Aquaculture* **2014**, *432*, 426–433. [CrossRef]
124. Li, Y.; Yang, B.; Zhang, X. Oral delivery of imatinib through galactosylated polymeric nanoparticles to explore the contribution of a saccharide ligand to absorption. *Int. J. Pharm.* **2019**, *568*, 118508. [CrossRef]
125. Nair, A.B.; Sreeharsha, N.; Al-Dhubiab, B.E.; Hiremath, J.G.; Shinu, P.; Attimarad, M.; Venugopala, K.N.; Mutahar, M. HPMC- and PLGA-based nanoparticles for the mucoadhesive delivery of Sitagliptin: Optimization and in vivo evaluation in rats. *Materials* **2019**, *12*, 4239. [CrossRef] [PubMed]
126. Wu, Y.; Rashidpour, A.; Almajano, M.P.; Metón, I. Chitosan-based drug delivery system: Applications in fish biotechnology. *Polymers* **2020**, *12*, 1177. [CrossRef]
127. Akhtar, B.; Muhammad, F.; Aslam, B.; Saleemi, M.K.; Sharif, A. Pharmacokinetic profile of chitosan modified poly lactic co-glycolic acid biodegradable nanoparticles following oral delivery of gentamicin in rabbits. *Int. J. Biol. Macromol.* **2020**, *164*, 1493–1500. [CrossRef] [PubMed]
128. de Sousa, R.V.; da Cunha Santo, A.M.; de Sousa, V.B.; de Araújo Neves, G.; Navarro de Lima Santana, L.; Rpdrigues Menezes, R. A review on chitosan's uses as biomaterial: Tissue engineering, drug delivery systems and cancer treatment. *Materials* **2020**, *13*, 4995. [CrossRef]
129. Makadia, H.K.; Siegel, S.J. Polylactic-co-glycolic acid (PLGA) as biodegradable controlled drug delivery carrier. *Polymers* **2011**, *3*, 1377–1397. [CrossRef] [PubMed]
130. Song, S.K.; Beck, B.R.; Kim, D.; Park, J.; Kim, J.; Kim, H.D.; Ringø, E. Prebiotics as immunostimulants in aquaculture: A review. *Fish Shellfish Immunol.* **2014**, *40*, 40–48. [CrossRef] [PubMed]
131. Dawood, M.A.O.; Koshio, S.; Esteban, M.Á. Beneficial roles of feed additives as immunostimulants in aquaculture: A review. *Rev. Aquac.* **2018**, *10*, 950–974. [CrossRef]
132. Nawaz, A.; Javaid Bakhsh, A.; Irshad, S.; Hoseinifar, H.S.; Xiong, H. The functionality of prebiotics as immunostimulant: Evidences from trials on terrestrial and aquatic animals. *Fish Shellfish Immunol.* **2018**, *76*, 272–278. [CrossRef] [PubMed]
133. Castro, R.; Piazzon, M.C.; Zarra, I.; Leiro, J.; Noya, M.; Lamas, J. Stimulation of turbot phagocytes by Ulva rigida C. Agardh polysaccharides. *Aquaculture* **2006**, *254*, 9–20. [CrossRef]
134. Fernández-Díaz, C.; Coste, O.; Malta, E.-J. Polymer chitosan nanoparticles functionalized with Ulva ohnoi extracts boost in vitro ulvan immunostimulant effect in Solea senegalensis macrophages. *Algal Res.* **2017**, *26*, 135–142. [CrossRef]
135. Tian, J.; Yu, J. Poly(lactic-co-glycolic acid) nanoparticles as candidate DNA vaccine carrier for oral immunization of Japanese flounder (Paralichthys olivaceus) against lymphocystis disease virus. *Fish Shellfish Immunol.* **2011**, *30*, 109–117. [CrossRef] [PubMed]
136. Shilakari Asthana, G.; Asthana, A.; Kohli, D.V.; Vyas, S.P. Mannosylated chitosan nanoparticles for delivery of antisense oligonucleotides for macrophage targeting. *BioMed Res. Int.* **2014**, *2014*, 526391. [CrossRef] [PubMed]

Article

Encapsulation of the Natural Product Tyrosol in Carbohydrate Nanosystems and Study of Their Binding with ctDNA

Antonella Rozaria Nefeli Pontillo [1], Evangelia Konstanteli [1,2], Maria M. Bairaktari [1] and Anastasia Detsi [1,*]

1. Laboratory of Organic Chemistry, Department of Chemical Sciences, School of Chemical Engineering, National Technical University of Athens, 15780 Zografou, Greece; nefelipontillo@gmail.com (A.R.N.P.); e.kwnst@gmail.com (E.K.); maro.bairakt@gmail.com (M.M.B.)
2. Institute of Chemical Biology, National Hellenic Research Foundation, 48 Vassileos Constantinou Avenue, 11635 Athens, Greece

* Correspondence: adetsi@chemeng.ntua.gr

Abstract: Tyrosol, a natural product present in olive oil and white wine, possesses a wide range of bioactivity. The aim of this study was to optimize the preparation of nanosystems encapsulating tyrosol in carbohydrate matrices and the investigation of their ability to bind with DNA. The first encapsulation matrix of choice was chitosan using the ionic gelation method. The second matrix was β-cyclodextrin (βCD) using the kneading method. Coating of the tyrosol-βCD ICs with chitosan resulted in a third nanosystem with very interesting properties. Optimal preparation parameters of each nanosystem were obtained through two three-factor, three-level Box-Behnken experimental designs and statistical analysis of the results. Thereafter, the nanoparticles were evaluated for their physical and thermal characteristics using several techniques (DLS, NMR, FT-IR, DSC, TGA). The study was completed with the investigation of the impact of the encapsulation on the ability of tyrosol to bind to calf thymus DNA. The results revealed that tyrosol and all the studied systems bind to the minor groove of ctDNA. Tyrosol interacts with ctDNA via hydrogen bond formation, as predicted via molecular modeling studies and corroborated by the experiments. The tyrosol-chitosan nanosystem does not show any binding to ctDNA whereas the βCD inclusion complex shows analogous interaction with that of free tyrosol.

Keywords: tyrosol; nanoparticles; Design of Experiment (DoE); chitosan; β cyclodextrin; DNA binding

1. Introduction

Tyrosol (2-(4-Hydroxyphenyl)ethanol) is a biophenol that is found in olive oil, white wine, beer and vermouth (Figure 1) [1]. Even though tyrosol does not exhibit strong antioxidant activity, it contributes to the cellular defences due to intracellular accumulation [2,3]. Moreover, numerous studies affirm that tyrosol offers neuroprotective and cardioprotective effect and enhances the regulation of the human LDL levels [4,5]. However, its hydrophilic nature impedes its incorporation in lipid substrates and limits its absorption and bioavailability [6].

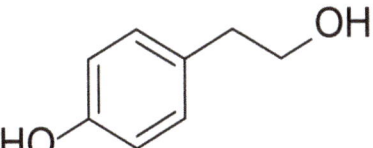

Figure 1. Structure of tyrosol.

Nanoencapsulation of bioactive compounds and pharmaceutical agents in suitable carriers is a very promising technology, as it offers protection and stabilisation of the encapsulated compound. Furthermore, the encapsulation of a compound may lead to a controllable and sustained release, thus enhancing its activity. Therefore, this technology is incorporated in a broad range of applications in different fields, such as in medicinal and pharmaceutical science, cosmetics, agrochemical and food industry [7–10].

β-cyclodextrin (βCD) is a truncated cone-shaped oligosaccharide, with a hydrophobic inner cavity and a hydrophilic outer surface [11]. Small, hydrophobic molecules can be entrapped in the cavity forming an inclusion complex (IC), increasing their solubility, while more hydrophilic compounds can be bound on the external surface [12–14].

Chitosan (CS) is a naturally occurring polymer widely used as a nanocarrier. It is nontoxic, biocompatible and biodegradable and is recognised as Generally Recognised as Safe (GRAS) by the Food and Drug Administration (FDA) [15,16]. The process of encapsulation in chitosan nanoparticles (NPs) has been extensively studied and various techniques have been reported. The nature of the polymer permits the encapsulation of small or larger molecules, natural products like plant extracts and essential oils, or even other nanosystems [17–19].

The properties of the particulate system are defined by the selected carrier and preparation process. Therefore, the ability to design and engineer the experimental process in order to obtain desirable results is an asset for any application. To that end, experimental design and statistical analysis are implemented. Box-Behnken design (BBD) is a Response Surface Methodology (RSM) that enables the multivariate optimisation of a quadratic model [20–22].

Intercalators and groove binders are a class of compounds that interact with the double-stranded DNA. Many anticancer drugs, such as anthracyclines, interact with the DNA through intercalation between adjacent base pairs perpendicularly to the axis of the helix. Many substituents in the intercalator molecule can greatly influence the binding mechanism, the geometry of the ligand–DNA complex and the selectivity of the sequence [23,24].

The interactions between the various cyclodextrins and DNA have yet to be completely identified; however, they are of utmost importance as there are many marketed formulations that contain cyclodextrins. Modified cationic cyclodextrins are known to interact with DNA for gene therapy applications while a strong interaction of a βCD complex was proved to be formed with DNA as the ribose and phosphate groups of the DNA exert a stabilizing effect by forming H-bonds with the outer surface of CD [25,26].

The aim of this study was to develop and optimize the encapsulation process of tyrosol in nanosystems using different matrices namely: chitosan (**TYR/CS**), βCD (**TYR-βCD**) as well as in the combined system of βCD/CS (**TYR-βCD/CS**). The kneading method was used for the preparation of the inclusion complex of tyrosol with βCD (**TYR-βCD**) and ionic gelation for the synthesis of the chitosan nanoparticles. The process optimisation was performed in both cases using a three-factor three-level BBD. The independent variables were set as the initial concentration of the polymer, the loading capacity of **TYR** or the **TYR-βCD** inclusion complex and the amount of the cross-linking agent. The examined range was elicited from literature data and data from preliminary experiments.

Complete characterisation of the systems was performed by various methods and techniques, such as Nuclear Magnetic Resonance Spectroscopy (NMR), Infrared Spectroscopy (FT-IR), antioxidant activity determination by the DPPH method, Dynamic Light Scattering (DLS) for the measurement of size, polydispersity index and ζ-potential, Thermogravimetric Analysis (TGA) and Scanning Electron Microscopy (SEM). Finally, the effect of the encapsulation matrix on the ability of tyrosol to interact with Calf-thymus DNA (ctDNA) was investigated.

2. Materials and Methods

2.1. Materials

Tyrosol was purchased from Fluorochem (Hadfield, Derbyshire, UK), β-Cyclodextrin in an assay ≥99% (HPLC) was obtained from Fluka (Buchs, Switzerland) and Chitosan (5–20 mPa·s, 0.5% in 0.5% Acetic Acid at 20 °C) from TCI (TCI (Shanghai, China). Tween 80,

Tris Base, rhodamine B and deoxyribonucleic acid sodium salt, calf thymus were purchased from Alfa Aesar (Ward Hill, MA, USA) and Sodium Tripolyphosphate from Acros Organics (Morris Plains, NJ, USA). For all the experiments double-deionised water was used.

2.2. Synthesis of Nanoparticles

2.2.1. Preparation of Inclusion Complexes of Tyrosol with βCD (**TYR-βCD**)

The kneading method was used for the preparation of the inclusion complex. Briefly, equimolar quantities of βCD (569 mg) and TYR (70 mg) were mixed in an agate mortar and pestle with minimum amount of the solvent (H_2O:EtOH 7:3 v/v) for 30 min. The paste was dried to constant weight in a high vacuum pump [27].

2.2.2. Encapsulation of Tyrosol in Chitosan Nanoparticles (**TYR/CS**)

In a 1% aqueous acetic acid solution, 0.05%, 0.2% or 0.35% CS was fully dissolved. Then, an equal to CS amount of emulsifier Tween 80 is added and left to stir at room temperature overnight until complete dissolution. In 5 mL of the occurring solution 2.5, 10 or 17.5 mg of tyrosol was added and the solution was left to stir until total dissolution. Then 1 mL of TPP solution of concentration 0.42, 1.67 or 0.35 mg/mL was added dropwise. The sample was left under magnetic stirring at 700 rpm, for 10 min at 25 °C. The nanoparticles were centrifuged at 30,000 rpm for 45 min. The sediment was dispersed and washed with two more consecutive centrifugations. Finally, the nanoparticles were freeze dried and stored in a desiccator.

2.2.3. Coating of the Tyrosol-βCD Inclusion Complexes with Chitosan (**TYR-βCD/CS**)

For the synthesis of **TYR-βCD/CS** nanoparticles, the procedure described in "Section 2.2.2" was followed, adding **TYR-βCD** instead of tyrosol. For preparation of blank chitosan nanoparticles, in 5 mL of 0.2% chitosan, 1 mL of TPP solution 1.67 mg/mL was added and the solution was left to stir for 10 min.

2.3. Design of Experiments (DoE)

Two experimental designs were conducted to optimize the processes for preparation of **TYR/CS** and **TYR-βCD/CS** nanoparticles. Design-Expert® in trial version (Version 12, Stat-Ease, Inc., Minneapolis, MN, USA) was used.

Three factors at three levels were selected to control the size (response R_1) and ζ-potential (response R_2) of the nanosystem. Factor A was the concentration of the chitosan solution, factor B was the mg of TPP and factor C the amount of tyrosol for the **TYR/CS** nanosystem or of the inclusion complex for the **TYR-βCD/CS** system.

Factors' levels and their normalised values are shown in Table 1. Central Point (0, 0, 0) was repeated three times, and a total of 15 runs were performed for each set.

Table 1. Factors, level and responses of the DoE studies.

	Factors	Levels			Responses		Constrains
		−1	0	+1			
A	CS (% w/v)	0.05	0.20	0.35	R_1	Size (nm)	Minimize
B	TPP (mg)	0.42	1.67	2.92			
C	(i) TYR (mg) (ii) TYR-βCD (mg)	2.5	10.0	17.5	R_2	ζ-potential (mV)	Maximize

The data obtained were analysed with Analysis of Variance (ANOVA). Linear, second-order and quadratic models were evaluated for all responses and in terms of statistical significance, R^2 values and the deviation of the predicted to the experimentally obtained results.

The confidence level was set at 95% and p-values ≤ 0.05 to determine statistically significant factors.

2.4. Characterisation of the Nanoparticles

2.4.1. Dynamic Light Scattering (DLS)

The measurements for size, polydispersity index (PDI) and ζ-potential were performed in a Zetasizer Nano ZS, Malvern Instruments Ltd. (Malvern, UK) using a cuvette DTS1070. The results were analysed with the Zetasizer 7.12, Malvern Instruments Ltd. Software.

For **TYR-βCD**, 1 mg of the dried sample was dissolved in 4 mL of water and was fully dispersed with 2 min ultrasound at a 2210 Ultrasonic Bath, Branson. DLS measurements of the **TYR/CS and TYR-βCD/CS** samples were conducted by diluting 1 mL of freshly made sample in 1 mL of water.

2.4.2. Encapsulation Efficiency (EE%) and Loading Capacity (LC%)

After ultracentrifugation, the quantification of free tyrosol in the supernatant was performed using UV-Vis spectroscopy.

$$EE\% = \frac{Total\ TYR\ (mg) - TYR\ in\ supernatant\ (mg)}{Total\ TYR\ (mg)} \times 100 \quad (1)$$

$$LC\% = [Total\ encapsulated\ (mg)/total\ nanoparticles\ weight\ (mg)] \times 100 \quad (2)$$

2.4.3. Release Study

The release profile of tyrosol was investigated by determining the quantity of tyrosol released form the nanosystem at given time intervals. For that reason, 50 mg of the each nanosystem were dissolved in a 12 mL solution of aqueous 1% acetic acid (6 mL) and 6 mL DMSO and were stirred at 37 °C at 100 rpm. At specific time intervals, 0.5 mL of sample was obtained and filtered through a 0.45 μm pore syringe filter and analysed by UV-Vis spectroscopy. Each time 0.5 mL of fresh solvent was added to the solution

2.4.4. Fourier Transform Infrared Spectroscopy (FTIR)

KBr pellets containing the dried sample were prepared with hydraulic pellet press. The FT-IR spectra were recorded with a JASCO FT/IR-4200 spectrometer (Japan Spectroscopic Company, Tokyo, Japan).

2.4.5. Nuclear Magnetic Resonance Spectroscopy (NMR)

^{1}H NMR spectrum of TYR-βCD IC was obtained to determine the host–guest interactions. The spectra were obtained on a Varian V 600 MHz instrument (National Hellenic Research Foundation, Institute of Chemical Biology, Athens, Greece). The inclusion complex was dissolved in deuterium oxide (D_2O).

2.4.6. Thermal Characterisation

Thermal characterisation of the dried samples was performed with Differential Scanning Calorimetry (DSC) with a DSC 1 STARe System device (Mettler Toledo, Columbus, OH, USA) at temperature range from 20 °C to 350 °C with heating rate 10 °C/min, under nitrogen gas flow 20 mL/min and Thermogravimetric Analysis (TGA) the TGA/DSC 1 STARe System Thermobalance (Mettler Toledo, Columbus, OH, USA) at 25 °C–600 °C, with heating rate 10 °C/min, under nitrogen gas flow 10 mL/min.

2.5. Molecular Docking

The study of the interaction mode and binding affinity docking studies has been performed with the crystal structure of the DNA (PDB ID: 1bna), was obtained from the RSCB protein Data Bank. The optimisation of the docking parameters was performed using AutoDock Vina software (The Scripps Research Institute, La Jolla, CA, USA) and implemented empirical free energy function. Only polar hydrogens were added to the DNA in AutoDock Tools [28]. Finally, the image has been generated using PyMol software. The name and the number of nucleotides were designed according to PyMol software.

2.6. DNA Binding Studies Using UV-Vis Spectroscopy

Lyophilised Calf-thymus DNA (ctDNA) was dissolved in Tris-HCl buffer solution of concentration 10 mM and pH 7.4, and left overnight at 4 °C. Then, 1 mg ctDNA was dissolved in 1 mL buffer and the concentration was determined from the absorbance at 260 nm using an extinction coefficient of 6600 M^{-1} cm^{-1} [29]. Tyrosol, βCD, CS, TYR-βCD, TYR-βCD/CS, TYR/CS and Rhodamine B were dissolved in the buffer to a concentration of 10 mM for tyrosol or Rhodamine B or 2 mg/mL for the two carriers and all the nanosystems, which were then used as the stock solution for the preparation of the concentration of 100 µM. Afterwards, various concentrations (0–100 µM) of ctDNA were added to the prepared solutions which were incubated for 5 min and 30 min at 37 °C. Absorption spectra were measured using a JASCO double beam V-770 UV-Vis/NIR spectrophotometer in range of 230–400 nm.

3. Results

3.1. DoE for the TYR/CS Preparation Process

The measured responses for the **TYR/CS** nanoparticles are presented in Table 2.

Table 2. Experimental data of **TYR/CS** nanosystem and obtained results.

No	Coded Name	Factors			Responses	
		A	B	C	R1 Size (nm)	R2 ζ-Potential (mV)
1	TYR/CS 54	1	0	−1	168.8	41.6
2	TYR/CS 52	1	−1	0	574.6	46.4
3	TYR/CS 49	−1	1	0	precipitated	4.3
4	TYR/CS 50	0	−1	−1	679.6	41.7
5	TYR/CS 77	0	0	0	127.2	24.8
6	TYR/CS 57	0	1	−1	139.4	32.0
7	TYR/CS 45	0	−1	1	474.8	36.4
8	TYR/CS 53	1	0	1	199.3	43.3
9	TYR/CS 42	0	0	0	139.8	36.4
10	TYR/CS 51	0	0	0	115.3	35.5
11	TYR/CS 56	0	1	1	148.9	37.5
12	TYR/CS 47	−1	−1	0	393.4	28.6
13	TYR/CS 43	−1	0	−1	precipitated	8.7
14	TYR/CS 46	−1	0	1	precipitated	12.7
15	TYR/CS 55	1	1	0	206.1	37.0

From the table above, it can be observed that the size of the occurring nanoparticles ranged from 115.3 nm at central point to 679.6 nm except for points (−1, 1, 0), (−1, 0, 1) and (−1, 0, −1) in which the particles were over 3 µm and precipitated.

ζ-potential was positive in all cases and ranged from 4.3 to 46.4 mV. As chitosan is a cationic polymer at acidic environments, highly positive ζ-potential is expected, and is indicative of its stability in dispersion. However, the presence of the polyanion TPP manages to reduce the value, by interacting with the protonated amino groups.

Particles in the micro scale were observed only in three points in all of which the concentration of chitosan was at the low level while the concentration of TPP was on its medium or high level. Hence, it can be deduced that when the ratio of chitosan to TPP is low, there is accumulation of the cross-linker in the particles surface, which can also be confirmed by the low ζ-potential. On the other hand, increase in the particles' size is also observed when the ratio of chitosan to TPP is high, as it occurs from points (1, −1, 0), (0, −1, −1) and (0, −1, 1). This could be attributed to an excess of chitosan in the particles surface resulting in insufficient cross-linking of the polymer and low cross-linking density between the polymer and the cross-linking agent. The high ζ-potential of those points confirms the existence of protonated amino groups on the surface of the NPs. Moreover, increased concentration of chitosan leads in decreased intermolecular distance and increased intermolecular hydrogen bonding between the polymeric chains [30–34].

3D surface plots of R_1 were designed (Figure 2), and statistical analysis of the results was performed. A Reduced Quadratic Model better described the results (Equation (3)) and was found to be statistically significant (p-value 0.0012). Factors A, B, AB, A^2 were found to be statistically significant, verifying the observation that the size of the nanoparticles depends both on the amount of chitosan and the interaction with the crosslinking agent TPP. The Model F-value was calculated to be 9.43, indicating a significant model.

The coded equation that describes the size response is:

$$R_1 = -85.74 - 2666.41\,A + 1450.25\,B + 162.06\,C - 3301.28\,AB + 2674.75\,A^2 + 729.22\,B^2 \qquad (3)$$

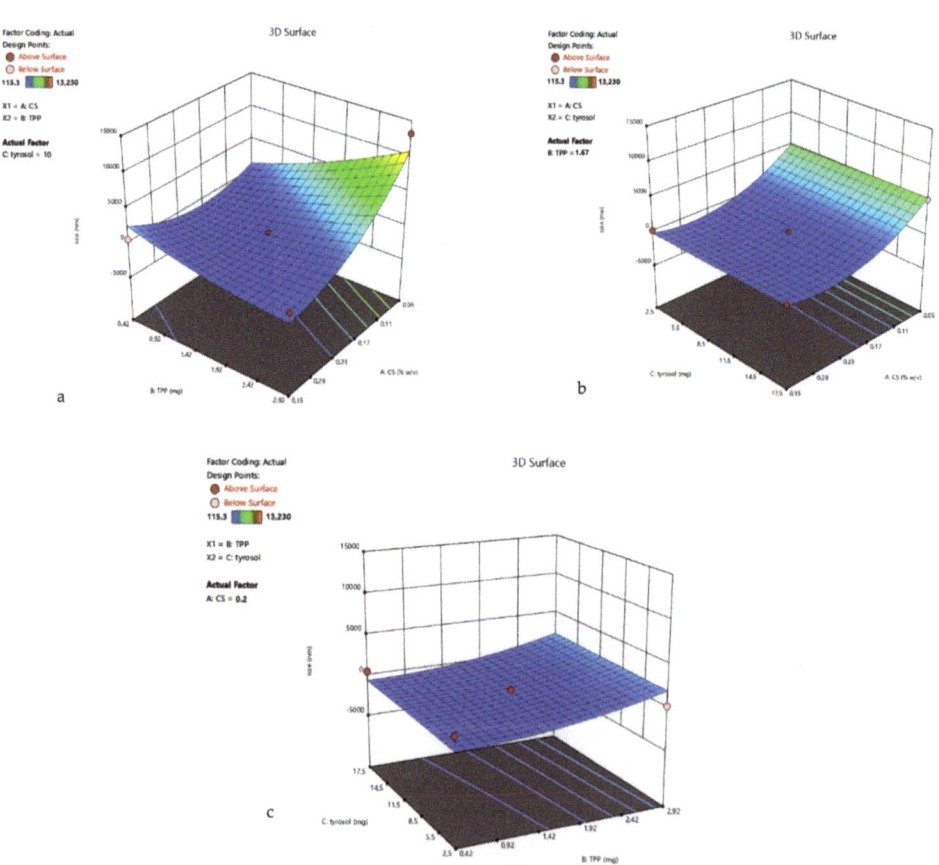

Figure 2. 3D surface plot of response R1 of TYR/CS system (**a**) CS Vs TPP (tyrosol: 10 mg) (**b**) CS Vs TYR (TPP: 1.67 mg) (**c**) TPP Vs TYR (CS: 0.2%).

For the ζ-potential response, the linear model (Equation (4)) best described the relation between the experiment data of R_2 and the model F-value was 9.64. The 3D surface plots of R_2 are presented in Figure 3.

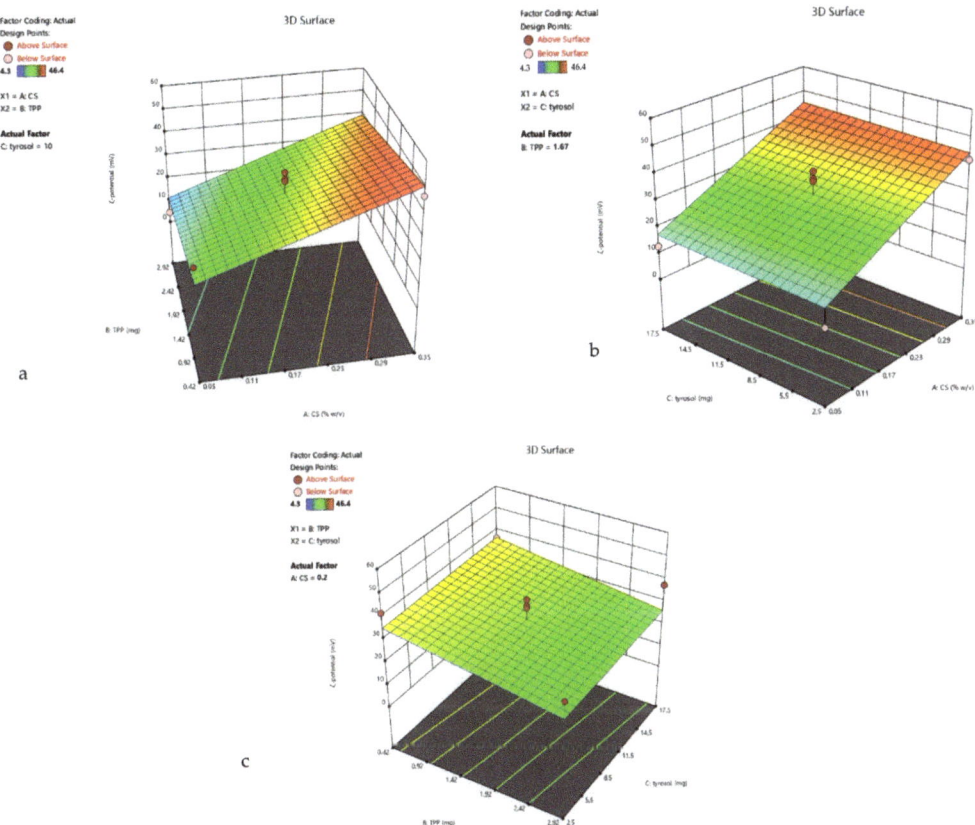

Figure 3. 3D surface plot of response R_2 of TYR/CS system (a) CS Vs TPP (tyrosol: 10 mg) (b) CS Vs TYR (TPP: 1.67 mg) (c) TPP Vs TYR (CS: 0.2%).

The coded equation that describes the ζ-potential-response is:

$$R_2 = 30.64 + 14.25\,A - 5.29\,B + 0.74\,C \tag{4}$$

Table 3 summarises the significance of each factor for the responses R_1 and R_2 of the TYR/CS nanosystem.

Optimal preparation conditions for TYR/CS nanoparticles are found to be close to the Central Point (CP) values of factors B and C and in the range 0.2–0.35% for the CS concentration.

Those results are in accordance with literature. Shah et al. [31], prepared CS nanoparticles loaded with quetiapine fumarate sized between 140 and 487 nm. The optimal preparation conditions were CS concentration 0.1% and CS:TPP ratio 4.8:1, with stirring time 15 min at 700 rpm and resulted in nanoparticles of size 131.08 nm with ζ-potential 34.4 mV. Delan et al. [30] used BBD for the optimisation of the synthesis of chitosan nanoparticles loaded with the anionic and lipophilic drug simvastatin. It was found that the best CS concentration was 0.34% and the CS:TPP ratio was 3:1, leading to nanoparticles of size

106 nm and ζ-potential 43.3 mv. Sharma et al. [34] ran a four-factor, three-level BBD to assess the process of synthesis of Carvedilol loaded CS nanoparticles. In their research, optimum CS concentration was found to be 0.262% and the nanoparticle size was measured with TEM 102.12 nm.

Table 3. Significance of each factor equation model terms of the TYR/CS system.

Factor	p-Value	
	R1	R2
Model	0.0012	0.0013
A	0.0010	0.0002
B	0.0309	0.0843
C	0.7848	0.7984
AB	0.0024	-
A^2	0.0072	-
B^2	0.3807	-
R^2	0.850	0.6899
Adjusted R^2	0.760	0.6184
Adeq Precision	11.385	10.0663

3.2. DoE for TYR-βCD/CS

The aqueous dispersion of the inclusion complex of tyrosol with βCD is formed by nanoparticles of size 478.1 nm with a negative, almost neutral ζ-potential of −7.18 mV. The entrapment of the inclusion complex into the chains of chitosan reversed the ζ-potential, due to the presence of chitosan in the outer layer of the inclusion complexes. Furthermore, in most of the 15 runs of the experimental design, the obtained particles were significantly smaller than the inclusion complex. This could be attributed to the strong electrostatic interaction between the oppositely charged carriers, which could separate agglomerations.

From the data in Table 4, it can be observed that the measured sizes ranged from 132.6 nm to over 3 µm, which precipitate, while the ζ-potential ranged from 4.8 to 46.5 mV.

The lowest result for R_1 is at the point (0, +1, −1), but at the CP, the size is also very close (average is 190.5 nm). In this design, it can also be observed that the chitosan-to-TPP ratio has a strong impact on the particles' size and ζ-potential. Therefore, for points (−1, 1, 0), (−1, 0, −1) and (−1, 0, 1), the particles precipitate, and the ζ-potential is very low. Moreover, from the 3D surface plot of the response R_1 (Figure 4), the size tends to decrease when the concentration of CS increases.

According to the mathematical analysis, the Reduced Quadratic Model was the most suitable for describing this response (F = 12.08). The coded equation (Equation (5)) that describes the system is:

$$R_1 = -185.97 - 3350.14\,A + 1093.59\,B - 226.53\,C - 2506.92\,AB + 3392.22\,A^2 + 906.57\,C^2, \quad (5)$$

The ζ-potential values range from 4.8 to 46.5 mV for the TYR-βCD/CS system.

Response R_2 is best described by the quadratic model, with F-value = 24.02. The coded equation (Equation (6)) of the model is as follows:

$$R_2 = 25.04 + 13.11\,A - 4.30\,B + 2.54\,C + 2.18\,AB + 1.50\,AC - 1.03\,BC - 6.72\,A^2 + 7.21\,B^2 + 5.98\,C^2, \quad (6)$$

The 3D surface plots of R_2 are presented in Figure 5.

Table 5 summarises the significance of each factor for the responses R_1 and R_2 of the TYR-βCD/CS nanosystem.

Table 4. Experimental data of TYR-βCD/CS nanosystem and obtained results.

Run	Coded Name	Factors			Responses	
		A	B	C	R_1	R_2
1	TYR-βCD/CS 67	0	−1	−1	364.8	36.9
2	TYR-βCD/CS 78	0	0	0	159.0	28.7
3	TYR-βCD/CS 66	−1	+1	0	precipitated	4.8
4	TYR-βCD/CS 73	0	+1	+1	142.2	37.5
5	TYR-βCD/CS 70	+1	0	+1	231.0	39.9
6	TYR-βCD/CS 68	0	0	0	162.5	21.2
7	TYR-βCD/CS 71	+1	0	−1	190.3	34.3
8	TYR-βCD/CS 64	−1	−1	0	276.7	19.4
9	TYR-βCD/CS 72	+1	+1	0	221.3	36.0
10	TYR-βCD/CS 69	+1	−1	0	595.0	41.9
11	TYR-βCD/CS 79	0	0	0	141.5	27.8
12	TYR-βCD/CS 62	0	−1	+1	441.5	46.5
13	TYR-βCD/CS 61	−1	0	−1	precipitated	11.7
14	TYR-βCD/CS 63	−1	0	+1	precipitated	11.3
15	TYR-βCD/CS 74	0	+1	−1	132.6	32.0

Table 5. Significance of each factor equation model terms of the TYR-βCD/CS system.

Factor	p-Value	
	R_1	R_2
Model	0.0028	0.0016
A	0.0006	<0.0001
B	0.1131	0.0130
C	0.7221	0.0766
AB	0.0204	0.2353
AC	-	0.3950
BC	-	0.5530
A^2	0.0046	0.0080
B^2	-	0.0100
C^2	0.2881	0.0213
R^2	0.8773	0.9760
Adjusted R^2	0.7853	0.9327
Adeq Precision	9.8620	15.2455

For this system, the optimal preparation formula was found to be the CP, giving the smallest nanoparticles and a ζ-potential of 25.9 mV. This point was chosen for further experiments.

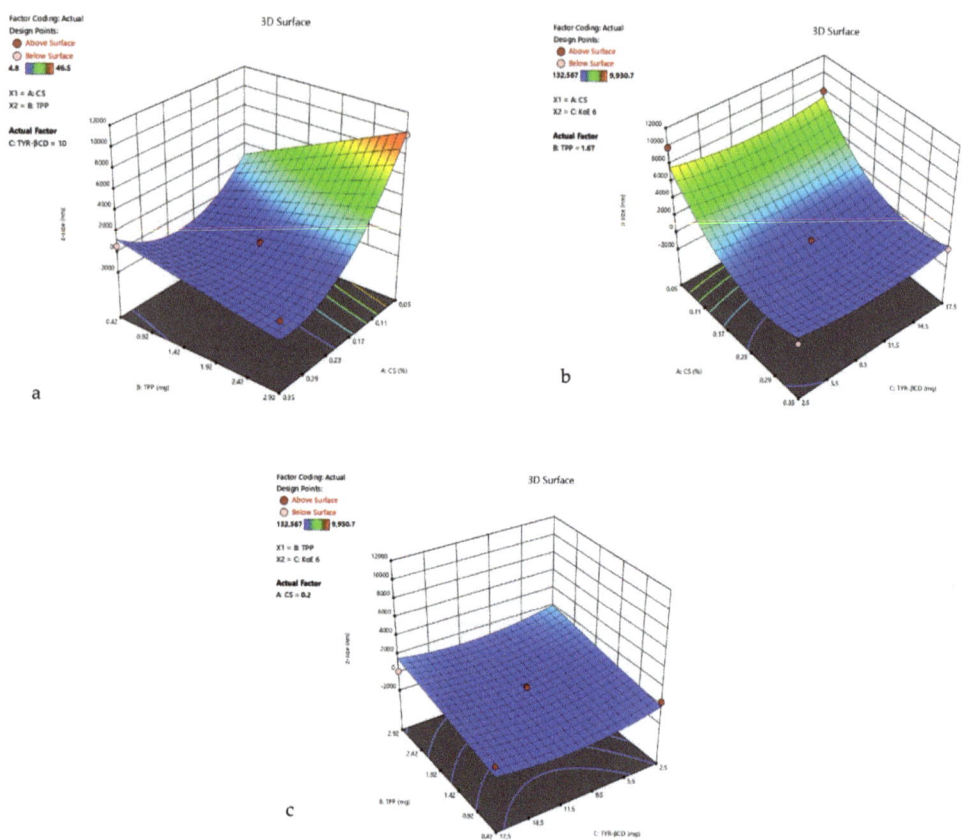

Figure 4. 3D surface plot of response R_1 of TYR-βCD/CS system (**a**) CS Vs TPP (TYR-βCD: 10 mg) (**b**) CS Vs TYR-βCD (TPP: 1.67 mg) (**c**) TPP Vs TYR-βCD (CS: 0.2%).

Comparing the experimental results, the lowest size values were given at the point (0, +1, −1) and the CP. In both cases, the ζ-potential was highly positive; hence, the central point was chosen for comparative reasons. This result is in accordance with the software's predictions of the optimal points.

Comparing the two nanosystems, many similarities can be observed. First, the initial concentration of chitosan in the nanoparticle-forming solution plays a significant role in the properties of the particles. Increased concentration of chitosan results in the agglomeration of particles and a very high ζ-potential, attributed to the presence of many protonated amino groups. On the other hand, chitosan to TPP ratio is also important and strongly affects both responses. Moreover, the range of both responses does not differ significantly for the two systems, suggesting that the polymeric chains form a matrix entrapping the molecule or the inclusion complex and form the nanoparticles.

The main difference between the two systems is the impact of the **TYR-βCD** to the ζ-potential. The reason for this difference could be that the inclusion complex is negatively charged and that a strong electrostatic interaction between the oligosaccharide and chitosan exists.

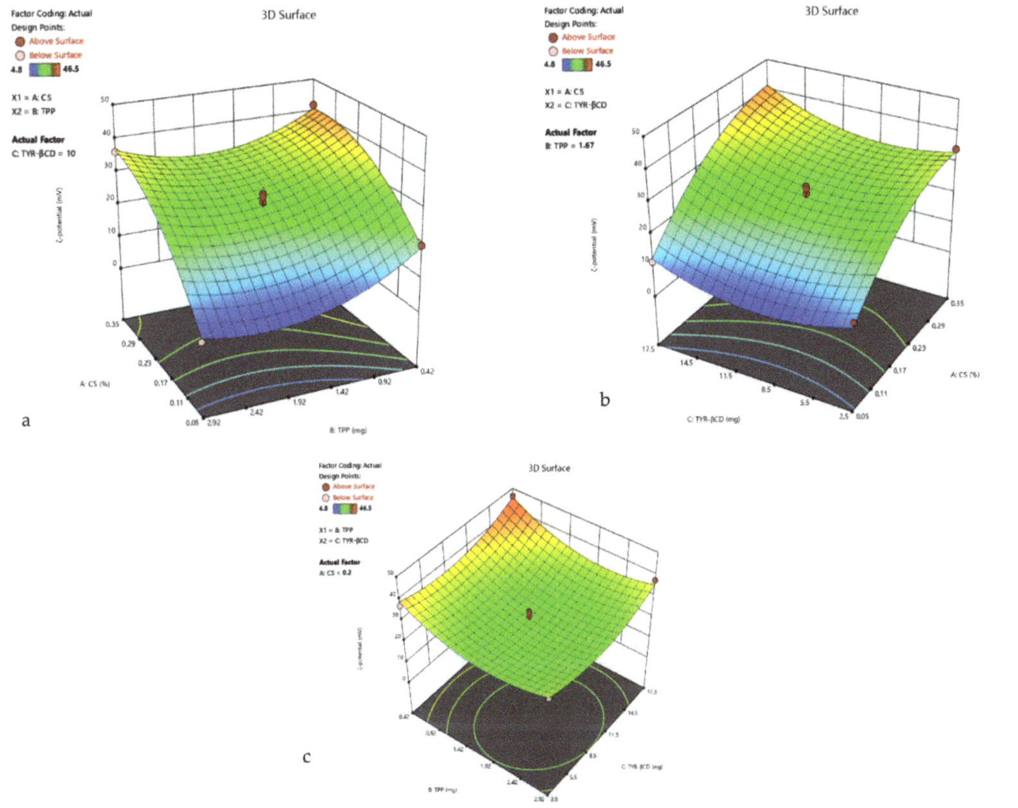

Figure 5. 3D surface plot of response R_2 of the TYR-βCD/CS system (a) CS Vs TPP (TYR-βCD: 10 mg) (b) CS Vs TYR-βCD (TPP: 1.67 mg) (c) TPP Vs TYR-βCD (CS: 0.2%).

3.3. Encapsulation Efficiency and Loading Capacity Calculation

After identifying the optimal conditions for obtaining nanoparticles, the encapsulation efficiency (EE%) and loading capacities (LC%) were determined for the three nanosystems.

For the inclusion complex of tyrosol with the βCD, it was found that the EE% was 98%. The high encapsulation efficiency is expected for this system, as the kneading technique was implemented, and no washing of the tyrosol was performed. The structure of βCd is presented in Figure 6.

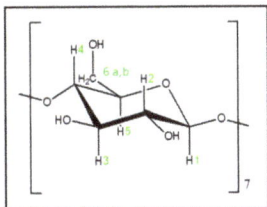

Figure 6. The glucose monomer in βCD.

For the **TYR/CS** nanosystem, the EE% was found to be 46% and the LC 12%, while for the **TYR-βCD/CS** nanosystem, the corresponding values were 12 and 4.2%, respectively.

3.4. Structural Identification of TYR-βCD Using ^{1}H NMR Spectroscopy

The analysis of the ^{1}H NMR spectrum of the inclusion complexes with βCD provides important evidence regarding the host–guest interactions. The ^{1}H NMR spectrum of **TYR-βCD**, as well as those of tyrosol and βCD, are presented in Figure 7. In Table 6, the chemical shift changes of ^{1}H-NMR signals of the protons of βCD before and after the formation of the **TYR-βCD** inclusion complexes are shown.

Table 6. Chemical shift changes of ^{1}H-NMR signals of βCD before and after the formation of the TYR-βCD IC.

Proton	Chemical Shifts (δ1) of βCD Protons (ppm)	Chemical Shifts (δ2) of βCD Protons (ppm) in TYR-βCD IC	Δδ = (δ2 − δ1) ppm
1	5.081	5.077	−0.004
2	3.661	3.658	−0.003
3	3.975	3.945	−0.030
4	3.596	3.595	−0.001
5	3.867–3.913	3.813	−0.100
6a,b		3.874	−0.039

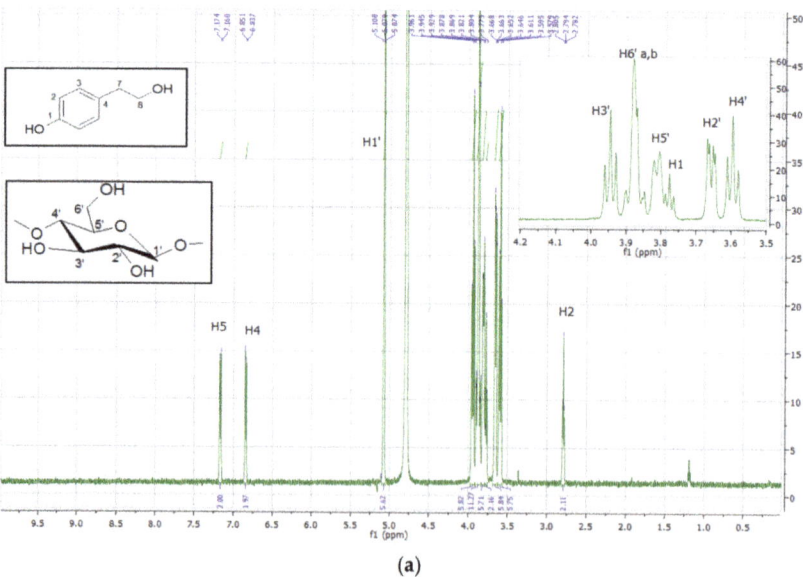

(a)

Figure 7. Cont.

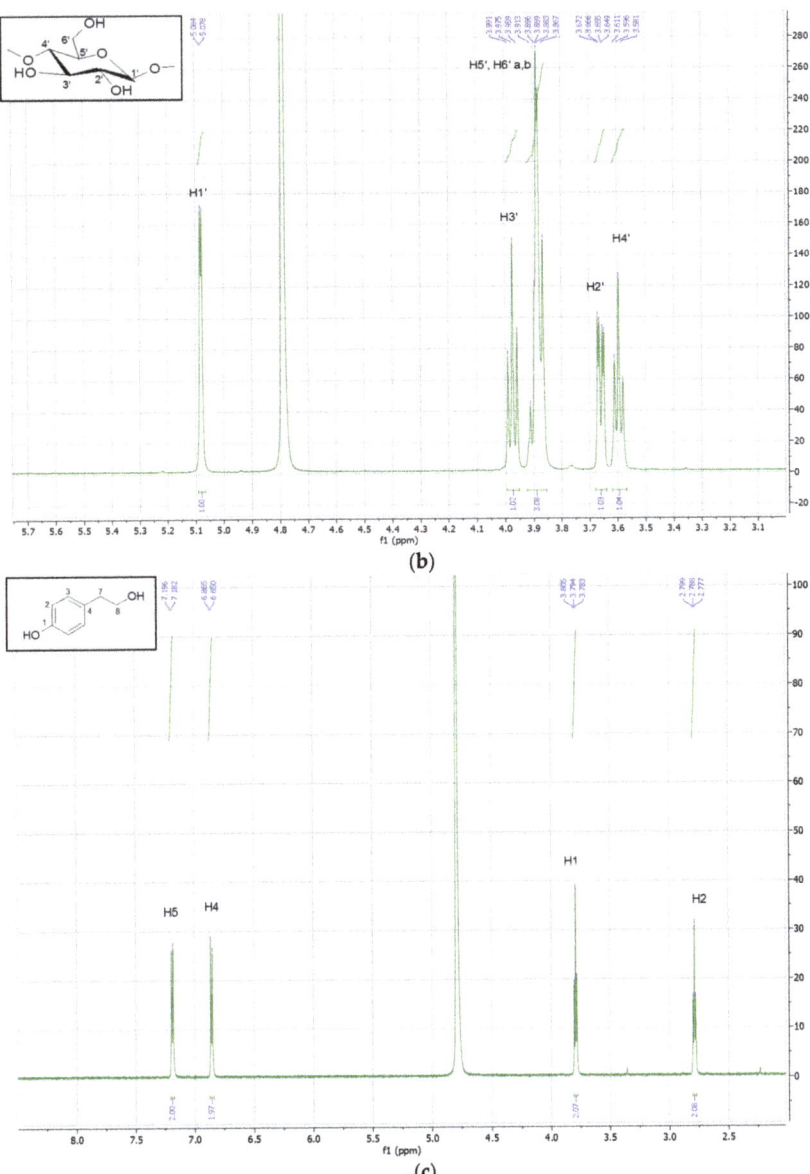

Figure 7. ^{1}H NMR spectrum (600MHz, D$_2$O) of (a) **TYR-βCD** and expansion of the region 3.4–4.1 ppm, (b) βCD, and (c) tyrosol.

The peaks of the H-3 and H-5 protons of βCD (which are located inside the cavity) present significant upfield shift (−0.030 and −0.100 ppm, respectively), indicative of strong hydrophobic interactions between tyrosol and βCD inside the cavity. The upfield shift of the H-6 (−0.039 ppm), which lie at the primary face of the cyclodextrin cone, implies that there is also a strong interaction with tyrosol, probably between the aliphatic OH of the tyrosol molecule and the 6-OH of βCD. These results lead us to believe that the tyrosol molecule enters the cyclodextrin cone in such a way that the aromatic ring lies well inside the hydrophobic cavity whereas the aliphatic hydroxyethyl moiety points towards the primary face of the cone and strong hydrogen bonds are formed between the OH groups. These observations are in accordance with the results of Rescifina et al. [35] and Lopez-Garcia et al. [36], who studied the structure of the inclusion complexes of tyrosol and hydroxytyrosol with βCD.

3.5. FT-IR Analysis of the TYR-β-CD IC

The analysis of the FT-IR spectra of pure βCD, tyrosol and the inclusion complex can serve as further proof of the formation of the inclusion complex. The most characteristic peaks in the FT-IR spectra of the above-mentioned compounds are shown in Table 7 and the spectra in Figure 8.

Table 7. Characteristic FT-IR absorption bands of β-CD, tyrosol and the tyrosol-β-CD inclusion complex.

	Characteristic Absorption Bands (cm^{-1})		
	βCD	Tyrosol	Inclusion Complex
OH stretching	3382	3389	3376
C–H stretching (aromatic compounds)	-	3148	-
C–H stretching	2925	-	2924
C–H asymmetric stretching (CH$_2$)	1643	-	1643
C=C stretching (Aromatic compounds)	-	1512	1541
O-H bending (alcohol)	1415	1452	1423
C–O stretching (secondary alcohols)	1029	-	1029
C–H stretching (aromatic, para substituted)	-	818	-

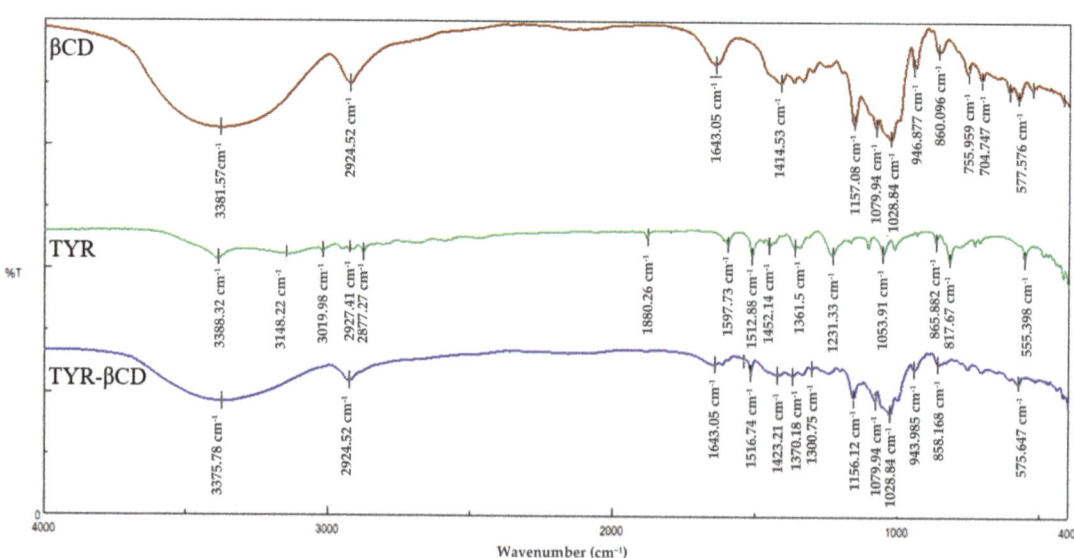

Figure 8. FT-IR spectra of β-CD (red), tyrosol (green) and tyrosol-β-CD inclusion complex (blue).

The FT-IR spectrum of the inclusion complex of tyrosol with βCD (TYR-βCD) shows a broad absorption band at 3376 cm^{-1} owed to the OH stretching vibration. This band is shifted compared to the corresponding band at the spectra of pure βCD (3382 cm^{-1}) and tyrosol (3389 cm^{-1}). This shift is indicative of strong interactive forces between the host and guest in the inclusion complex. The band at 1543 cm^{-1} present in the FT-IR spectrum of the inclusion complex is attributed to the C=C stretching vibration in aromatic compounds and is shifted by 29 cm^{-1} from the value of the same band at the spectrum of pure tyrosol. This large shift is further evidence of the efficient inclusion of tyrosol in the cyclodextrin cone with the aromatic part of the molecule lying inside the cone, as concluded also by the NMR spectra. The absorption band of the OH bending vibration appears at 1423 cm^{-1} in the spectrum of the inclusion complex and is shifted from the value of the same absorbance at the spectra of β-CD and tyrosol by 9 cm^{-1} and 29 cm^{-1}, respectively. Again, these shifts corroborate the strong interaction between tyrosol and β-CD.

3.6. Differential Scanning Calorimetry Analysis (DSC)

The DSC thermograms of tyrosol, the two carbohydrate carriers and the corresponding nanosystems are presented in Figure 9.

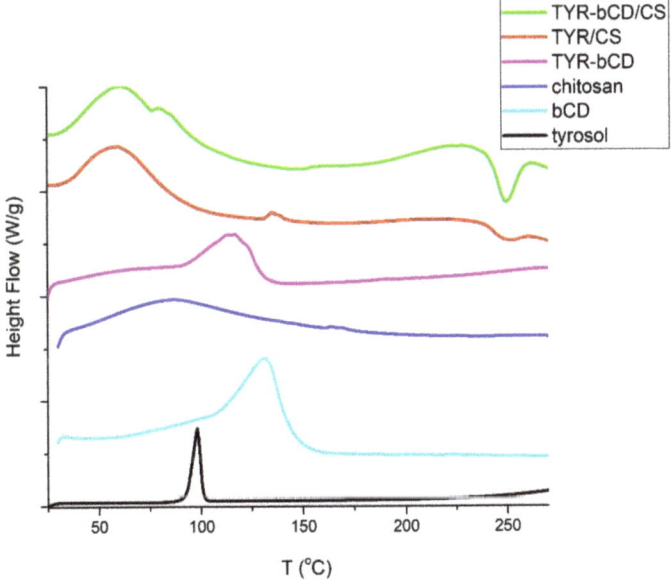

Figure 9. DSC thermograms of tyrosol (black), β-cyclodextrin (cyan), chitosan (blue), TYR-βCD inclusion complex (magenta), TYR/CS (red), TYR-βCD/CS (green).

In the DSC thermogram of tyrosol, the sharp endothermic process at 85–113 °C with peak at 95 °C, corresponds to the melting point of the compound [6]. In the chitosan DSC thermogram, in the studied range, only the loss of water is observed at the temperature range 85–107 °C, with a peak at 94 °C corresponding to the loss of water. The loss of water from cyclodextrin occurs in the range 98–155 °C with an endothermic peak at 131 °C [37–39].

The DSC thermogram of the inclusion complex undergoes an endothermic thermal transition from 87 to 133 °C with peak observed at 116 °C, ascribed to the water loss. The melting of tyrosol cannot be observed in this curve, yet the decrease of the temperature of water loss, compared to the one of the pure βCD, may be evidence of the protection that the carrier offers to the molecule. This is in accordance with what has previously been reported when encapsulating a molecule in βCD [37,39].

In the **TYR/CS** nanosystem, water loss occurs at 60 °C. Similarly, in the **TYR-βCD/CS** nanosystem, the water loss is observed with an endothermic peak at 61 °C. Moreover, in this system a second endothermic process takes place at the temperature range 75–87 °C with peak at 79 °C and could correspond to the loss of the water bound inside the cavity of the βCD.

The endothermic peak owed to the melting of tyrosol is not present in any of the nanosystems studied, and can be considered as further evidence of the successful encapsulation of the compound in the different matrices.

3.7. Thermogravimetric Analysis (TGA)

The thermal stability of the samples can be determined using TGA. As can be seen in Figure 10, the degradation temperature (T_d) of TYR is 220 °C.

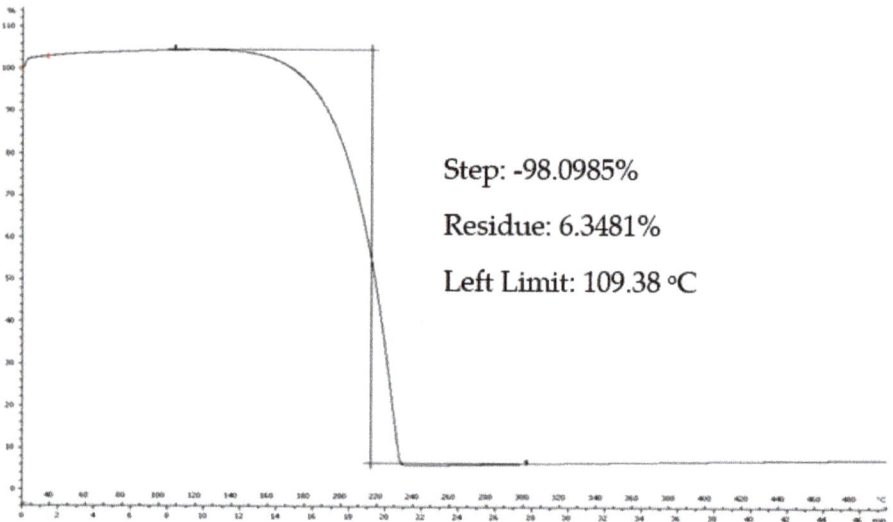

Figure 10. The TGA thermogram of tyrosol.

The thermal degradation of chitosan (Figure 11) occurs in two stages: water loss occurs at 92 °C, resulting in an 8% mass loss; whereas the decomposition temperature is 300 °C. The residue of the polymer at 500 °C is approximately 44%.

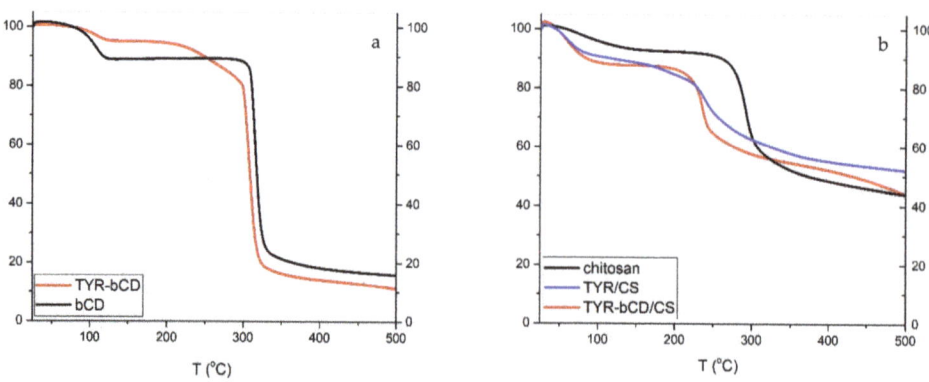

Figure 11. TGA thermograms of (**a**) β-cyclodextrin and TYR-βCD and (**b**) chitosan, TYR/CS and TYR-βCD.

For the **TYR/CS** nanosystem, the TGA profile presents three stages of degradation ((Figure 11). First, there is the water loss of at 67 °C then there is a mass loss of 5.0% at 187 °C that could be ascribed to tyrosol and, finally, the decomposition of chitosan occurs at 270 °C, leaving a residue of 52% at 500 °C. The slight decrease in the T_d of chitosan is attributed to the formation of nanoparticles through anionic gelation and, hence, the synthesis of a new material [40].

In the TG curve of βCD, first there is the loss of water molecules, externally and internally bound, at 103 °C, resulting in a decrease of 11% of the total mass [41]. The decomposition of βCD happens at 323.8 °C and the mass loss is 72.4%. The degradation profile of **TYR-βCD** presents three stages: the water loss at 105 °C with 4.7% mass loss, then the decomposition of tyrosol at 266 °C, resulting in further mass loss of 9.5%, and finally, at 316 °C, the decomposition of the βCD. The decomposition of tyrosol happens at a significantly higher temperature compared to the decomposition of the free molecule, confirming that the formation of the inclusion complex protects it from thermal degradation. The residue of the inclusion complex at 500 °C is 11%.

For the double-encapsulation system **TYR-βCD/CS**, TGA reveals an improved thermal stability of the system. This system degrades in two stages: at 64 °C, the dehydration of the sample happens, losing approximately 14% of the mass, and at the temperature range 161–438 °C, there is another mass loss of 37%, which corresponds to the degradation of the nanosystem. The residue of the sample at 500 °C is 44%.

Therefore, it seems that tyrosol is better protected in the double encapsulated system than in the chitosan matrix.

3.8. Molecular Docking

In Figure 12, the binding architecture of tyrosol in the crystal structure of DNA (source: PDB:1bna) is presented, depicting its stabilisation in the binding cavity of minor groove of DNA. The docked complex between tyrosol and DNA is illustrated as cartoon (Figure 12a,c,d) and in the form of spheres (Figure 12b) showing the interaction of tyrosol in the binding cavity of minor groove of DNA. The minor groove is smaller in size than the major groove and has the benefit that it is available for attack from small molecules such as tyrosol. Most of the anticancer and antibiotic drugs that have been reported are small molecules, so the minor groove is important as their main binding site.

The stabilisation of the complex is achieved by the formation of hydrogen bond (Figure 12), polar and hydrophobic interactions. From the five hydrogen bonds between DC-11, DG-10, DG-14 and DG-16 nucleotides, three hydrogen bonds are formed between the aliphatic hydroxyl group of tyrosol and the purines of DG-10 and DG-16 base pairs. One more hydrogen bond is formed between the hydroxyl group and the pentose of DG-11 nucleotide and another hydrogen bond between the phenolic hydroxyl group of tyrosol and the purines of DG-14 base pair.

In addition, Table 8 illustrates the nucleotides, the number of hydrogen bonds, and the binding energy that are formed with tyrosol.

Table 8. Binding scores of the docked tyrosol on the active site of DNA.

Ligand	Binding Energy (kcal/mol)	No. of Hydrogen Bonds	Nucleotides
(tyrosol structure: HO-C6H4-CH2CH2-OH)	−5.0	5	DC-11, DG-10, DG-14, DG-16

The above results are in accordance with the results obtained from the DNA-binding studies using UV spectroscopy, which indicated an external interaction between TYR and ctDNA.

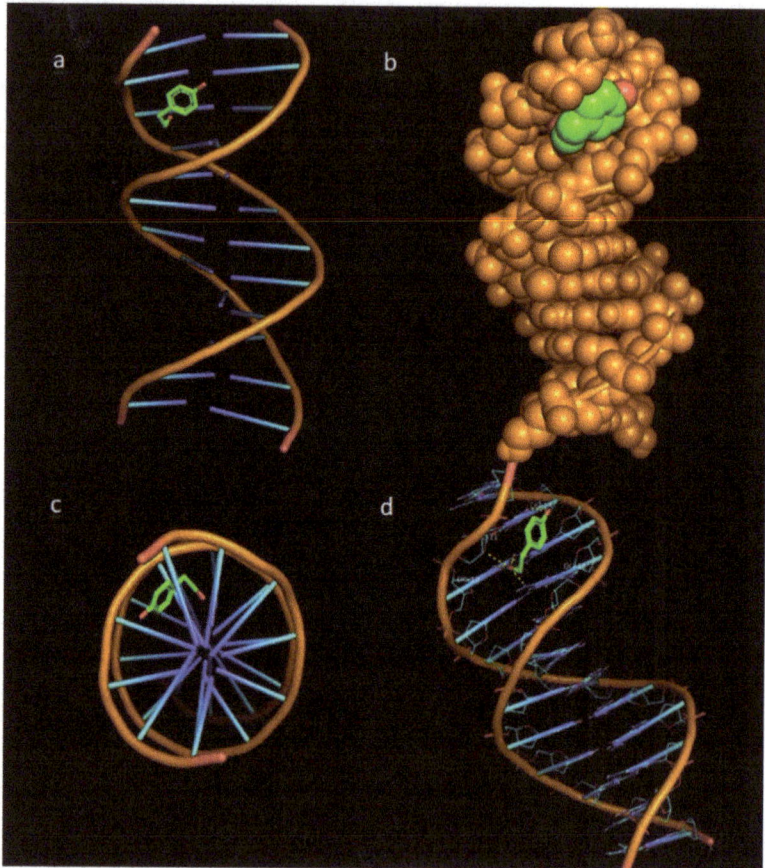

Figure 12. Binding architecture of tyrosol in the crystal structure of CT DNA (PDB:1bna) depicting its stabilisation in the binding cavity of minor groove of DNA. (**a**) DNA structure and tyrosol are illustrated as cartoons, (**b**) DNA structure and tyrosol are formed as spheres, (**c**) The docking pose from a view above the axis of the helix, (**d**) Nucleotides are rendered in line mode and the yellow dotted lines indicate hydrogen bonds between the docked molecule and the nucleotides of the binding pocket in the minor groove of DNA.

3.9. DNA Binding Studies with ctDNA Using UV Spectroscopy

The interaction of the compounds and nanosystems with ctDNA was studied by UV spectroscopy to obtain information on the existence of any interaction and to further calculate the DNA-binding constants of the compounds (k_b). An interaction between a chemical entity and DNA can disrupt the ctDNA band located at 260–280 nm in the presence of increasing amounts of ctDNA.

In absorption spectroscopy, hyperchromism and hypochromism are significant spectral features to study the changes of the double helical structure of DNA. Due to the strong interactions between a molecule and DNA bases, a change in absorption is observed, showing the proximity of the molecule to the DNA bases. On the basis of the interaction of compounds with DNA, the binding constant k_b for ligand–DNA binding was determined in the present work, using the Benesi–Hildebrand plot [42]. For the sake of comparison, the binding of a well-known xanthene dye, namely Rhodamine B, which has been shown to have a nonintercalative ctDNA binding in the DNA minor groove [43,44], was also studied.

It is worth noting that, in the present work, no ctDNA binding was observed for any of the nanosystems at 5 min; thus, the measurements were repeated after 30 min. This phenomenon is attributed to the nature of the nanosystems: at 5 min, no significant amount of tyrosol was released from the matrix of the nanosystem; whereas after 30 min, the released tyrosol was able to bind to ctDNA. On the other hand, the binding of rhodamine B and tyrosol to ctDNA at 5 min and 30 min did not show specific change. The UV-Vis data are summarised in Table 9.

Table 9. UV-Vis absorption data for rhodamine B, tyrosol and nanosystems in the absence and presence of ctDNA.

Compound or Nanosystem	λ_{max} Absent (nm)	λ_{max} Present (nm)	$\Delta\lambda$ (nm)	Hypochromicity (%)	Hyperchromicity * (%)	K_b 10^4 (M^{-1})
Rhodamine B	554	554	0	26.90	-	10.92 ± 0.20
Tyrosol	275.8	274	1.8	-	43.44	2.09 ± 0.07
βCD	261	259.2	1.8	-	93.43	0.78 ± 0.05
TYR-βCD	276.4	272.6	3.8	-	76.61	2.40 ± 0.16
TYR-βCD/CS	260.2	258.8	1.4	-	92.97	1.30 ± 0.08

* Hyperchromicity for complexes formed by compounds and nanosystems and 100 μM of ctDNA in comparison to free ligands.

As shown in Figure 13a, TYR (100 μM, pH = 7.4) showed absorption maxima at 275.8 nm. With incremental addition of ctDNA to the solution of TYR, an increase in the absorption intensity at 275.8 nm was observed with concomitant blue shift of the λ_{max} at 274 nm. This hyperchromism suggests that TYR binds to ctDNA by groove binding mode. The binding constant (k) of TYR was calculated from the ratio of the intercept to the slope and was found to be k = 2.09 ± 0.07 × 10^4 M^{-1}. The results indicate the binding of TYR in the minor groove of ctDNA, as predicted by the molecular modeling studies.

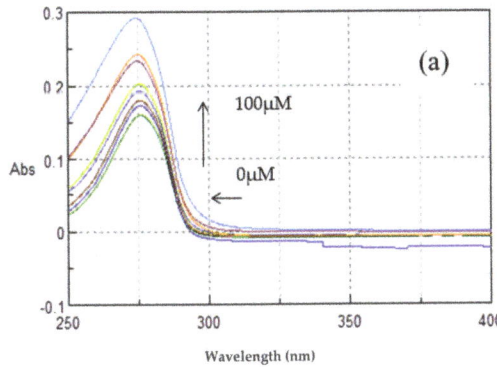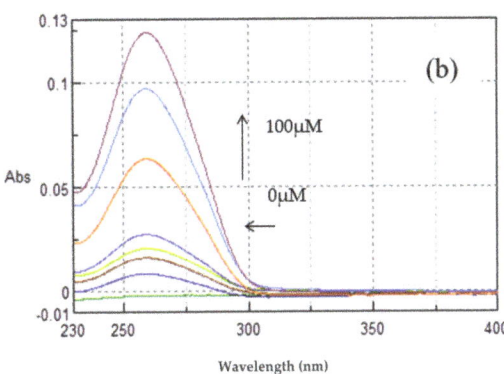

Figure 13. Absorption spectra of (a) TYR and (b) βCD (100 μM, pH 7.4) in Tris-HCl buffer with increasing concentrations of ctDNA (0–100 μM). Arrows (↑) and (←) refer to hyperchromic and hypochromic (blue shift) effects, respectively.

CDs are known to interact with nucleic acids. Recently, it was reported that when a complex with βCD is formed, the deoxyribose or ribose and the phosphate groups stabilize the docked complex by hydrogen bonds with the outer rim of the CD molecules [26,45]. In this context, the interaction of β-CD and the **TYR-β-CD** inclusion complex with ctDNA was investigated.

As shown in Figure 13b, βCD showed absorption maxima at 261 nm. With incremental addition of ctDNA to the βCD solution, an increase in the absorption intensity at 261 nm was observed with concomitant shift of the λ_{max} at 259.2 nm (blue shift). The binding constant (k) of βCD was calculated from the ratio of the intercept to the slope and was found to be k = 0.78 ± 0.05 × 10^4 M^{-1}. Therefore, it seems that βCD binds to ctDNA in a non-intercalative mode, yet much less strongly than tyrosol.

As shown in Figure 14 the complex **TYR-βCD** showed absorption maxima at 276.4 nm. With incremental addition of ctDNA to the solution of **TYR-βCD**, an increase in the absorption intensity at 276.4 nm was observed with a blue shift at λ_{max} 272.6 nm. The binding constant was found to be k = 2.40 ± 0.16 × 10^4 M^{-1}. The complex **TYR-βCD** showed enhanced interaction with ctDNA in comparison to the free tyrosol, which could be explained by the enhanced aqueous solubility of the inclusion complex and by the mode of insertion of tyrosol in the βCD cavity. As previously mentioned, if the aromatic ring lies well inside the hydrophobic cavity, while the aliphatic hydroxyethyl group points towards the primary face of the cone, it means that hydrogen bonds can be formed and stabilize the interaction between the complex and ctDNA.

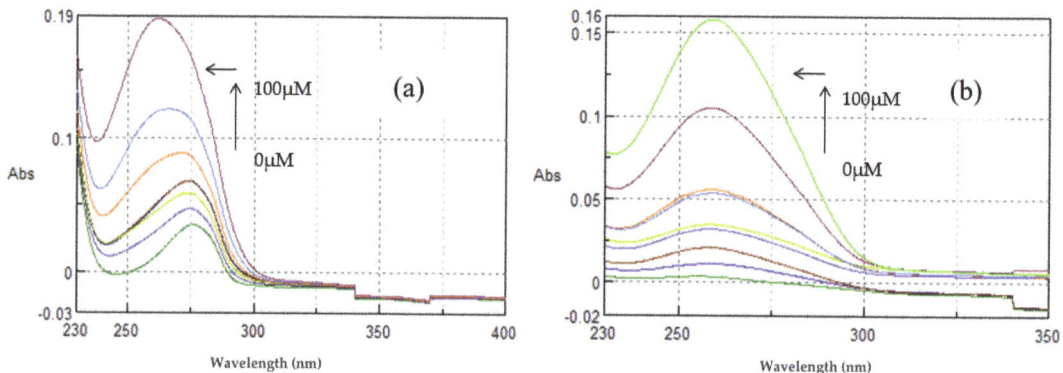

Figure 14. Absorption spectra of (**a**) **TYR-βCD** and (**b**) **TYR-βCD/CS** (100 μM, pH 7.4) in Tris-HCl buffer with increasing concentrations of ctDNA (0–100 μM). Arrows (↑) and (←) refer to hyperchromic and hypsochromic (blue shift) effects respectively.

The complex **TYR-βCD/CS** showed absorption maxima at 260.2 nm (Figure 14). With incremental addition of ctDNA to the solution of **TYR-βCD/CS**, an increase in the absorption intensity at 260.2 nm was observed, with a blue shift of the λ_{max} 258.8 nm. The binding constant was found to be k = 1.30 ± 0.08 × 10^4 M^{-1}.

Rhodamine B showed a hypochromic behaviour and the binding constant was determined to be 10.92 ± 0.20 × 10^4 M^{-1} (Figure 15). This is consistent with the minor groove binding of Rhodamine B in ctDNA, in accordance with the work of Islam et al. [44].

All the tested compounds and nanosystems, with the exception of Rhodamine B, showed hyperchromism and a blue-shift upon increasing DNA concentration, indicating that they all interact with the DNA helix. However, as no significant changes in the spectra could be observed, the results indicate a nonintercalative mode of binding. CS and **TYR/CS** were also studied, but no binding with ctDNA was observed.

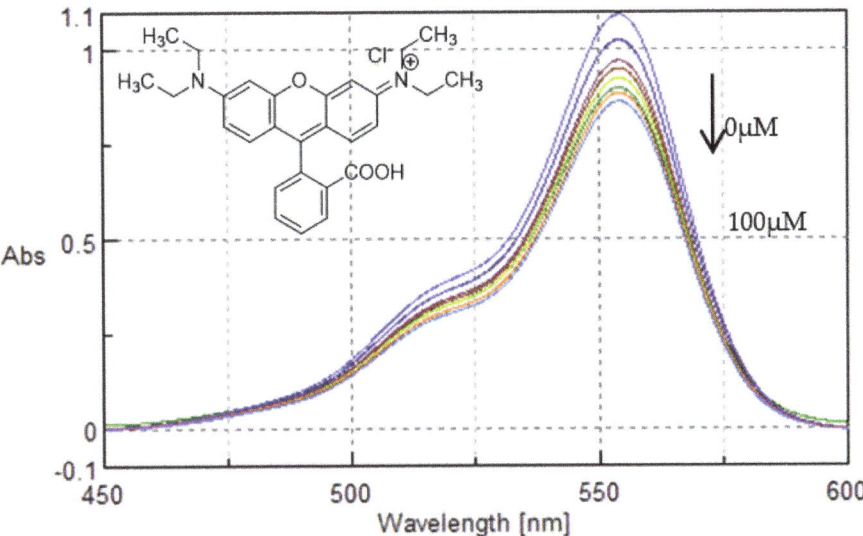

Figure 15. UV-Vis spectra of Rhodamine B with increasing concentrations of ctDNA (0–100 µM). Arrow refers to hypochromic effect.

3.10. Release Profiles

The release profile of tyrosol was investigated for all nanosystems prepared, and the results are presented in Figure 16. For all systems, the release study was carried out for 45 h in a solution of pH 3.4.

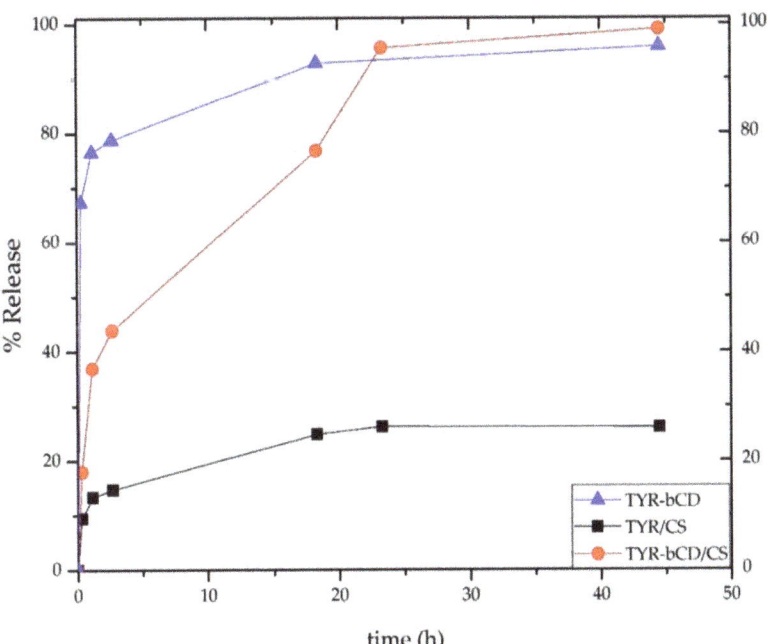

Figure 16. The release profile of tyrosol from **TYR-βCD** (blue line), **TYR/CS** (black line) and **TYR-βCD/CS** (red line).

All nanosystems showed a "burst release" of tyrosol during the first hour. For the **TYR-βCD** system, approximately 77% of the encapsulated tyrosol was released during that time. Thereafter, the release rate significantly slowed down, reaching 93% after 18 h.

On the other hand, the release profile of tyrosol from the **TYR/CS** system was significantly different. During the first hour of the study, 13% of tyrosol was released reaching the maximum release after 23 h (26%).

For the **TYR-βCD/CS** system, 37% of the encapsulated tyrosol was released during the first hour, reaching the 99% after 45 h. The slower rate presented by this system as compared to the **TYR-βCD** system can be attributed to the existence of chitosan as a coating and provides proof that this "double" encapsulation system can offer a more sustained release than the TYR/βCD by slowing down the initial burst effect observed in the case of the inclusion complex.

4. Discussion

The comparative study of the optimisation of the encapsulation parameters of tyrosol and the inclusion complex of β-cyclodextrin:tyrosol in chitosan revealed that the concentration of chitosan plays a significant role on the properties of the particles that are formed. Moreover, the ratio between the polymer and the cross-linking agent or the molecule to be encapsulated also affects the responses of the particulate system. In both systems studied, it was observed that the smaller sized particles presented a good ζ-potential, which is indicative of a very stable formulation.

The interaction of tyrosol, the inclusion complex of tyrosol with βCD and the inclusion complex coated with chitosan with ctDNA was studied in an effort to elucidate the potential of the prepared nanosystems to be exploited as pharmacologically active agents. Tyrosol and all the examined nanosystems showed a nonintercalative mode of binding to ctDNA. The complexation of tyrosol with cyclodextrin resulted in slightly better interaction with ctDNA as compared to the interaction exhibited by the **TYR-βCD/CS** system. This difference is attributed to the small, yet existent, interaction of the oligosaccharide with ctDNA whereas the polysaccharide chitosan did not interact with ctDNA in the performed experiments. Moreover, from the NMR analysis of the inclusion complex, it was deduced that the aromatic ring of the molecule lies inside the cavity, while the aliphatic hydroxyethyl moiety points towards the primary face of the cone. Therefore, it could be suggested that βCD carries the molecule close to ctDNA and an interaction can be observed. On the other hand, no interaction was observed when tyrosol was encapsulated in chitosan nanoparticles, presumably due to the strong positive charge on the surface of the **TYR/CS** nanosystem which impedes it from approaching DNA.

Finally, the release profile of tyrosol is different for each nanosystem. A burst effect is observed for all systems. The release of tyrosol from the inclusion complex formed with βCD is completed in approximately 20 h hours, but when chitosan coating is added, the release rate is delayed significantly. On the other hand, when tyrosol is encapsulated into chitosan nanoparticles, the release of only the 26% of encapsulated tyrosol occurs during the same time.

5. Conclusions

In the present study, the impact of two different carriers for the encapsulation of tyrosol is investigated: the oligosaccharide β-cyclodextrin and the polysaccharide chitosan. The effect of coating the tyrosol-βCD inclusion complex with chitosan was also investigated as a potential tool to modify the release profile of the bioactive compound. We were gratified to find that this was true for the systems studied in this work: the coating resulted in a sustained release of tyrosol and slowed down the initial burst effect observed from the inclusion complex.

For the formation of the tyrosol-βCD inclusion complex a well-known methodology was implemented, whereas for the two chitosan-containing systems, the formation of the nanoparticles was succeeded via the ionic gelation method with sodium tripolyphosphate. The latter processes were optimised using experimental design.

Moreover, the interaction of tyrosol and the corresponding nanosystems with ctDNA was investigated. The results show that tyrosol is a ctDNA groove binder and this was confirmed by molecular modeling studies. The same mode of binding was found for the **TYR/βCD** and **TYR/βCD/CS** nanosystems.

Author Contributions: Conceptualisation, A.D. and A.R.N.P.; Methodology, A.R.N.P., E.K., M.M.B.; Software, A.R.N.P., E.K., M.M.B.; Resources, A.D.; Data Curation, A.R.N.P., E.K., M.M.B.; Writing—Original Draft Preparation, A.D., A.R.N.P., E.K., M.M.B.; Writing—Review & Editing, A.D., A.R.N.P., E.K., M.M.B.; Supervision, A.D.; Project Administration, A.D. All authors have read and agreed to the published version of the manuscript.

Funding: A R N Pontillo would like to thank the General Secretariat for Research and Technology (GSRT) and the Hellenic Foundation for Research and Innovation (HFRI) for funding her PhD research. M Bairaktari gratefully acknowledges financial support from State Scholarships Foundation (IKY): This research is co-financed by Greece and the European Union (European Social Fund- ESF) through the Operational Programme "Human Resources Development, Education and Lifelong Learning" in the context of the project "Strengthening Human Resources Research Potential via Doctorate Research" (MIS-5000432), implemented by the State Scholarships Foundation (IKY).

Institutional Review Board Statement: Not applicable.

Informed Consent Statement: Not applicable.

Data Availability Statement: Data sharing not applicable.

Conflicts of Interest: The authors declare no conflict of interest.

References

1. Covas, M.I.; EMiro-Casas, M.; Fito, M.; Farre-Albadalejo, E.; Gimeno, J.; Marrugat, J.; De La Torre, R. Bioavailability of tyrosol, an antioxidant phenolic compound present in wine and olive oil, in humans. *Drugs Exp. Clin. Res.* **2003**, *29*, 203. [PubMed]
2. Di Benedetto, R.; Varì, R.; Scazzocchio, B.; Filesi, C.; Santangelo, C.; Giovannini, C.; Matarrese, P.; D'Archivio, M.; Masella, R. Tyrosol, the major extra virgin olive oil compound, restored intracellular antioxidant defences in spite of its weak antioxidative effectiveness. *Nutr. Metab. Cardiovasc. Dis.* **2007**, *17*, 535–545. [CrossRef] [PubMed]
3. San Miguel-Chávez, R. Phenolic antioxidant capacity: A review of the state of the art. *Phenolic Compd. Biol. Act.* **2017**, 59–74. [CrossRef]
4. St-Laurent-Thibault, C.; Arseneault, M.; Longpre, F.; Ramassamy, C. Tyrosol and hydroxytyrosol two main components of olive oil, protect N2a cells against amyloid-β-induced toxicity involvement of the NF-κB signaling. *Curr. Alzheimer Res.* **2011**, *8*, 543–551. [CrossRef]
5. Gris, E.F.; Mattivi, F.; Ferreira, E.A.; Vrhovsek, U.; Filho, D.W.; Pedrosa, R.C.; Bordignon-Luiz, M.T. Stilbenes and tyrosol as target compounds in the assessment of antioxidant and hypolipidemic activity of Vitis vinifera red wines from Southern Brazil. *J. Agric. Food Chem.* **2011**, *59*, 7954–7961. [CrossRef]
6. Paulo, F.; Santos, L. Encapsulation of the Antioxidant Tyrosol and Characterization of Loaded Microparticles: An Integrative Approach on the Study of the Polymer-Carriers and Loading Contents. *Food Bioproc. Tech.* **2020**, 1–22. [CrossRef]
7. Suganya, V.; Anuradha, V. Microencapsulation and nanoencapsulation: A review. *Int. J. Pharm. Clin. Res.* **2017**, *9*, 233–239. [CrossRef]
8. Shahidi, F.; Han, X.Q. Encapsulation of food ingredients. *Crit. Rev. Food Sci. Nutr.* **1993**, *33*, 501–547. [CrossRef]
9. Singh, R.; Lillard, J.W., Jr. Nanoparticle-based targeted drug delivery. *Exp. Mol. Pathol.* **2009**, *86*, 215–223. [CrossRef]
10. Kedare, S.B.; Singh, R.P. Genesis and development of DPPH method of antioxidant assay. *Int. J. Food Sci. Technol.* **2011**, *48*, 412–422. [CrossRef]
11. Del Valle, E.M. Cyclodextrins and their uses: A review. *Process. Biochem.* **2004**, *39*, 1033–1046. [CrossRef]
12. Tiwari, G.; Tiwari, R.; Rai, A.K. Cyclodextrins in delivery systems: Applications. *J. Pharm. Bioallied. Sci.* **2010**, *2*, 72. [CrossRef] [PubMed]
13. Villalonga, R.; Cao, R.; Fragoso, A. Supramolecular chemistry of cyclodextrins in enzyme technology. *Chem. Rev.* **2007**, *107*, 3088–3116. [CrossRef] [PubMed]
14. Ramos, M.A.D.S.; Da Silva, P.B.; Spósito, L.; De Toledo, L.G.; Bonifacio, B.V.; Rodero, C.F.; Dos Santos, K.C.; Chorilli MBauab, T.M. Nanotechnology-based drug delivery systems for control of microbial biofilms: A review. *Int. J. Nanomed.* **2018**, *13*, 1179. [CrossRef]
15. Bellich, B.; D'Agostino, I.; Semeraro, S.; Gamini, A.; Cesàro, A. The good, the bad and the ugly of chitosans. *Mar. Drugs* **2016**, *14*, 99. [CrossRef]

16. Muxika, A.; Etxabide, A.; Uranga, J.; Guerrero, P.; De La Caba, K. Chitosan as a bioactive polymer: Processing, properties and applications. *Int. J. Biol. Macromol.* **2017**, *105*, 1358–1368. [CrossRef]
17. Pontillo, A.R.N.; Detsi, A. Nanoparticles for ocular drug delivery: Modified and non-modified chitosan as a promising biocompatible carrier. *Nanomedicine* **2019**, *14*, 1889–1909. [CrossRef]
18. Cheung, R.C.F.; Ng, T.B.; Wong, J.H.; Chan, W.Y. Chitosan: An update on potential biomedical and pharmaceutical applications. *Mar. Drugs* **2015**, *13*, 5156–5186. [CrossRef]
19. Detsi, A.; Kavetsou, E.; Kostopoulou, I.; Pitterou, I.; Pontillo, A.R.N.; Tzani, A.; PChristodoulou ASiliachli Zoumpoulakis, P. Nanosystems for the encapsulation of natural products: The case of chitosan biopolymer as a matrix. *Pharmaceutics* **2020**, *12*, 669. [CrossRef]
20. Koukouzelis, D.; Pontillo, A.R.N.; Koutsoukos, S.; Pavlatou, E.; Detsi, A. Ionic liquid–Assisted synthesis of silver mesoparticles as efficient surface enhanced Raman scattering substrates. *J. Mol. Liq.* **2020**, *306*, 112929. [CrossRef]
21. Koutsoukos, S.; Tsiaka, T.; Tzani, A.; Zoumpoulakis, P.; Detsi, A. Choline chloride and tartaric acid, a Natural Deep Eutectic Solvent for the efficient extraction of phenolic and carotenoid compounds. *J. Clean. Prod.* **2019**, *241*, 118384. [CrossRef]
22. Tsintavi, E.; Rekkas, D.M.; Bettini, R. Partial tablet coating by 3D printing. *Int. J. Pharm.* **2020**, *581*, 119298. [CrossRef] [PubMed]
23. Zhao, J.; Li, W.; Ma, R.; Chen, S.; Ren, S.; Jiang, T. Design, synthesis and DNA interaction study of new potential DNA bis-intercalators based on glucuronic acid. *Int. J. Mol. Sci.* **2013**, *14*, 16851–16865. [CrossRef] [PubMed]
24. Pasolli, M.; Dafnopoulos, K.; Andreou, N.P.; Gritzapis, P.S.; Koffa, M.; Koumbis, A.E.; Psomas, G.; Fylaktakidou, K.C. Pyridine and p-nitrophenyl oxime esters with possible photochemotherapeutic activity: Synthesis, DNA photocleavage and DNA binding studies. *Molecules* **2016**, *21*, 864. [CrossRef]
25. Alves, P.S.; Mesquita, O.N.; Rocha, M.S. Controlling Cooperativity in β-Cyclodextrin–DNA Binding Reactions. *J. Phys. Chem. Lett.* **2015**, *6*, 3549–3554. [CrossRef]
26. Leclercq, L. Interactions between cyclodextrins and cellular components: Towards greener medical applications? *Beilstein J. Org. Chem.* **2016**, *12*, 2644–2662. [CrossRef]
27. García-Padial, M.; Martínez-Ohárriz, M.C.; Isasi, J.R.; Vélaz, I.; Zornoza, A. Complexation of tyrosol with cyclodextrins. *J. Incl. Phenom. Macrocycl. Chem.* **2013**, *75*, 241–246. [CrossRef]
28. Tang, B.; Shen, F.; Wan, D.; Guo, B.H.; Wang, Y.J.; Yi, Q.Y.; Liu, Y.J. DNA-binding, molecular docking studies and biological activity studies of ruthenium (II) polypyridyl complexes. *RSC Adv.* **2017**, *7*, 34945–34958. [CrossRef]
29. Singla, P.; Luxami, V.; Paul, K. Quinazolinone-benzimidazole conjugates: Synthesis, characterization, dihydrofolate reductase inhibition, DNA and protein binding properties. *J. Photochem. Photobiol. B Biol.* **2017**, *168*, 156–164. [CrossRef]
30. Delan, W.K.; Zakaria, M.; Elsaadany, B.; ElMeshad, A.N.; Mamdouh, W.; Fares, A.R. Formulation of simvastatin chitosan nanoparticles for controlled delivery in bone regeneration: Optimization using Box-Behnken design, stability and in vivo study. *Int. J. Pharm.* **2020**, *577*, 119038. [CrossRef]
31. Shah, B.; Khunt, D.; Misra, M.; Padh, H. Application of Box-Behnken design for optimization and development of quetiapine fumarate loaded chitosan nanoparticles for brain delivery via intranasal route*. *Int. J. Biol. Macromol.* **2016**, *89*, 206–218. [CrossRef] [PubMed]
32. Kalam, M.A.; Khan, A.A.; Khan, S.; Almalik, A.; Alshamsan, A. Optimizing indomethacin-loaded chitosan nanoparticle size, encapsulation, and release using Box–Behnken experimental design. *Int. J. Biol. Macromol.* **2016**, *87*, 329–340. [CrossRef] [PubMed]
33. Yan, J.; Guan, Z.Y.; Zhu, W.F.; Zhong, L.Y.; Qiu, Z.Q.; Yue, P.F.; Wu, W.T.; Liu, J.; Huang, X. Preparation of Puerarin Chitosan Oral Nanoparticles by Ionic Gelation Method and Its Related Kinetics. *Pharmaceutics* **2020**, *12*, 216. [CrossRef] [PubMed]
34. Sharma, M.; Sharma, R.; Jain, D.K.; Saraf, A. Enhancement of oral bioavailability of poorly water soluble carvedilol by chitosan nanoparticles: Optimization and pharmacokinetic study. *Int. J. Biol. Macromol.* **2019**, *135*, 246–260. [CrossRef] [PubMed]
35. Rescifina, A.; Chiacchio, U.; Iannazzo, D.; Piperno, A.; Romeo, G. β-cyclodextrin and caffeine complexes with natural polyphenols from olive and olive oils: NMR, thermodynamic, and molecular modeling studies. *J. Agric. Food Chem.* **2010**, *58*, 11876–11882. [CrossRef] [PubMed]
36. López-García, M.Á.; López, Ó.; Maya, I.; Fernández-Bolaños, J.G. Complexation of hydroxytyrosol with β-cyclodextrins. An efficient photoprotection. *Tetrahedron* **2010**, *66*, 8006–8011. [CrossRef]
37. Kotronia, M.; Kavetsou, E.; Loupassaki, S.; Kikionis, S.; Vouyiouka, S.; Detsi, A. Encapsulation of Oregano (Origanum onites L.) essential oil in β-cyclodextrin (β-CD): Synthesis and characterization of the inclusion complexes. *Bioengineering* **2017**, *4*, 74. [CrossRef]
38. Guinesi, L.S.; Cavalheiro, É.T.G. The use of DSC curves to determine the acetylation degree of chitin/chitosan samples. *Thermochim. Acta* **2006**, *444*, 128–133. [CrossRef]
39. Chatzidaki, M.; Kostopoulou, I.; Kourtesi, C.; Pitterou, I.; Avramiotis, S.; Xenakis, A.; Detsi, A. β-Cyclodextrin as carrier of novel antioxidants: A structural and efficacy study. *Colloids and Surfaces A: Physicochem. Eng. Asp.* **2020**, *603*, 125262. [CrossRef]
40. De Pinho Neves, A.L.; Milioli, C.C.; Müller, L.; Riella, H.G.; Kuhnen, N.C.; Stulzer, H.K. Factorial design as tool in chitosan nanoparticles development by ionic gelation technique. *Colloids Surf. A Physicochem. Eng. Asp.* **2014**, *445*, 34–39. [CrossRef]
41. Abarca, R.L.; Rodríguez, F.J.; Guarda, A.; Galotto, M.J.; Bruna, J.E. Characterization of beta-cyclodextrin inclusion complexes containing an essential oil component. *Food Chem.* **2016**, *196*, 968–975. [CrossRef] [PubMed]
42. Benesi, H.A.; Hildebrand, J.H.J. A spectrophotometric investigation of the interaction of iodine with aromatic hydrocarbons. *J. Am. Chem. Soc.* **1949**, *71*, 2703–2707. [CrossRef]

43. Islam, M.M.; Chakraborty, M.; Pandya, P.; Al Masum, A.; Gupta, N.; Mukhopadhyay, S. Binding of DNA with Rhodamine B: Spectroscopic and molecular modeling studies. *Dye. Pigment.* **2013**, *99*, 412–422. [CrossRef]
44. Al Masum, A.; Chakraborty, M.; Ghosh, S.; Laha, D.; Karmakar, P.; Islam, M.M.; Mukhopadhyay, S. Biochemical activity of a fluorescent dye rhodamine 6G: Molecular modeling, electrochemical, spectroscopic and thermodynamic studies. *J. Photochem. Photobiol. B Biol.* **2016**, *164*, 369–379. [CrossRef]
45. Sayed, M.; Gubbala, G.K.; Pal, H. Contrasting interactions of DNA-intercalating dye acridine orange with hydroxypropyl derivatives of β-cyclodextrin and γ-cyclodextrin hosts. *New J. Chem.* **2019**, *43*, 724–736. [CrossRef]

Article

A Statistical Study on the Development of Metronidazole-Chitosan-Alginate Nanocomposite Formulation Using the Full Factorial Design

Hazem Abdul Kader Sabbagh [1], Samer Hasan Hussein-Al-Ali [1,2,*], Mohd Zobir Hussein [3,*], Zead Abudayeh [1], Rami Ayoub [1] and Suha Mujahed Abudoleh [1]

1. Department of Basic Pharmaceutical Science, Faculty of Pharmacy, Isra University, Amman 11622, Jordan; hazem.sabbagh@yahoo.com (H.A.K.S.); Zead.abudayeh@iu.edu.jo (Z.A.); rami.ayoub@iu.edu.jo (R.A.); abudoleh81@gmail.com (S.M.A.)
2. Department of Chemistry, Faculty of Science, Isra University, Amman 11622, Jordan
3. Materials Synthesis and Characterization Laboratory, Institute of Advanced Technology (ITMA), Universiti Putra Malaysia, 43400UPM Serdang, Selangor, Malaysia
* Correspondence: samer.alali@iu.edu.jo (S.H.H.-A.-A.); mzobir@upm.edu.my (M.Z.H.)

Received: 27 February 2020; Accepted: 22 March 2020; Published: 1 April 2020

Abstract: The goal of this study was to develop and statistically optimize the metronidazole (MET), chitosan (CS) and alginate (Alg) nanoparticles (NP) nanocomposites (MET-CS-AlgNPs) using a $(2^1 \times 3^1 \times 2^1) \times 3 = 36$ full factorial design (FFD) to investigate the effect of chitosan and alginate polymer concentrations and calcium chloride ($CaCl_2$) concentration on drug loading efficiency(LE), particle size and zeta potential. The concentration of CS, Alg and $CaCl_2$ were taken as independent variables, while drug loading, particle size and zeta potential were taken as dependent variables. The study showed that the loading efficiency and particle size depend on the CS, Alg and $CaCl_2$ concentrations, whereas zeta potential depends only on the Alg and $CaCl_2$ concentrations. The MET-CS-AlgNPs nanocomposites were characterized by X-ray diffraction (XRD), Fourier-transform infrared spectroscopy (FTIR), thermal gravimetric analysis (TGA), scanning electron microscopy (SEM) and in vitro drug release studies. XRD datashowed that the crystalline properties of MET changed to an amorphous-like pattern when the nanocomposites were formed.The XRD pattern of MET-CS-AlgNPs showed reflections at $2\theta = 14.2°$ and $22.1°$, indicating that the formation of the nanocompositesprepared at the optimum conditions havea mean diameter of (165 ± 20) nm, with a MET loading of $(46.0 \pm 2.1)\%$ and a zeta potential of (-9.2 ± 0.5) mV.The FTIR data of MET-CS-AlgNPs showed some bands of MET, such as 3283, 1585 and 1413 cm^{-1}, confirming the presence of the drug in the MET-CS-AlgNPs nanocomposites. The TGA for the optimized sample of MET-CS-AlgNPs showed a 70.2% weight loss compared to 55.3% for CS-AlgNPs, and the difference is due to the incorporation of MET in the CS-AlgNPs for the formation of MET-CS-AlgNPs nanocomposites. The release of MET from the nanocomposite showed sustained-release properties, indicating the presence of an interaction between MET and the polymer. The nanocomposite shows a smooth surface and spherical shape. The release profile of MET from its MET-CS-AlgNPs nanocomposites was found to be governed by the second kinetic model (R^2 between 0.956–0.990) with more than 90% release during the first 50 h, which suggests that the release of the MET drug can be extended or prolonged via the nanocomposite formulation.

Keywords: full factorial design; optimization; metronidazole; nanocomposites; sodium alginate; chitosan

1. Introduction

Design of experiment (DE) is a systematic method in research used to determine the relationship between independent variables and response variables [1,2]. There are different types of DE, which include factorial designs [3–5], fractional factorial designs [6,7], full factorial designs (FFD) [8,9], Plackett–Burman designs [10,11], central composite designs (Box–Wilson designs) [12,13], Box–Behnken designs (BBD) [14,15], Taguchi designs (TD) [16,17] and response surface designs (RSD) [18,19].

In pharmacy, the term optimization can be defined as the process of discovering the best way of using the existing resources while taking into account all the parameters that influence the decisions of any experiment [20]. Modern pharmaceutical optimization involves a systematic design of experiments to improve drug formulation. The process begins with predicting and evaluating the independent variables that affect the formulation response and selecting the best response values. With optimization, the formulation steps and preparation that fulfill the desired characteristics of the final product could be minimized.

Polynomial is one form of regression analysis. It is a non-linear analysis that correlates between the independent variable (x) and the dependent variable (y) as an nth degree polynomial in x. Different ways can be used in the fitting of the regression analysis for establishing approximate mathematical models. One of these fitting methods is called the stepwise method [21,22].It involves choosing the predictive variables by an automatic procedure [23,24]. In each step, a variable is added to or subtracted from the set of explanatory variables based on some pre-specified criteria.

After decades of basic nanosciences research, nanotechnology applications offer a wide range of opportunities in the fields of agriculture [24,25], food [26], environment [27,28] and drug delivery [29–31]. Nano-formulation technology has produced many new innovative drug delivery systems. Smart drug delivery, as well as polymeric nano-formulation as solid colloidal particles with diameters ranging from 1 to 1000 nm, preserves drugs against chemical decomposition and modifies drug release profiles in a controlled manner [25,26]. Polymeric nanoparticles are one example of nano-formulation. Research on polymeric nanoparticles has been especially focused on their role in drug delivery and drug targeting owing to their particle size and long circulation in the blood [27,28].They can be used therapeutically in vaccines, or as drug carriers, in which the drug can be encapsulated, entrapped, chemically attached, adsorbed or dissolved [29].

Chitosan and alginate, which are polymeric materials, were widely used in the development of nano-formulation products [30]. Both are non-toxic, stable hydrophilic polymers [31,32]. Chitosan-alginate has been used as a sustained release polymer matrix in different dosage forms [30,31,33,34]. Drug side effects may occur when administered in large quantities and sustained release formulations in nanocomposites by a single dose might be a suitable way to decrease drug complications due to its high concentration and increased patient compliance [35–38].

The drug used in this research, metronidazole (MET), is an antibiotic drug usually used to treat bacterial infections of the vagina, stomach or intestines, liver, skin, joints, brain, heart and respiratory tract. However, it is ineffective for viral infections (such as the common cold and flu).

There have been many attempts by researchers to load MET using nano-formulations such as nanostructure lipid carriers (NLCs) [39], nano-emulsions [40], MET loaded into niosomes and then coated on dental implants using a layer-by-layer dip-coating technique with poly(lactic acid) (PLA) [41] and magnetic nanocomposites [42].

In the present study, the incorporation of MET into polymer nanoparticles was achieved. To the best of our knowledge, this work is reported for the first time, where MET as a guest drug was encapsulated into CS for the formation of CS-Algnanoparticleswith optimized preparation parameters.Traditionally, optimization is done by evaluating each factor independently. However, in a single-factor experiment, the interactions between important parameters are ignored. Response surface methodology (RSM) can be used to analyze the interactions between the different variables. The experimental data is input as a quadratic equation and the response is predicted. Stepwise regression analysis is one of the methodsthat can derive the best equation that can describe the data via surface or contour plots. Thus, in this work, Minitab softwareversion 18.1 and full factorial design were used to examine the effect of three independent variables (concentration of CS, Alg, and $CaCl_2$) on three dependent variables (loading efficiency, particle size and zeta potential) for the synthesis of CS-Alg nanoparticles.

2. Materials and Methods

2.1. Materials

The chemicals used in this study are metronidazole ($C_6H_9N_3O_3$ (99% purity), Sigma-Aldrich, Taufkirchen, Germany), low molecular weight chitosan (10–120 kDa, 90% deacetylation, Sigma-Aldrich), and low viscosity sodium alginate (10–100 kDa, AZ Chem., Sigma Aldrich). All HPLCsolutions were used from VWR (West Chester, PA, USA). All other chemicals, including acetic acid and calcium chloride, were purchased from Chem CO (Port Louis, Mauritius).

2.2. Preparation of CS-Alg Nanoparticles and MET-CS-AlgNPs Nanocomposites

The method used was a modification of what is called the ionotropic pregelation method [43,44]. Solutions of CS, Alg and $CaCl_2$ were first prepared. The pH of the CS and Alg solutions was adjusted to 5.5 and 5.0, respectively. The first step was the formation of the AlgNPs pre-gel, which was achieved by adding 6 mL of different concentrations of aqueous $CaCl_2$ solution to 10 mL of Alg, followed by 30 min of stirring. The second step was the addition of 4 mL of CS solution to the AlgNPs pre-gel with stirring for another 30 min. The resultant solution was stirred overnight at room temperature to form uniform nanoparticles. The same procedure was used to form MET-CS-AlgNPs nanocomposites using only 100 milligrams of MET mixed with the Alg solution.

2.3. Methodology

First, the modeling of the responses (loading efficiency, particle size and zeta potential) was presented. Secondly, the FFD was built to perform the experiments. This was followed by the use of multiple regressions to develop the loading efficiency, particle size and zeta potential model responses. Finally, the analysis of concentration variance was used to analyze the experimental data to predict the effects and contribution of parameters on responses.

2.3.1. Modeling of Different Responses

The loading efficiency percentis the first response, which was taken as a parameter and was defined as the amount of total entrapped drug divided by the total weight of the nanoparticles. The second and third responses measured were particle size and zeta potential. Table 1 shows the three independent parameters and their levels.

Table 1. Independent parameters and their levels.

Parameter		Levels (mg)		
		Low	Medium	High
A	Alg	200	-	400
B	CS	50	100	200
C	$CaCl_2$	30	-	60

2.3.2. Full Factorial Design

Full factorial design (FFD) is a method used by researchersto design experiments that consist of several factors with separate possible levels. With FFD, the experiment takes all possible combinations of the levels across all such factors. FFD allows researchers to study the effect of each factor, as well as their interactions, on the response variable [45]. In this study, the FFDwas used to conduct the experiments. Therefore, $(2^1 \times 3^1 \times 2^1) \times 3 = 36$ combinations were used, corresponding to $n = 3$ parameters or factors (CS, Alg and CaCl$_2$) (Table 2).

Table 2. Composition of formulations.

Std Order	Run Order	Sample Code	Alg	CS	CaCl$_2$
17	1	MAC1	200	200	30
24	2	MAC2	400	200	60
10	3	MAC3	400	100	60
2	4	MAC4	200	50	60
35	5	MAC5	400	200	30
20	6	MAC6	400	50	60
32	7	MAC7	400	50	60
6	8	MAC8	200	200	60
22	9	MAC9	400	100	60
29	10	MAC10	200	200	30
36	11	MAC11	400	200	60
14	12	MAC12	200	50	60
25	13	MAC13	200	50	30
5	14	MAC14	200	200	30
9	15	MAC15	400	100	30
1	16	MAC16	200	50	30
31	17	MAC17	400	50	30
26	18	MAC18	200	50	60
3	19	MAC19	200	100	30
7	20	MAC20	400	50	30
16	21	MAC21	200	100	60
11	22	MAC22	400	200	30
28	23	MAC23	200	100	60
27	24	MAC24	200	100	30
13	25	MAC25	200	50	30
23	26	MAC26	400	200	30
30	27	MAC27	200	200	60
15	28	MAC28	200	100	30
34	29	MAC29	400	100	60
18	30	MAC30	200	200	60
12	31	MAC31	400	200	60
19	32	MAC32	400	50	30
4	33	MAC33	200	100	60
8	34	MAC34	400	50	60
33	35	MAC35	400	100	30
21	36	MAC36	400	100	30

2.4. MET Loading Efficiency

High-speed centrifugation was used to determine the loading efficiency (LE) of MET in the prepared nanocomposites, in which 2.0 mL of suspension were centrifuged (Hettich Universal 30 RF) at 10,000 rpm for 10 min, and the drug loading was measured by high-performance liquid chromatography (HPLC, Shimadzu, Japan), using a Venusil C18 column (4.6 mm × 250 mm, 5 µm) at 25°C. The UV detection wavelength was 323 nm, and the mobile phase was prepared by mixing acetonitrile/0.1% with phosphoric acid (5:95, v/v). The flow rate was 1.0 mL/min. The LE was calculated as follows:

$$\% \text{ Loading Effciency (LE)} = \frac{T_p - T_f}{\text{mass of nanoparticles}} \times 100 \qquad (1)$$

where T_p is the total mass of MET used to prepare the nanocomposites, and T_f is the free mass of MET in the supernatant.

2.5. Particle Size and Zeta Potential of Nanocomposites

Particle size and zeta potential of the nanocompositeswere analyzed through a dynamic light scattering (DLS) method using a Zetasizer Nano S (Malvern, UK) at the Arab Pharmaceutical Manufacturing. The analysis was performed in triplicates at a temperature of 25 °C.

For the particle size analysis, the samples were dispersed in distilled water, the cells were filled and capped and checked for the absence of any bubbles.

The samples were prepared for zeta potential analysis by dispersing themin the distilled water and measuring the zeta values at 25 °C.

2.6. Controlled Release Study of the MET from the Nanocomposites

The in vitro release of MET from the nanocomposites wasdetermined in a solution at pH 1.2, using a Perkin Elmer UV–Vis spectrophotometer with λ_{max} of 323 nm. A suitable amount of each nanocomposite was added to 2 mL of the media. The cumulative amount of drug released into the solution was measured at preset time intervals at corresponding λ_{max}.

The percentage release of MET into the release media was calculated according to the formula:

$$\%Release = \frac{\text{Concentration of MET at time t (ppm)}}{\text{Concentration corresponding to 100\% release of MET (ppm)}} \times 100 \quad (2)$$

The concentration corresponding to 100% release was obtained by adding a known amount of the nanocomposites into 2 mL HCl followed by sonicating and heating the nanocomposites at 37 °C.

2.7. Instrumentation

Powder X-ray diffraction (XRD) patterns were used to determine the crystal structure of the samples in the range of 2–70 degrees on an XRD-6000 diffractometer (Shimadzu, Tokyo, Japan) using CuK$_\alpha$ radiation (λ=1.5406 Å) at 30 kV and 30 mA at Universiti Putra Malaysia. Fourier transform infrared spectroscopy (FTIR) spectra of the materials were recorded over the range of 400–4000 cm^{-1} on a Perkin Elmer (model Smart UAIR-two). The thermogravimetric analysis was carried out using a Metter-Toledo 851e instrument (Switzerland) with a heating rate of 10 °C min^{-1}, in 150 µL alumina crucibles and in the range of 30–900 °C. The zeta potential was measured at 25 °C by the dynamic light scattering (DLS) method using a Malvern Zetasizer Nano ZS (Malvern Instruments, Malvern, UK). UV–Vis spectra were measured to determine the release profiles using a Shimadzu UV-1601 spectrophotometer.

3. Results and Discussion

3.1. MultipleLinear Regression Analysis

Multiple regression analysis is a statistical method that is used to estimate the correlation between dependent and independent variables. The term correlation coefficient (R^2) indicates how well the data fit the multiple regression models. It provides a measure of how well-observed outcomes are replicated by the model, as the proportion of total variation of outcomes explained by the model. An R^2 close to 1 indicates that the regression model perfectly fits the data; the higher the R^2, the more the dependent variations are explained by input variables and therefore the better the model.

However, the demerit with R^2 is that it will stay the same or increase with the addition of more variables, even if they do not have any relationship with the output variables. This can be solved by using the "adjusted R square", which is sensitive for adding variables that do not improve the model.

The linear (CS, Alg, and CaCl$_2$), linear-square (CS*CS, Alg*Alg, and CaCl$_2$*CaCl$_2$) and linear-interaction equations (CS*Alg, CS*CaCl$_2$ and Alg*CaCl$_2$) have been fitted using a Minitab software for LE, particle size and zeta potential variables. The data was analyzed using stepwise regression which is a way to build a model by adding or removing independent variables, usually via a series of F-tests or T-tests. The variables to be added or removed are chosen based on the test statistics of the estimated coefficients.

Table 3 shows the ANOVA values for LE, particle size and zeta potential given in the suggested models. The P-value is less than 0.05, showing the model which is significant at a 95% confidence level. These LE, particle size and zeta potential models show that lack-of-fit error value is insignificant (0.283, 0.821 and 0.432, respectively) indicating that the fitted model is accurate enough to predict the response. The mathematical models were developed to determine the optimal values of the MET-CS-AlgNPs formulation conditions leading to maximum values of LE, minimumvalues of particle size and a negative value (~20 mV) of zeta potential.

Table 3. ANOVA values for loading efficiency (LE), particle size and zeta potential.

	DF	Adj SS	Adj MS	F value	Coef	T Value	VIF	P value
LE model								
Model	7	9585.39	1369.34	337.95	47.908	67.86	-	0.000
Alg	1	1771.00	1771.00	437.07	−7.385	−20.91	1.04	0.000
CS	1	431.85	431.85	106.58	4.361	10.32	1.04	0.000
$CaCl_2$	1	208.24	208.24	51.39	−2.532	−7.17	1.04	0.000
CS*CS	1	2.09	2.09	0.51	−0.605	−0.72	1.03	0.480
Alg*CS	1	236.52	136.52	9.01	1.252	3.00	1.04	0.006
Alg*$CaCl_2$	1	6545.51	6545.51	1615.40	−13.998	−40.19	1.01	0.000
CS*$CaCl_2$	1	22.71	22.71	5.60	0.987	2.37	1.04	0.026
Lack-of-fit	4	20.77	5.19	1.35	-	-	-	0.283
Particle size model								
Model	7	141548	20221.1	202.86	185.00	51.90	-	0.000
Alg	1	45889	45889.3	460.35	−43.50	−21.46	1.10	0.000
CS	1	30270	30270	303.67	44.42	17.43	1.12	0.000
$CaCl_2$	1	19575	19574.9	196.37	−28.53	−14.01	1.12	0.000
CS*CS	1	6104	6103.7	61.23	−36.54	−7.83	1.13	0.000
Alg*CS	1	1963	1962.6	19.69	11.64	4.44	1.19	0.000
Alg*$CaCl_2$	1	700	700.2	7.02	−5.43	−2.65	1.06	0.016
CS*$CaCl_2$	1	11146	11145.6	111.81	−27.64	−10.57	1.17	0.000
Lack-of-fit	4	173	43.4	0.38	-	-	-	0.821
Zeta potential model								
Model	7	399.875	57.125	303.51	−10.501	−44.19	-	0.000
Alg	1	85.093	85.093	452.11	2.119	21.26	1.07	0.000
CS	1	0.256	0.256	1.36	0.133	1.17	1.19	0.263
$CaCl_2$	1	191.991	191.991	1020.07	3.308	31.94	1.25	0.000
CS*CS	1	30.172	30.172	160.31	3.347	12.66	1.13	0.000
Alg*CS	1	0.404	0.404	2.15	0.164	1.47	1.17	0.165
Alg*$CaCl_2$	1	181.922	181.922	966.57	-3.182	−31.09	1.22	0.000
CS*$CaCl_2$	1	1.855	1.855	9.85	0.344	3.14	1.05	0.007
Lack-of-fit	3	0.562	0.187	0.99	-	-	-	0.432

DF: degrees of freedom, SS: sum of squares, F: F-test value and P: error variance.

The equations can be given in terms of the coded values of the independent variables as shown in the following Table 4.

Table 4. Regression model for dependent variables.

Regression Model	R-sq (%)	R-sq (adj)%
LE= −46.07 + 0.3252Alg − 0.0045 CS +2.5210CaCl$_2$ 0.000108 CS*CS+0.000167Alg*CS − 0.009332Alg*CaCl$_2$+ 0.000878 CS*CaCl$_2$	98.91	98.62
Size= 96.7 − 0.4660Alg + 2.856CS + 2.256CaCl2 − 0.006495CS*CS + 0.001552Alg*CS − 0.00362 Alg*CaCl2 − 0.02457CS*CaCl2	98.68	98.19
Potential= −43.80 + 0.11391Alg − 0.1673CS + 0.8187CaCl2 + 0.000595CS*CS + 0.000022Alg*CS − 0.002121Alg*CaCl2 + 0.000305CS*CaCl2	99.35	99.02

Table 4 shows the regression model for three dependent variables for LE, particle size and zeta potential. The LE model showed that R-square values were found to be 98.91%, 98.68% and 99.35%, respectively. Moreover, the Adj-R-square values were found to be 98.62%, 98.19% and 99.02%, respectively.

3.2. Evaluation of the Models

3.2.1. Pareto Chart of Responses of Standardized Effects and Normal Plot of the Standardized Effects

A Pareto chart (Figure 1) is a graphical overview of the process factors and/orinteractions of influence, in ranking order from the most influencialto theleast influencial. A threshold line (P-value 0.05) indicates the minimum magnitude of statistically significant effects, considering the statistical significance of 95%.

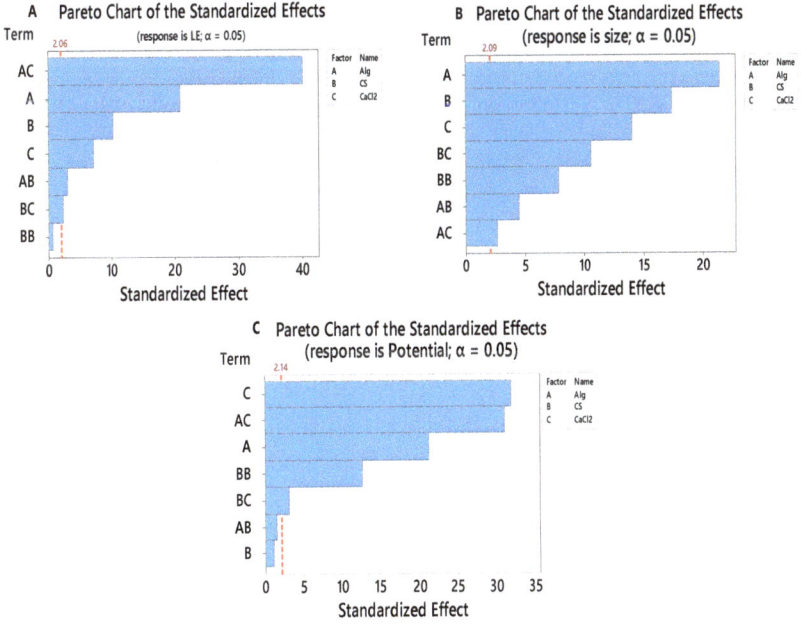

Figure 1. Pareto Chart of the standardized effects toward LE (A), particle size (B) and zeta potential (C).

Figure 1A indicates that the effect of BB i.e., CS × CS is statistically insignificant toward LE. The effect of AC (Alg × CaCl$_2$) has the highest standardized effect on the LE followed by A, B, C, AB and BC. Hence, the term BB should not be considered for the empirical relation. The insignificance of BB can also be reasserted from the normal plot (Figure 1A), in which the points that do not fall near the fitted line are important. The factors having a negligible effect on the output response tend to be smaller and are centeredaround zero.

Figure 1B represents the effect of different parameters on particle size. The results indicate that all the effects are statistically significant. Factor A (Alg) has the highest standardized effect on the particle size followed by B, C, BC, BB, AB and AC. The significance of factors can be shown in the normal plot (Figure 1B).

Figure 1C shows the effect of different factors on the zeta potential response. The main factors (A, and C), square factors (B*B) and 2-way interaction (A*C and B*C) have a statistically significant effect on the response. C (CaCl$_2$) has the highest standardized effect on the zeta potential followed by AC, A, BB, and BC. Hence, the terms AB and B should not be considered for the empirical relation.

3.2.2. Contour Plot and Surface Plot of LE, Particle Size and Zeta Potential Against Selected Independent Variables

The effect of the formulation and process variables on LE responsecan be evaluated by studying thecontour and response surface plots. Figure 2A-1,A-2 shows the response plots of LE as a function of CS and Alg concentrations, and it is seen to display a stationary ridge pattern. As the color gets darker, the LE response increases. The stationary ridge has a flat shape. Increasing the concentration of CS and decreasing the Alg can afford more space for LE (>55%). In Figure 2B-1,B-2, the contour and response surface plots show minimax patterns, with the stationary point (saddle point) being near the center of the design. From the stationary point (saddle point), increasing CaCl$_2$ concentration while decreasing the Alg concentrationled to an increase in the LE response. Figure 2C-1,C-2 showsa flat shaped stationary ridge, and increasing the concentration of CS while decreasing CaCl$_2$ concentration led to an increase of the LE by more than 52%.

Figure 3A-1,A-2 shows a rising ridge pattern. As the color gets lighter, the particle size decreases. The minimum particle size was achieved using high concentrations of Alg and the lowest concentration of CS. From Figure 3A-2 it can be seen that the particle size below 50 nm can be prepared using 50 mg of CS and 400 mg of Alg. Figure 3B-1,B-2 shows that the particle size below 120 nm can be prepared by using 400 mg of Alg and 60 mg of CaCl$_2$. In the case of CS and CaCl$_2$ variables in Figure 3C-1,C-2, rising ridge pattern can also be seen. The particle size lower than 120 nm can be obtained using CS concentrations of 200 mg and CaCl$_2$ concentrations ranging between 30 and 60 mg.

Figure 4 shows the3Dresponse surface and contourplots of the combined effect of CS, Alg and CaCl$_2$on the zeta potential charge. The plots show that all the variables affect the zeta potential with rising ridge patterns. Figure 4A-1,A-2 shows the combined effect of Alg and CS concentrations; when the color gets lighter, the zeta potential becomesgreater than −12.5 mV, whereas when the color gets darker, the zeta potential becomesless than −5.0 mV. The zeta potential was higher than −5.0 mV when the Alg concentration was higherthan 300 mg and CS concentration was between 50–75 mg and 160–200 mg, whereas the zeta potential was lower than −12.5 mV when the concentration of Alg was less than 300 mg and the CS concentration was between 60–185 mg.

Figure 4B-1,B-2 shows the contour plots of the effect of Alg and CaCl$_2$ on the zeta potential. The zeta potential was between −8 and −18 mV; it was around −18 mV at low concentrations of both Alg and CaCl$_2$, and around −8 mV at Alg concentrations between 200–350 mg with concentrations of CaCl$_2$ between 55–60 mg.

Figure 4C-1,C-2 shows that the 3D surface and contour plots represent a rising ridge pattern. As the color gets darker, the zeta potential response reaches −4 mV; this occurs at high concentrations of CaCl$_2$ of 55–60 mg and CS concentrations below 50 mg and higher than 200 mg. The zeta potential response at −4 mV can be achieved at a low concentration of CaCl$_2$ of below 30 mg and a CS concentration between 75–175 mg.

3.2.3. Main effects plot for LE, particle size and zeta potential

Figure 5 shows a plot of the main effects (CS, Alg and CaCl$_2$) used to examine differences between level means for LE, particle size and zeta potential factors. All factors seem to affect the LE, particle size and zeta potential because the line is not horizontal. Figure 5A shows that Alg at a concentration of 200 mg gave a higher LE (55%) compared to 400 mg (40%). A CaCl$_2$ concentration of 30 mg had a higher LE mean (50%) than the one at 60 mg (45%). The CS also affected the LE, with 200 mg of CS having had a higher LE mean (51%) than at 60 mg (43%). It is evident from Figure 5B that particle size is minimal (≈150 nm) at the highest level of Alg (400 mg) and CaCl$_2$ (60 mg). In addition, the minimal particle size of approximately 100 nm can be obtained with the lowest level of CS (60 mg).

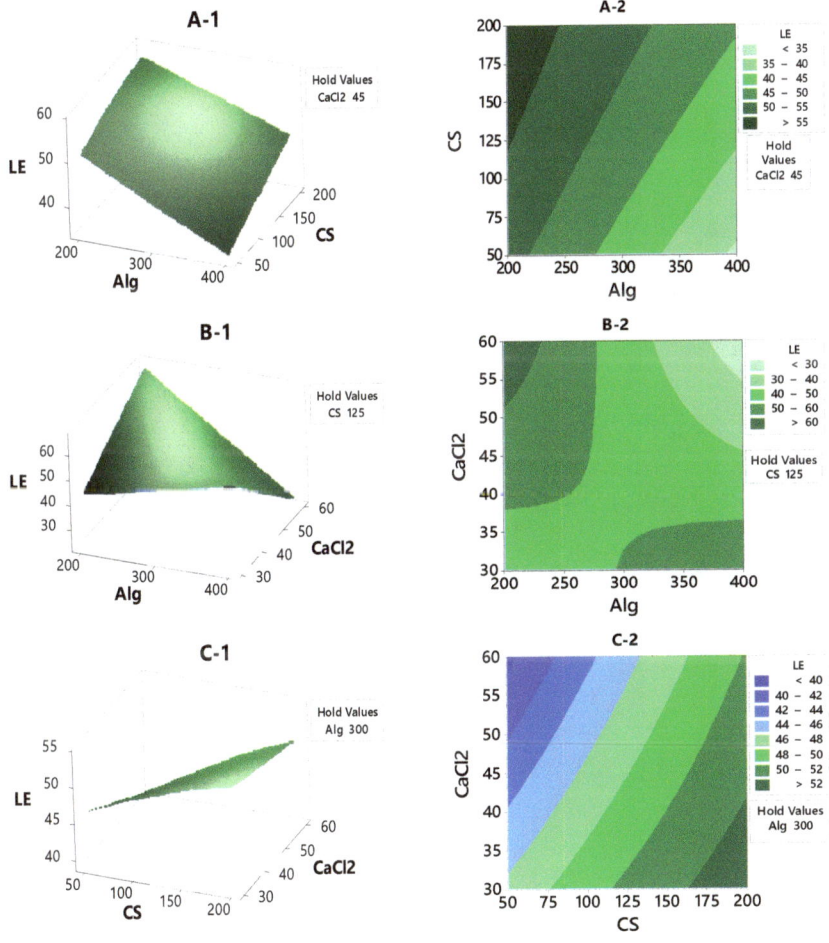

Figure 2. The contour plot and response surface of the LE with variances of CaCl$_2$, Alg and CS concentrations.

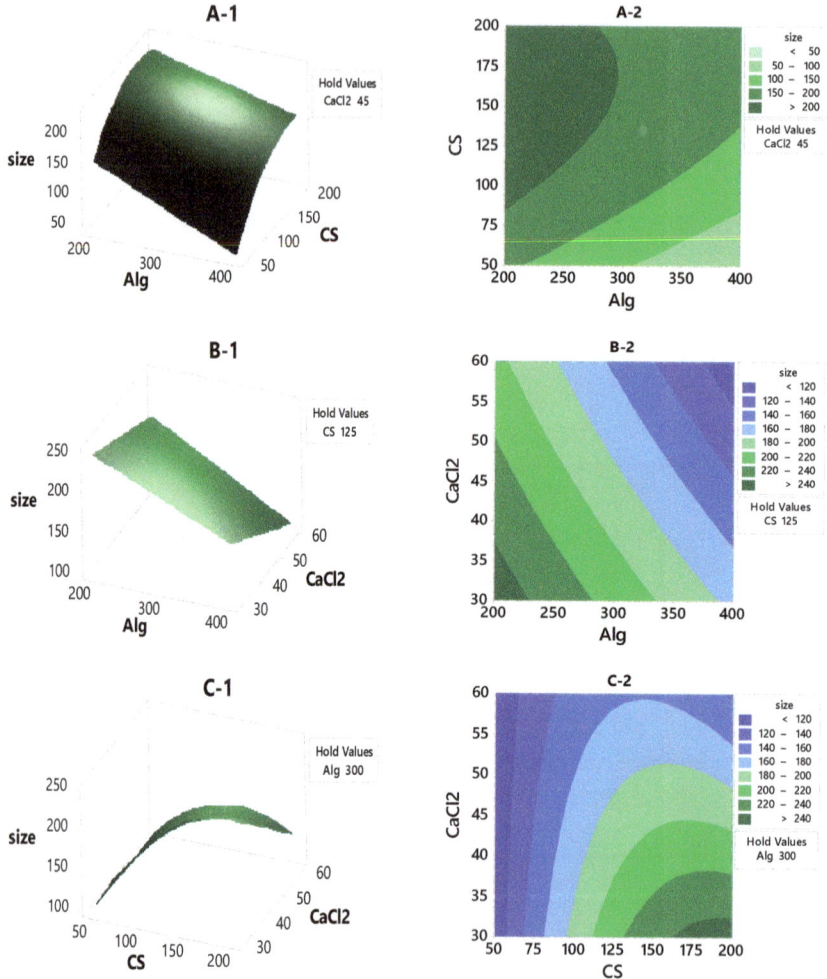

Figure 3. The contour plot and response surface of the particle size with variances of $CaCl_2$, Alg and CS concentrations.

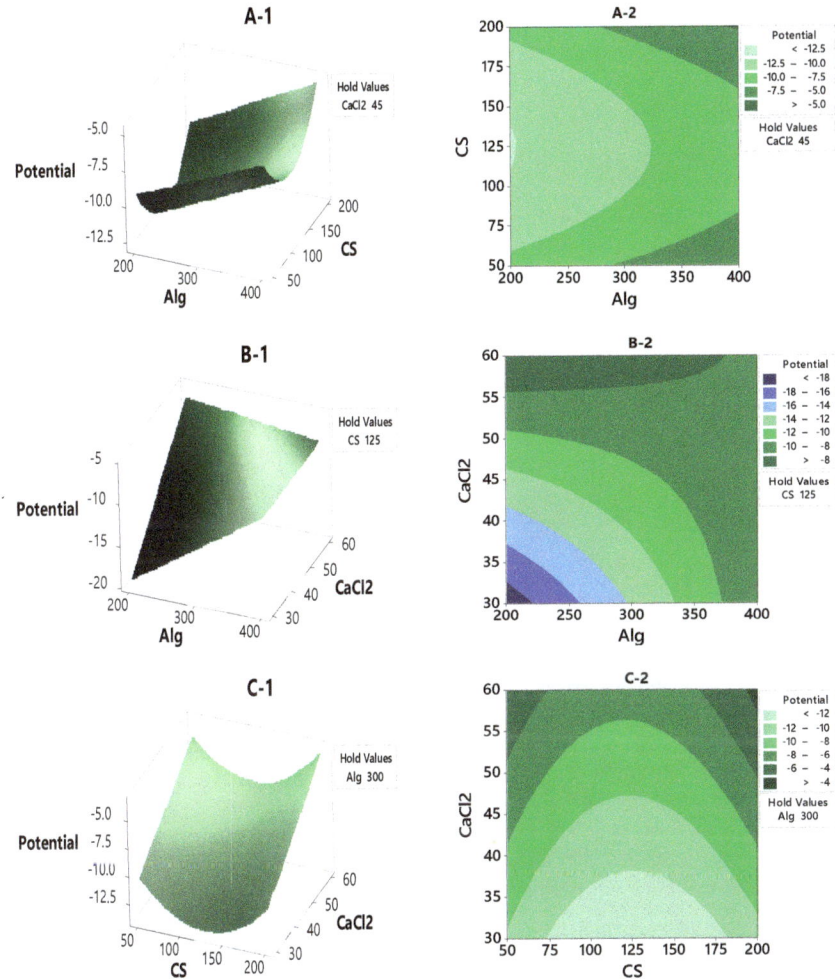

Figure 4. The contour plot and response surface of the zeta potential with variances of CaCl$_2$, Alg and CS concentrations.

Based on the main effect plots in Figure 5C, the zeta potential was found to be the lowest at all of the highest values of Alg, CS and CaCl$_2$ parameters tested. Both the parameters of Alg and CaCl$_2$ concentrations show a linear potential pattern with an increase in their levels. However, CS concentration shows otherwise; although the highest level of CS concentration tested resulted in −7 mV potential, its mid-point shows a downward curvature in its response. The −11, −10 and −14 mV values of the mean zeta potential are observed at 200 mg of Alg, 120 mg of CS and 30 mg of CaCl$_2$. From our studies, based on their potential data, the prepared nanocomposites were stable.

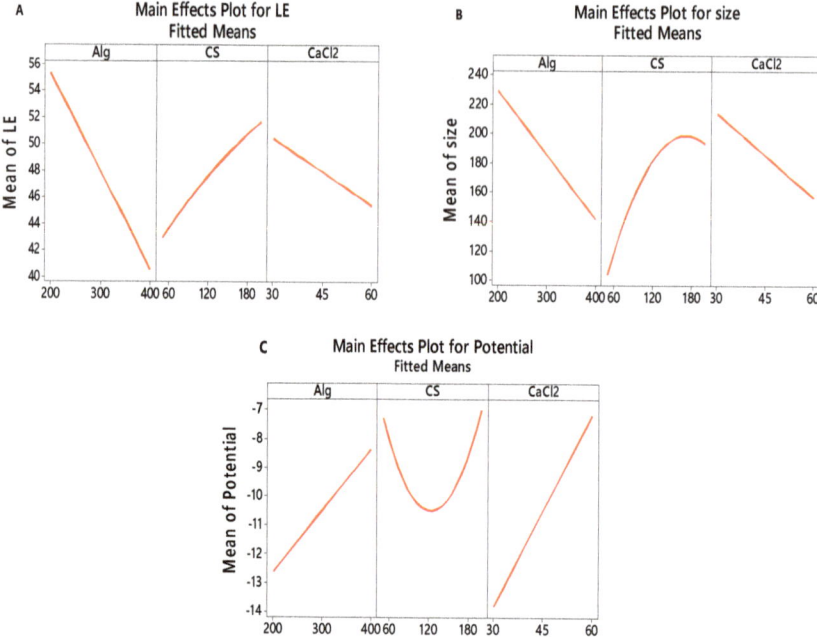

Figure 5. Main effects plot for LE, particle size and zeta potential.

3.2.4. The Interaction between the Factors thatAffects the LE, Particle Size and Zeta Potential

The interaction plots in Figures 6–8 show how the relationship between one independent factor and a continuous response depends on the value of the second independent factor. The plot displays mean values for the levels of one factor on the x-axis and a separate line for each level of the other factor. The parallel lines in the interaction indicate that there is no relationship between the variables. When an interaction occurs, the lines are less parallel, and the strength of the interaction becomes greater.

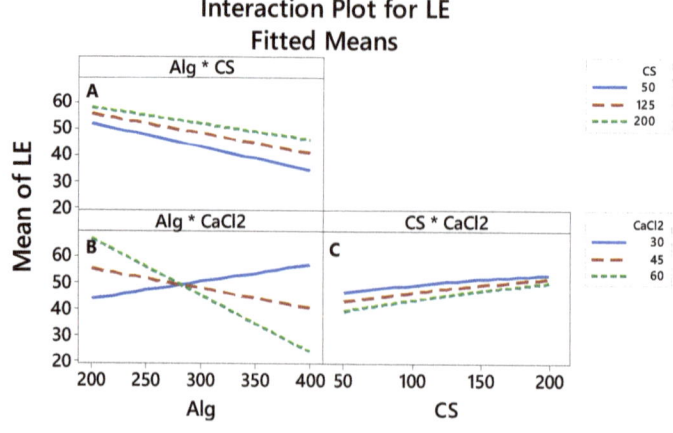

Figure 6. Interaction effects of factors on the loading efficiency.

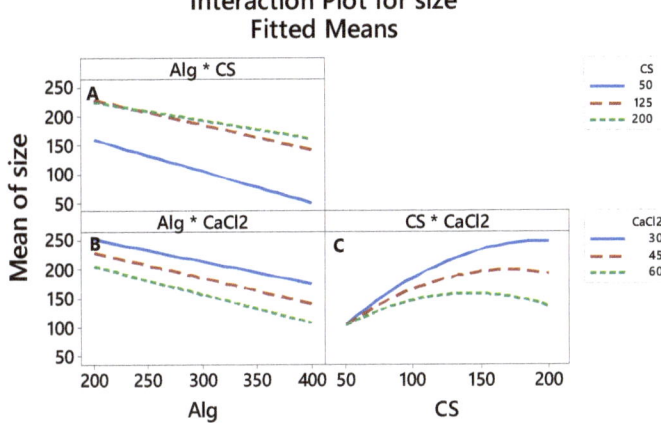

Figure 7. Interaction effects of factors on the particle size.

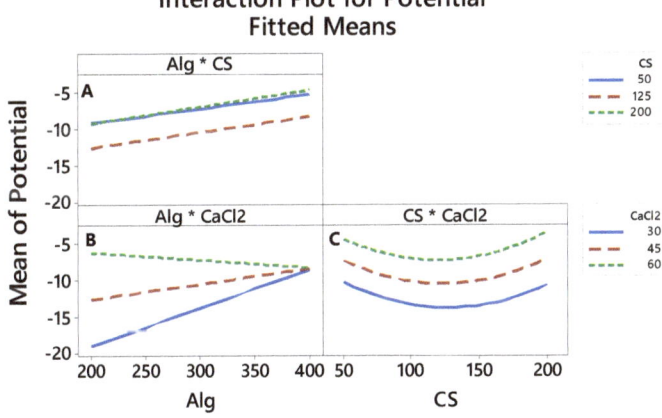

Figure 8. Interaction effects of factors on the zeta potential.

In this interaction plot, the lines in Figure 6A are parallel, which indicates that there is a relationship between the variables. The interaction in Figure 6B has a nonparallel line, indicating that the relationship between Alg and LE depends on the value of $CaCl_2$. For example, if 300 mg of Alg is used, then 30, 45 and 60 mg of $CaCl_2$ are associated with the 50 % LE means, similar to that in Figure 6C.

Figure 7 shows that there is an interaction between the Alg*CS (Figure 7A) and CS*$CaCl_2$ (Figure 7C). Figure 7A shows that there is a significant interaction between Alg and CS. The green and redlines (200 and 125 mg CS, respectively) show that the mean size response decreases when the Alg factor level is low, while in Figure 7C, the green, red and blue lines, which correspond to 60, 45 and 30 mg $CaCl_2$, respectively, show that the particle size mean response decreases when the CS factor level is low.

The interaction plotsshown in Figure 8A,B show the lines are not parallel, indicating that the relationship between Alg concentration and zeta potential depends on the value of CS (Figure 8A) and $CaCl_2$ (Figure 8B). For example, when Algwas used at concentrations of 200 mg, then $CaCl_2$ at 30 mg was associated with the −20 mV mean zeta potential (Figure 8B). However, when Alg with concentrationsof 200 mg was used, then CS at 50 and 125 mg was associated with −10 mV mean zeta potential (Figure 8A).

The Normal Plot of the Standardized effects, Normal probability plots, Residuals versus fitted value and Residuals versus observation toward LE, Particle Size and Zeta Potential (Figures S1–S4 in Supplementary Materials).

3.3. Optimization of LE, Particle Size and Zeta Potential

In this study, the data was used to build a mathematical model such as linear, linear interaction, linear square and second-order model. Table 5 shows the selected mathematical model used to optimize the conditions of 46.05% for LE, minimizing the particle size to a 164 nm value and achieving a −9.25 mV zeta potential, using 350 mg Alg, 150 mg CS and 40 mg CaCl$_2$ (Figure 9).

Table 5. Response optimization plot for different responses.

Value	Alg (350 mg)	CS (150 mg)	CaCl$_2$ (40 mg)
	Optimization Responses		
LE	46.0 ± 2.1%		
Minimum Size	164.71 ± 20.03 nm		
Zeta potential	−9.25 ± 0.51 mV		

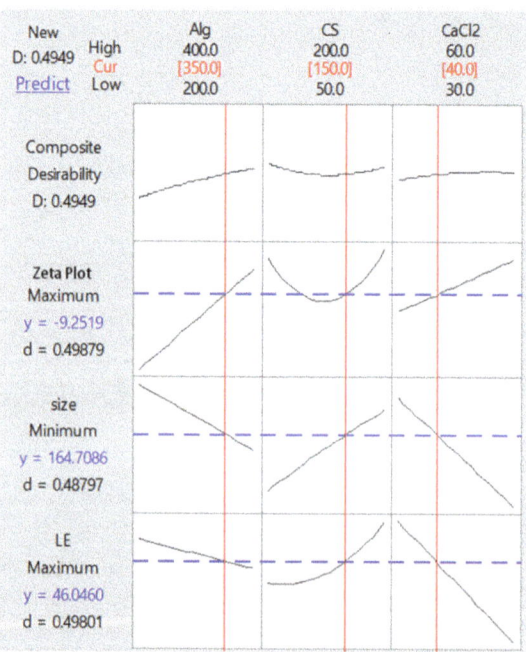

Figure 9. The optimization plot for metronidazole (MET), chitosan (CS) and alginate (Alg) nanoparticles (NP) (MET-CS-AlgNPs) nanocomposites.

3.4. Validation Test for Building Model

The comparison of experimental results with predicted values is shown in Table 6. From the table, the theoretical values for response were close to the experimentally obtained values. This result indicates that the mathematical models can be successfully used to predict the LE, particle size and zeta potential values for any combination of the Alg, CS and CaCl$_2$ within the range of the performed experimentation.

Table 6. Response optimization for LE, particle size and zeta potential.

No.	Alg	CS	CaCl$_2$	%LE			Particle Size (nm)			Zeta Potential (mV)		
				Exp	Theo	Error %	Exp	Theo	Error %	Exp	Theo	Error %
1	300	100	50	45.0	43.0	4.7	115	126	8.7	−9.5	−8.9	6.7
2	200	200	30	43.3	45.5	4.8	285	277	2.9	−14.5	−16.2	10.5
3	350	150	40	48.8	46.0	6.1	150	165	9.1	−10.8	−11.5	6.1

3.5. X-Ray Diffraction of MET-CS-AlgNPs Nanocomposites

From the literature, the XRD diffractogram of CS shows crystalline properties with an intense peak at 2θ = 19.7°. At the same time, the XRD diffractogram of Alg shows semi-crystalline properties with a peak at 2θ = 13.6° [46].

XRD patterns of pure MET, CS-AlgNPs and MET-CS-AlgNPs nanocomposite formulations are illustrated in Figure 10A–C. The MET powder shows two sharp single peaks at 2θ = 11.0° and 22.3°, whereas the blank CS-AlgNPs nanoparticles gave a peak at 2θ = 14.9° and 21.6°, which indicates there is an amorphous pattern. The intensity of the diffraction peak of the CS-AlgNPs nanoparticles at 21.6° 2θ decreased after loading of MET and the peak for MET at 2θ = 11.0 and 22.3° disappeared in the MET-CS-AlgNPs nanocomposite. This might be due to the loading of MET inside the amorphous region of the nanocomposite matrix.

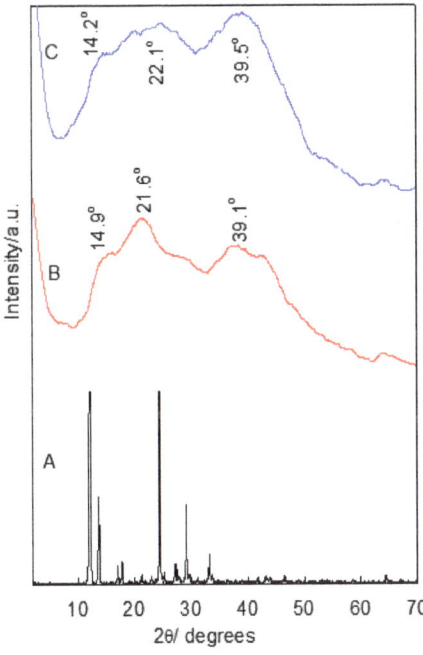

Figure 10. XRD diffraction spectra of MET (**A**), CS-AlgNPs (**B**) and MET-CS-AlgNPs (**C**).

3.6. FTIR Spectroscopic Analysis of CS-AlgNPs and MET-CS-AlgNPs

FTIR spectra of MET, CS-AlgNPs and MET-CS-AlgNPs are presented in Figure 11A–C. The FTIR spectra of pure MET (Figure 11A) show characteristic peaks at 3457 cm^{-1} (Hydroxyl –OH), 3100 cm^{-1} (C–C stretching), 1534 and 1366 cm^{-1})nitroso N–O stretch), and 1075, 875 cm^{-1} (C–N stretch) [47].

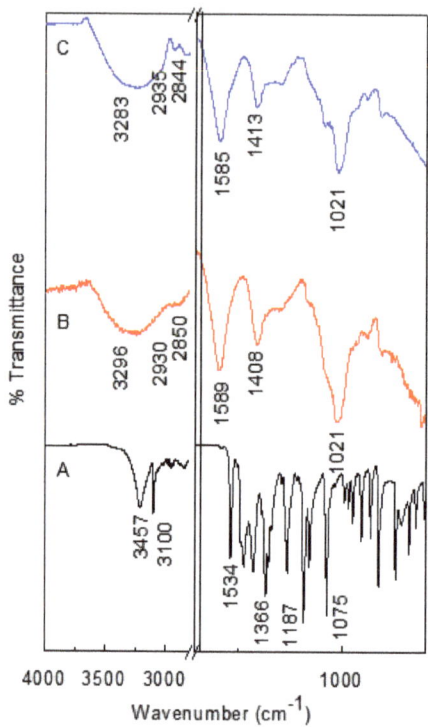

Figure 11. FTIR spectra of MET (**A**), CS-AlgNPs (**B**) and MET-CS-AlgNPs (**C**).

For CS-AlgNPs (Figure 11B), a band at 3296 cm^{-1} was observed due to O–H and N–H stretching. Absorptions due to vibration asymmetry CH$_2$ and symmetry CH$_2$ were located at 2930 and 2850 cm^{-1}, respectively. A strong band near 1589 cm^{-1} corresponds to the C=O, C–N and N–H bending of amide I. Asymmetric stretching band of the COO$^-$ group was centered near to 1420 cm^{-1} [48].

For MET-CS-AlgNPs (Figure 11C), some bands were downshifted; for example, from 3296to 3283 cm^{-1}, from 1589 to 1585 cm^{-1} and from 1408 to 1413 cm^{-1}. This can be explained due to the interaction between MET and CS-AlgNPs.

3.7. Thermogravimetric Analysis of MET-CS-AlgNPs Nanocomposites

The thermal decomposition process of MET-CS-AlgNPs nanocomposites and its pure counterpart CS-AlgNPs was evaluated by TGA/DTG analyses. These analysis curves give thepercentage weight loss due to the thermal decomposition (Figure 12). The results show that a pure MET sample undergoes a one-stage thermal degradation process, whileCS-AlgNPs and MET-CS-AlgNPs samples are degraded in a three-stage process. For the MET sample, the decomposition process occurred between 137–288 °C and with a mineral residue of 0.9% [49], which was due to the vaporization of volatile components [50].

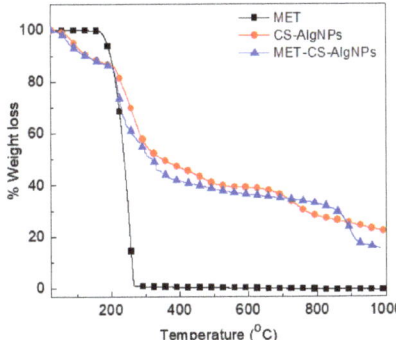

Figure 12. TGA curves of MET, CS-AlgNPs and MET-CS-AlgNPs nanocomposites.

The CS-AlgNPs show three main thermal stages; the first stage of the decomposition process occurred between 60–200 °C, which was due to the vaporization of volatile components, such as water molecules immobilized between chitosan chains during the coating process [51]. Based on the structure of CS and Alg, H_2O molecules can be bounded by the hydroxyl group [52].

The second stage of weight loss, which occurred between 200–520 °C, is due to the release of water bound to the functional groups of CS and Alg polymers, which was not completely removed in the first step of the dehydration, and to the degradation of both polymers.

A third inflection point occurred between 520–800 °C, which may be associated with the decomposition of functional groups of both polymers which were not completely removed by the previous stages.

The TGA of MET-CS-AlgNPs (Figure 12) also shows three weight loss steps similar to CS-AlgNPs. The MET-CS-AlgNPs shows 70.2% weight loss compared to 55.3% for CS-AlgNPs. The extra weight loss is due to the incorporation of MET in the CS-AlgNPs.

3.8. Scanning Electron Microscopy

The CS-AlgNPs and MET-CS-AlgNPs were morphologically characterized using the SEM (Figure 13). The micrographs of CS-AlgNPs (Figure 13A) show that the nanoparticles have a smooth surface with a spherical shape which is in agreement with previous studies [53]. Figure 13B shows that MET-CS-AlgNPsnanocomposites also have a spherical shape.

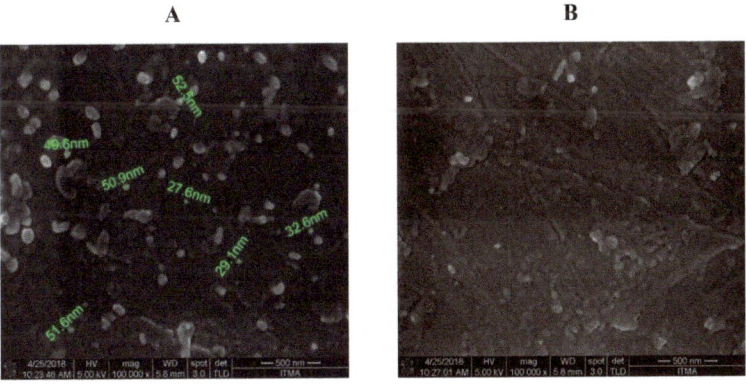

Figure 13. SEM micrographs of CS-AlgNPs (**A**) and MET-CS-AlgNPs (**B**) (100,000×).

3.9. Transmission Electron Microscopy

The MET-CS-AlgNPs nanocomposites were also examined using the transmission electron microscope (TEM), and the structure is as shown in Figure 14. From the Figure, it can be seen that the nanocomposites have irregular spherical shapes with agglomerate behaviors. The size of the main individual nanocomposites is around 80–110 nm.

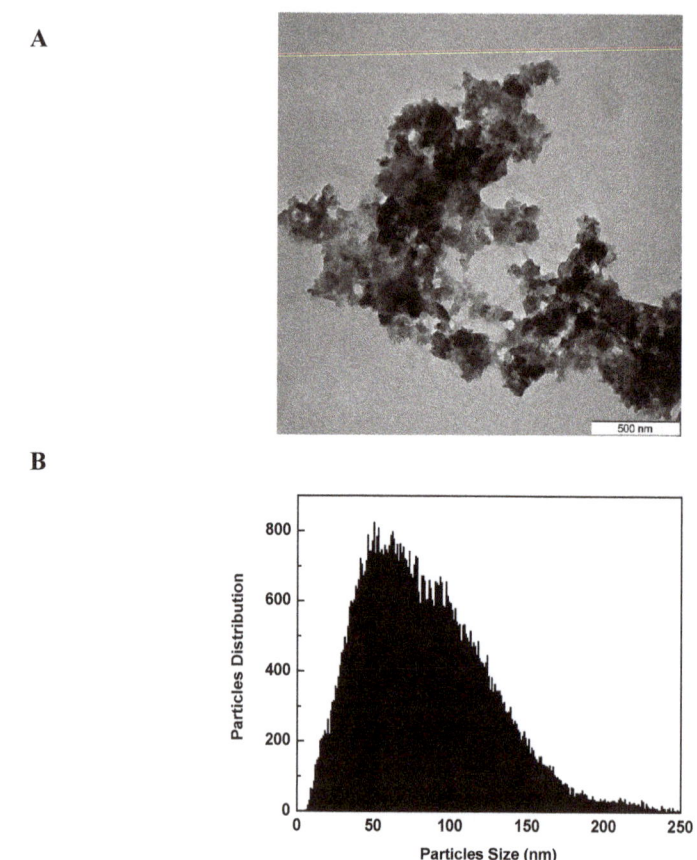

Figure 14. TEM images of MET-CS-AlgNPs (**A**) and their particle size distribution (**B**).

3.10. Interactions between Chemical Components of MET-CS-AlgNPs Nanocomposites

Possible interaction between the components of the nanocomposites is shown in Figure 15. From the Figure, it can be seen that CS and Alg chains polymers are electrostatically held between positive charges of CS (protonated by acetic acid) and negative charges of Alg [54]. Moreover, calcium cations interact with negative charges of Alg. The structure of MET contains hydroxide (OH–) and nitro (NO_2–) groups, which led to the formation of different hydrogen bonds with CS and Alg polymers (Figure 15).

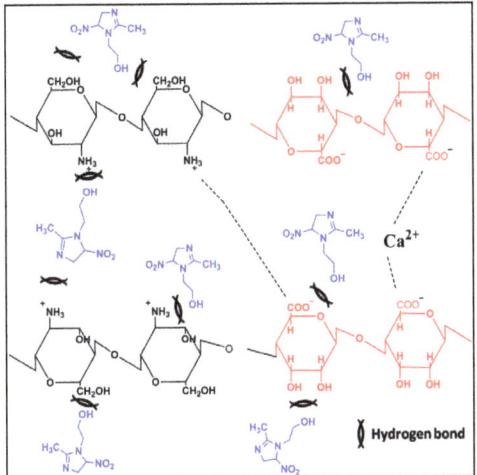

Figure 15. Possible interactions between components of MET-CS-AlgNPs nanocomposites.

3.11. Release Properties of MET from MET-CS-AlgNPs Nanocomposites

The release profiles of MET-loaded CS-AlgNPs were obtained at 0.1M HCl (pH 1.2, to simulate physiological environments in the stomach). As shown in Figure 16, free MET was initially released very rapidly and almost 95% was released within 3.3 h for MAC 5 nanocomposite. This phenomenon is called the burst effect, and it may be due to the presence of the free drug in the nanocomposite. The MET release process from MAC 8, MAC 21 and MAC 19 nanocomposites was observed in two stages with sustained release properties. After 23 h, 90% of the MET was released from the MAC 19, whereas, after 40 hours, 90%, of the MET was released from the MAC 8 and MAC 21. The MAC 8 nanocomposite reached 97% release after 63 h. The MET release at 0.1M HCl could be explained by the enhanced solubility of CS at lower pH (1.2), which in turn promoted the diffusion of the MET through the pores of the AlgNPs matrix into the media [55,56]. These results suggest that the MET-CS-AlgNPs nanocomposites can be used in oral or intravenous administration.

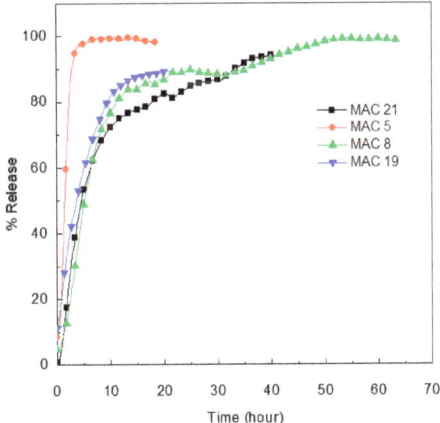

Figure 16. In vitrorelease behaviors of MET from MET-CS-AlgNPs nanocomposites in the 0.1 M HCl solutions.

The release kinetics of MET from MAC 8, MAC 21, MAC 19, and MAC 5 nanocomposites in 0.1M HCl were evaluated by fitting the data to various kinetic models (Table 7). Based on the highest adjusted R^2, the best fitted model for all MAC 8, MAC 21, MAC 19, and MAC 5 nanocomposites was the second kinetic model with R^2 values of 0.988, 0.956, 0.990 and 0.977, respectively.

Table 7. The correlation coefficients (R^2) obtained by fitting the MET release data from MET-CS-AlgNPs nanocomposites in aqueous solutions at 0.1M HCl [57–59].

Samples	R^2			
	Pseudo-First Order	Pseudo-Second Order	Hixson-Crowell Model	Korsmeyer-Peppas Model
MAC 8	0.917	0.988	0.781	0.877
MAC 21	0.903	0.956	0.734	0.882
MAC 19	0.930	0.990	0.822	0.891
MAC 5	0.664	0.977	0.787	0.856
Equation	$\ln(q_e - q_t) = \ln q_e - k_1 t$	$t/q_t = 1/k_2 q_{e2} + t/q_e$	$\sqrt[3]{M_o} - \sqrt[3]{q_t} = Kt$	$\frac{q_t}{q_\infty} = Kt^n$

q_e is the quantity released at equilibrium, q_t is the quantity released at the time (t), M_o is the initial quantity of drug in the nanocomposite, q_∞ is the release at the infinite time and k is the rate constant of the release kinetics

4. Conclusions

For the multiple linear regression analysis, the mathematical models for LE, particle size and zeta potential were developed using the responsesurface methodology to formulate the input parameters, which were Alg, CS and $CaCl_2$ concentrations. Selected mathematical models showed that the developed response surface methodology models werestatistically significant and suitable for all conditions to have higher R^2 and adjusted R^2 values. High correlation values were determined between the experimental data and predicted ones.The concentrations of Alg, CS and $CaCl_2$ with values of 350, 150 and 40 mg, respectively, were determined as the optimum conditions, resulting in the maximum LE (46.04%), the minimum particle size (164 nm) and the optimum zeta potential (−9.25 mV). The verification experiment was carried out to check the validity of the developed mathematical model that predicted LE, particle size and zeta potential within the range of 10% error limit and the prepared nanocomposites were generally stable.

In vitro MET release study of selected formulations; MAC 8, MAC 21, MAC 19, and MAC 5 showed 97%, 90%, 90% and 99% release in 60, 40, 20 and 10h, respectively. These results indicate that the nanocomposites could be effective in sustaining the MET release for a prolonged period.

Supplementary Materials: The following are available online at http://www.mdpi.com/2073-4360/12/4/772/s1, Figure S1: Normal Plot of the Standardized effects toward LE (A), particles size (B) and zeta potential (C); Figure S2: Normal probability plots for LE (A), particles size (B), and zeta potential (C); Figure S3: Residuals versus fitted value for LE (A), particles size (B), and zeta potential (C); Figure S4: Residuals versus observation order for LE (A), particles size (B) and zeta potential (C).

Author Contributions: H.A.K.S., Methodology, Formal analysis, Writing—original draft; S.H.H.-A.-A., Supervision, Funding acquisition; M.Z.H., Funding acquisition; Z.A.; R.A. and S.M.A., review & editing. All authors have read and agreed to the published version of the manuscript.

Funding: The author would like to thank the Faculty of Pharmacy, Isra University and Universiti Putra Malaysia for Grant Putra Berimpak, UPM/800-3/3/1/GPB/2019/9678800, Vot. No. 9678800 for providing funding for part of this research. This study was also supported by Hikma Pharmaceuticals Research and Development Department.

Conflicts of Interest: The authors report no conflict of interest in this work.

References

1. Abdallah, I.; Ibrahim, A.; Ibrahim, N.; Rizk, M.; Tawakkol, S. Simultaneous determination of atenolol and nifedipine by using spectrophotometric method with multivariate calibration and HPLC method implementing "design of experiment". *Pharm. Anal. Acta* **2015**, *6*, 2.
2. Agrahari, V.; Meng, J.; Zhang, T.; Youan, B.-B.C. Application of design of experiment and simulation methods to liquid chromatography analysis of topical HIV microbicides stampidine and HI443. *J. Anal. Bioanal. Tech.* **2014**, *5*, 180. [PubMed]
3. Lewis, G.A.; Mathieu, D.; Phan-Tan-Luu, R. *Pharmaceutical Experimental Design*; CRC Press: Boca Raton, FL, USA, 1998.
4. Rani, A.P.; Hema, V. Full factorial design in formulation of lamotrigine suspension using locust bean gum. *Int. J. Chem. Sci.* **2013**, *11*, 751–760.
5. Kumar, R.S.; Yagnesh, T.N.S.; Kumar, V.G. Optimisation of ibuprofen fast dissolving tablets employing starch xanthate using 23 factorial design. *Int. J. Appl. Pharm.* **2017**, *9*, 51–59. [CrossRef]
6. Box, G.E.; Hunter, J.S. The 2 k—P fractional factorial designs. *Technometrics* **1961**, *3*, 311–351. [CrossRef]
7. Gunst, R.F.; Mason, R.L. Fractional factorial design. *Wiley Interdiscip. Rev. Comput. Stat.* **2009**, *1*, 234–244. [CrossRef]
8. Shahabadi, S.M.S.; Reyhani, A. Optimization of operating conditions in ultrafiltration process for produced water treatment via the full factorial design methodology. *Sep. Purif. Technol.* **2014**, *132*, 50–61. [CrossRef]
9. Salea, R.; Widjojokusumo, E.; Hartanti, A.W.; Veriansyah, B.; Tjandrawinata, R.R. Supercritical fluid carbon dioxide extraction of Nigella sativa (black cumin) seeds using taguchi method and full factorial design. *Biochem. Compd.* **2013**, *1*, 1. [CrossRef]
10. Jacques, P.; Hbid, C.; Destain, J.; Razafindralambo, H.; Paquot, M.; De Pauw, E.; Thonart, P. Optimization of biosurfactant lipopeptide production from Bacillus subtilis S499 by Plackett-Burman design. *Appl. Biothem. Biotechnol.* **1999**, *77*, 223–233. [CrossRef]
11. Ahuja, S.; Ferreira, G.; Moreira, A. Application of plackett-Burman design and response surface methodology to achieve exponential growth for aggregated shipworm bacterium. *Biotechnol. Bioeng.* **2004**, *85*, 666–675. [CrossRef]
12. Sarlak, N.; Nejad, M.A.F.; Shakhesi, S.; Shabani, K. Effects of electrospinning parameters on titanium dioxide nanofibers diameter and morphology: An investigation by Box–Wilson central composite design (CCD). *Chem. Eng. J.* **2012**, *210*, 410–416. [CrossRef]
13. Wsól, V.; Fell, A.F. Central composite design as a powerful optimisation technique for enantioresolution of the rac-11-dihydrooracin—The principal metabolite of the potential cytostatic drug oracin. *J. Biochem. Biophys. Methods* **2002**, *54*, 377–390. [CrossRef]
14. Ferreira, S.C.; Bruns, R.; Ferreira, H.; Matos, G.; David, J.; Brandao, G.; da Silva, E.P.; Portugal, L.; Dos Reis, P.; Souza, A. Box-Behnken design: An alternative for the optimization of analytical methods. *Anal. Chim. Acta* **2007**, *597*, 179–186. [CrossRef] [PubMed]
15. Tak, B.-Y.; Tak, B.-S.; Kim, Y.-J.; Park, Y.-j.; Yoon, Y.-H.; Min, G.-H. Optimization of color and COD removal from livestock wastewater by electrocoagulation process: Application of Box–Behnken design (BBD). *J. Ind. Eng. Chem.* **2015**, *28*, 307–315. [CrossRef]
16. Thirugnanasambandham, K.; Sivakumar, V.; Shine, K. Optimization of reverse osmosis treatment process to reuse the distillery wastewater using Taguchi design. *Desalin. Water Treat.* **2016**, *57*, 24222–24230. [CrossRef]
17. Luo, W.; Pla-Roca, M.; Juncker, D. Taguchi design-based optimization of sandwich immunoassay microarrays for detecting breast cancer biomarkers. *Anal. Chem.* **2011**, *83*, 5767–5774. [CrossRef]
18. O'shea, N.; Rößle, C.; Arendt, E.; Gallagher, E. Modelling the effects of orange pomace using response surface design for gluten-free bread baking. *Food Chem.* **2015**, *166*, 223–230. [CrossRef]
19. Lesch, S. Sensor-directed response surface sampling designs for characterizing spatial variation in soil properties. *Comput. Electron. Agric.* **2005**, *46*, 153–179. [CrossRef]
20. Chowdary, K.; Shankar, K.R. Optimization of pharmaceutical product formulation by factorial designs: Case studies. *J. Pharm.Res.* **2016**, *15*, 105–109. [CrossRef]
21. Miller, A.J. Selection of subsets of regression variables. *J. R. Stat. Soc. Ser. A Gen.* **1984**, *147*, 389–410. [CrossRef]

22. Wagner, J.M.; Shimshak, D.G. Stepwise selection of variables in data envelopment analysis: Procedures and managerial perspectives. *Eur. J. Oper. Res.* **2007**, *180*, 57–67. [CrossRef]
23. Effroymson, M. Multiple regression analysis. In *Mathematical Methods for Digital Computers*; Ralson, A., Wilf, H.S., Eds.; John Wiley and Sons, Inc.: New York, NY, USA, 1960.
24. Draper, N.R.; Smith, H. *Applied Regression Analysis*; John Wiley & Sons: Hoboken, NJ, USA, 1998; Volume 326.
25. Gref, R.; Minamitake, Y.; Peracchia, M.T.; Trubetskoy, V.; Torchilin, V.; Langer, R. Biodegradable long-circulating polymeric nanospheres. *Science* **1994**, *263*, 1600–1603. [CrossRef] [PubMed]
26. Jawahar, N.; Meyyanathan, S. Polymeric nanoparticles for drug delivery and targeting: A comprehensive review. *Int. J. HealthAllied Sci.* **2012**, *1*, 217. [CrossRef]
27. Allemann, E.; Gurny, R.; Doelker, E. Drug-loaded nanoparticles: Preparation methods and drug targeting issues. *Eur. J. Pharm. Biopharm.* **1993**, *39*, 173–191.
28. Yih, T.; Al-Fandi, M. Engineered nanoparticles as precise drug delivery systems. *J. Cell. Biochem.* **2006**, *97*, 1184–1190. [CrossRef]
29. Meyer, M. Bioavailability of drugs and bioequivalence. *Encycl. Pharm. Technol.* **1998**, *1*. [CrossRef]
30. Li, T.; Shi, X.W.; Du, Y.M.; Tang, Y.F. Quaternized chitosan/alginate nanoparticles for protein delivery. *J. Biomed. Mater. Res. Part A Off. J. Soc. Biomater. Jpn. Soc. Biomater. Aust. Soc. Biomater. Korean Soc. Biomater.* **2007**, *83*, 383–390. [CrossRef]
31. Azhar, F.F.; Olad, A. A study on sustained release formulations for oral delivery of 5-fluorouracil based on alginate–chitosan/montmorillonite nanocomposite systems. *Appl. Clay Sci.* **2014**, *101*, 288–296. [CrossRef]
32. Dai, Y.N.; Li, P.; Zhang, J.P.; Wang, A.Q.; Wei, Q. Swelling characteristics and drug delivery properties of nifedipine-loaded pH sensitive alginate–chitosan hydrogel beads. *J. Biomed. Mater. Res. Part B Appl. Biomater. Off. J. Soc. Biomater. Jpn. Soc. Biomater. Aust. Soc. Biomater. Korean Soc. Biomater.* **2008**, *86*, 493–500. [CrossRef]
33. Taleb, M.F.A.; Alkahtani, A.; Mohamed, S.K. Radiation synthesis and characterization of sodium alginate/chitosan/hydroxyapatite nanocomposite hydrogels: A drug delivery system for liver cancer. *Polym. Bull.* **2015**, *72*, 725–742. [CrossRef]
34. Dubnika, A.; Loca, D.; Berzina-Cimdina, L. Functionalized hydroxyapatite scaffolds coated with sodium alginate and chitosan for controlled drug delivery. *Proc. Est. Acad. Sci.* **2012**, *61*, 193. [CrossRef]
35. Freiberg, S.; Zhu, X.X. Polymer microspheres for controlled drug release. *Int. J. Pharm.* **2004**, *282*, 1–18. [CrossRef] [PubMed]
36. Mayol, L.; Borzacchiello, A.; Guarino, V.; Serri, C.; Biondi, M.; Ambrosio, L. Design of electrospayed non-spherical poly (l-lactide-co-glicolide) microdevices for sustained drug delivery. *J. Mater. Sci. Mater. Med.* **2014**, *25*, 383–390. [CrossRef] [PubMed]
37. Deshmukh, R.K.; Naik, J.B. Aceclofenac microspheres: Quality by design approach. *Mater. Sci. Eng. C* **2014**, *36*, 320–328. [CrossRef] [PubMed]
38. Jiang, F.; Wang, D.-P.; Ye, S.; Zhao, X. Strontium-substituted, luminescent and mesoporous hydroxyapatite microspheres for sustained drug release. *J. Mater. Sci. Mater. Med.* **2014**, *25*, 391–400. [CrossRef] [PubMed]
39. Shinde, U.A.; Parmar, S.J.; Easwaran, S. Metronidazole-loaded nanostructured lipid carriers to improve skin deposition and retention in the treatment of rosacea. *Drug Dev. Ind. Pharm.* **2019**, *45*, 1039–1051. [CrossRef]
40. Vazini, H. Anti-Trichomonas vaginalis activity of nano Micana cordifolia and Metronidazole: An in vitro study. *J. Parasit. Dis.* **2017**, *41*, 1034–1039. [CrossRef]
41. Nasongkla, N.; Tanesanukul, C.; Nilyok, S.; Wongsuwan, N.; Tancharoen, S.; Nilanont, S. Nano-coating of metronidazole on dental implants for antibacterial application. In Proceedings of the 2018 IEEE 12th International Conference on Nano/Molecular Medicine and Engineering (NANOMED), Waikiki Beach, HI, USA, 2–5 December 2018; pp. 59–62.
42. Nasseh, N.; Barikbin, B.; Taghavi, L.; Nasseri, M.A. Adsorption of metronidazole antibiotic using a new magnetic nanocomposite from simulated wastewater (isotherm, kinetic and thermodynamic studies). *Compos. Part B Eng.* **2019**, *159*, 146–156. [CrossRef]
43. Rajaonarivony, M.; Vauthier, C.; Couarraze, G.; Puisieux, F.; Couvreur, P. Development of a new drug carrier made from alginate. *J. Pharm. Sci.* **1993**, *82*, 912–917. [CrossRef]
44. Loquercio, A.; Castell-Perez, E.; Gomes, C.; Moreira, R.G. Preparation of chitosan-alginate nanoparticles for trans-cinnamaldehyde entrapment. *J. Food Sci.* **2015**, *80*, N2305–N2315. [CrossRef]
45. Montgomery, D.C. *Design and Analysis of Experiments*; John Wiley & Sons: Hoboken, NJ, USA, 2001; Volume 52, pp. 218–286.

46. Trivedi, M.K.; Branton, A.; Trivedi, D.; Nayak, G. Characterization of physicochemical and thermal properties of chitosan and sodium alginate after biofield treatment. *Pharm. Anal. Acta* **2015**, *6*. [CrossRef]
47. Khan, G.; Yadav, S.K.; Patel, R.R.; Nath, G.; Bansal, M.; Mishra, B. Development and evaluation of biodegradable chitosan films of metronidazole and levofloxacin for the management of periodontitis. *AAPS Pharm. Sci. Tech.* **2016**, *17*, 1312–1325. [CrossRef] [PubMed]
48. Kumari, S.D.C.; Tharani, C.; Narayanan, N.; Kumar, C.S. Formulation and characterization of Methotrexate loaded sodium alginate chitosan Nanoparticles. *Indian J. Res. Pharm. Biotechnol.* **2013**, *1*, 915.
49. De Souza, N.; de Souza, F.; Basílio, I.; Medeiros, A.; Oliveira, E.; Santos, A.; Macwdo, R.; Macędo, R. Thermal stability of metronidazole drug and tablets. *J. Therm. Anal. Calorim.* **2003**, *72*, 535–538. [CrossRef]
50. Anand, M.; Sathyapriya, P.; Maruthupandy, M.; Beevi, A.H. Synthesis of chitosan nanoparticles by TPP and their potential mosquito larvicidal application. *Front. Lab. Med.* **2018**, *2*, 72–78. [CrossRef]
51. Kulig, D.; Zimoch-Korzycka, A.; Jarmoluk, A.; Marycz, K. Study on alginate–chitosan complex formed with different polymers ratio. *Polymers* **2016**, *8*, 167. [CrossRef]
52. Neto, C.d.T.; Giacometti, J.; Job, A.; Ferreira, F.; Fonseca, J.; Pereira, M. Thermal analysis of chitosan based networks. *Carbohydr. Polym.* **2005**, *62*, 97–103. [CrossRef]
53. Mukhopadhyay, P.; Paban Kundu, P. Chitosan-graft-PAMAM/alginate core-shell nanoparticles: A safe and promising oral insulin carrier in Animal Model. *RSC Adv.* **2015**, *5*. [CrossRef]
54. Dubey, R.; Bajpai, J.; Bajpai, A. Chitosan-alginate nanoparticles (CANPs) as potential nanosorbent for removal of Hg (II) ions. *Environ. Nanotechnol. Monit. Manag.* **2016**, *6*, 32–44. [CrossRef]
55. Patel, B.K.; Parikh, R.H.; Aboti, P.S. Development of oral sustained release rifampicin loaded chitosan nanoparticles by design of experiment. *J. Drug Deliv.* **2013**, *2013*, 370938. [CrossRef]
56. Sorasitthiyanukarn, F.N.; Muangnoi, C.; Bhuket, P.R.N.; Rojsitthisak, P.; Rojsitthisak, P. Chitosan/alginate nanoparticles as a promising approach for oral delivery of curcumin diglutaric acid for cancer treatment. *Mater. Sci. Eng. C* **2018**, *93*, 178–190. [CrossRef] [PubMed]
57. Dong, L.; Yan, L.; Hou, W.-G.; Liu, S.-J. Synthesis and release behavior of composites of camptothecin and layered double hydroxide. *J. Solid State Chem.* **2010**, *183*, 1811–1816. [CrossRef]
58. Ho, Y.-S.; Ofomaja, A.E. Pseudo-second-order model for lead ion sorption from aqueous solutions onto palm kernel fiber. *J. Hazard. Mater.* **2006**, *129*, 137–142. [CrossRef] [PubMed]
59. Sakore, S.; Chakraborty, B. Formulation and evaluation of enalapril maleate sustained release matrix tablets. *Int. J. Pharm.* **2013**, *4*, 21–26.

© 2020 by the authors. Licensee MDPI, Basel, Switzerland. This article is an open access article distributed under the terms and conditions of the Creative Commons Attribution (CC BY) license (http://creativecommons.org/licenses/by/4.0/).

Article

Investigating Novel Syntheses of a Series of Unique Hybrid PLGA-Chitosan Polymers for Potential Therapeutic Delivery Applications

Jason Thomas Duskey [1,2], Cecilia Baraldi [3], Maria Cristina Gamberini [3], Ilaria Ottonelli [1,4], Federica Da Ros [1], Giovanni Tosi [1], Flavio Forni [1], Maria Angela Vandelli [1] and Barbara Ruozi [1,*]

1. Te.Far.T.I.-Nanotech Lab, Department of Life Sciences, University of Modena and Reggio Emilia, 41121 Modena, Italy; jasonthomas.duskey@unimore.it (J.T.D.); ilaria.ottonelli@unimore.it (I.O.); federica.daros93@gmail.com (F.D.R.); gtosi@unimore.it (G.T.); flavio.forni@unimore.it (F.F.); mariaangela.vandelli@unimore.it (M.A.V.)
2. Umberto Veronesi Foundation, 20121 Milano, Italy
3. Department of Life Sciences, University of Modena and Reggio Emilia, 41121 Modena, Italy; cecilia.baraldi@unimore.it (C.B.); mariacristina.gamberini@unimore.it (M.C.G.)
4. Clinical and Experimental Medicine PhD Program, University of Modena and Reggio Emilia, 41121 Modena, Italy
* Correspondence: barbara.ruozi@unimore.it

Received: 10 March 2020; Accepted: 30 March 2020; Published: 4 April 2020

Abstract: Discovering new materials to aid in the therapeutic delivery of drugs is in high demand. PLGA, a FDA approved polymer, is well known in the literature to form films or nanoparticles that can load, protect, and deliver drug molecules; however, its incompatibility with certain drugs (due to hydrophilicity or charge repulsion interactions) limits its use. Combining PLGA or other polymers such as polycaprolactone with other safe and positively-charged molecules, such as chitosan, has been sought after to make hybrid systems that are more flexible in terms of loading ability, but often the reactions for polymer coupling use harsh conditions, films, unpurified products, or create a single unoptimized product. In this work, we aimed to investigate possible innovative improvements regarding two synthetic procedures. Two methods were attempted and analytically compared using nuclear magnetic resonance (NMR), fourier-transform infrared spectroscopy (FT-IR), and dynamic scanning calorimetry (DSC) to furnish pure, homogenous, and tunable PLGA-chitosan hybrid polymers. These were fully characterized by analytical methods. A series of hybrids was produced that could be used to increase the suitability of PLGA with previously non-compatible drug molecules.

Keywords: PLGA; chitosan; hybrid polymers; chitosan-PLGA polymer; NMR; DSC; FT-IR

---

1. Introduction

The discovery of effective therapeutic drugs is becoming increasingly difficult as seen by the drastic decline of new therapeutics accepted for public use each year. This is seen even with advances in structure activity relationship (SAR) studies [1], computer simulations of target structures (specific binding sequences and shape elucidation) [2], and high throughput screening methodology [3]. Novel surfaces and delivery nanosystems have taken the spotlight as the leading hope to advance new drugs from research into and beyond clinical studies by overcoming factors such as: lack of solubility, poor stability, poor biodistribution, immune response activation, off-target affects, and poor accumulation at the target site. Polymeric and lipid formulations have been taken advantage of to create fine-tuned systems to include targeting [4–6], triggerable activation (heat, light, reactive oxygen species (ROS), pH) [7–9], and varied uptake mechanisms to deliver pharmaceutics against numerous diseases [10–12].

In this respect, poly(lactic-co-glycolic acid) (PLGA) is of high interest due to the fact that it is: (1) FDA approved; (2) chemico-physically tunable to match biodistribution or loading needs; (3) capable of producing both nanosystems or polymeric scaffolds; (4) chemically modifiable to include stealthing moieties (polyethylene glycol, PEG) and/or targeting ligands. All of these aspects have been widely exploited in production of PLGA nanoparticles (NPs) for the possible cure of a plethora of diseases [13–18].

While PLGA NPs display many advantages in drug formulation, in comparison with cationic bio/polymers, they can suffer poor encapsulation efficiency when loading negatively-charged molecules. For example, while cationic bio/polymers (i.e., chitosan, cationic lipids, poly-ethylenimine, etc.) [19] can ionically bind negatively-charged DNA and form polyplexes, repulsion between the negatively-charged gene material and PLGA leads to negligible loading efficiencies. In this view, production of a co-polymer including chemical features needed for controlled release, absence of charge repulsion, and stable loading within the protective hybrid polymer assembly could be the correct answer to these limitations.

Previously, attempts to overcome these limitations were investigated in various ways. First, by surface engineering negatively-charged NPs (such as PLGA) with cationic molecules in order to allow DNA absorption onto the surface [20–23]. While this approach could improve theoretical loading of gene material or other positively-charged molecules onto polymeric NPs, the stability of the exposed drugs in a biological environment and control of their release are still lacking. Secondly, a synthesis of chitosan on a PLGA film for adsorption of hydrophilic molecules of chitosan for protein loading [24]. While in this study loading was improved, the reaction was only monitored based on time and the film remained intact throughout all analysis and the presence of absorbed but not reacted chitosan could be present. By creating a controlled synthesis of hybrid polymers, it would be possible to include improved encapsulation of drugs into PLGA assemblies, improving encapsulation of the molecule as well as protecting it within the structure from desorption in the blood and degradation. Furthermore, systematically synthesizing series of hybrid polymers could allow for tunability to include controlled release kinetics and degradation kinetics of the molecules as well.

Therefore, in this research we attempt two different synthetic methods to create a pure hybrid PLGA-chitosan polymer series: solid phase synthesis on a film (adapted from Li et al. [24]), or in solution chemical reaction (adopted from a reaction to react chitosan to polycaprolactone [25]). This will allow for the synthesize of a unique series of PLGA-chitosan hybrid polymers with tailored and tunable physico-chemical characteristics that could be used to expand the use of PLGA delivery systems of currently incompatible drugs or environments and in a variety of drug delivery assemblies to treat a larger range of disease states.

2. Materials and Methods

2.1. Materials

Poly (D,L-lactide-co-glycolide) acid [PLGA RG-503H 50:50, inherent viscosity in 0.1% (w/v) chloroform (CHCl$_3$) at 25 °C = 0.38 dLg^{-1}] was used as received from the manufacturer (Boehringer-Ingelheim, Ingelheim am Rhein, Germany). According to the experimental titration results of the carboxylic end of the polymers (4.94 mg potassium hydroxide (KOH)/g polymer) the molecular weight of RG-503H was calculated to be 11,000 Da. Low Molecular Weight chitosan, (mw 14,000) was purchased from Sigma Aldrich (Sigma Aldrich, Milano, Italy). All the solvents were of analytical grade, and all other chemicals and media were used as received from the manufacturers, and unless otherwise indicated, obtained from Sigma-Aldrich (Sigma Aldrich, Milano, Italy).

2.2. Solid Phase Synthesis of PLGA-Chitosan Co-Polymer

The solid phase reaction of PLGA and chitosan was performed following the method of Ai.D. Li et al. with minor modifications (Scheme 1) [24]. Briefly, a PLGA solution (50 mg) was weighed into a round bottom flask and solubilized in 5 mL dicloromethane (DCM) and dried by rotary evaporation to create a thin film. The film was then washed for 1 h with 5 mL 6 w/v% NaOH (sodium hydroxide). This solution was discarded and the film was gently washed three times with 10 mL dilute HCl (hydrochloric acid 10%) followed by three more times with distilled water. The film was then completely covered in a solution containing N-Hydroxysuccinimide (NHS, 10 mgmL^{-1}) and 1-Ethyl-3-(3-dimethylaminopropyl)carbodiimide (EDC, 10 mgmL^{-1}) and reacted for another 6 h at room temperature in order to activate the acid group of PLGA with the NHS ester to promote the amide coupling with the amine of chitosan. This solution was discarded and the film was ultimately covered by a solution of 80 mL (reaction in round bottom flask) chitosan of 25 mgmL^{-1} (pH 3.5). Remarkably, to achieve this pH value in which the chitosan becomes more soluble with decreasing pH it becomes highly viscous, HCl (1N) was added dropwise and stirred vigorously for several minutes between each additional drop. Therefore, rigorous stirring for several minutes is needed in order to ensure the added HCl is dispersed uniformly throughout the solution and to avoid pockets of extreme acidity. After reacting for 48 h, the chitosan solution became much more transparent and less dense. The same procedure up to this point was also performed on a film on the surface of a glass petri dish (diameter 10 cm) with the following changes: the PLGA (150 mg) in 9 mL DCM evenly dispersed over the surface of the petri dish was left to evaporate at room temperature under a chemical hood overnight instead of on a rotary evaporator. The volumes required to cover the film with ~1 cm of each solution were decreased: NaOH 15 mL, EDC 15 mL (10 mgmL^{-1}), NHS (15 mL 15 mL (10 mgmL^{-1}), chitosan 15 mL (25 mgmL^{-1}, pH 3.5). This decrease in volume was possible because unlike in the round bottom flask where a large volume is needed to fill in the 3D spherical space, on a flat surface the volume needed to cover the film is a much smaller cylindrical cross-section of the round bottom flask (1 cm thick cylinder). All material from the round bottom or petri dish was poured into a separation funnel and the reaction vessel was washed 3 × with water followed by 3 × with DCM (10 mL each), in order to remove products and starting material that are soluble in aqueous or organic solvents, and added to the separatory funnel. After allowing the extraction to separate for 1 h at room temperature in the separation funnel, three distinct layers formed during separation: a clear DCM layer, a middle white emulsion, and the yellow chitosan solution. The three layers were separated into separate containers and lyophilized to calculate a percent yield and further characterization.

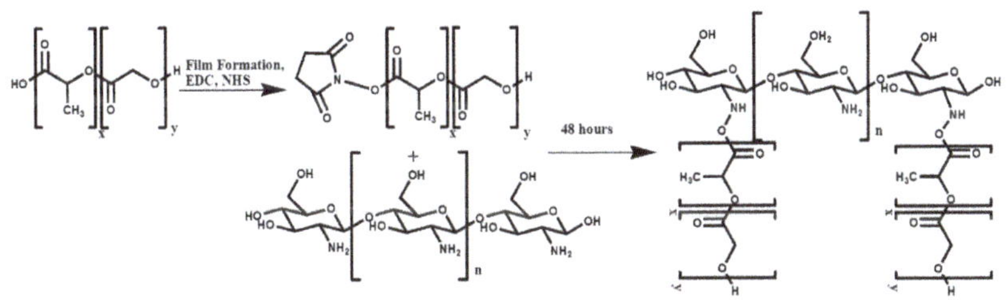

Scheme 1. Solid phase reaction of PLGA film with chitosan.

2.3. Characterization Protocols of PLGA-Chitosan Co-Polymer in Solid-Phase.

Characterization of chitosan-PLGA co-polymer was achieved by analysis in FTIR, and by NMR. The FT-IR spectra were recorded by a Vertex 70 (Bruker Optics, Ettlingen, Germany) FT-IR spectrophotometer, equipped with a deuterium triglycine sulphate (DTGS) detector (Bruker Optics, Ettlingen, Germany). Setting parameters are: resolution 4 cm^{-1}; apodization weak. The spectral range was 4000–600 cm^{-1} with 32 scans for each spectrum. The ATR spectra were recorded using the Golden-Gate accessory (Golden Gate™ Single Reflection Diamond ATR Series MkII).

1H NMR and 13C samples of the solid phase reactions were run on a Bruker 600 mHz NMR (Bruker, Milano, Italy). Simply, 4 mg of sample were dissolved in deuterated water with 1% v/v deuterated acetic acid added (Chitosan, 700 uL), or deuterated dimethylsulfoxide (DMSO) (PLGA-chitosan product, 700 uL), scanned 40 (1H) or 3000 (13C) times and analyzed by Bruker Top Spin software (Bruker, Milano, Italy).

2.4. Reaction of Chitosan and PLGA in Solution

To perform the conjugation between chitosan and PLGA in organic solution, an organic soluble SDS-chitosan salt was formed (Scheme 2). In particular, adapting a protocol published by Cai et al. [25], a solution of chitosan (200 mg) in 2% v/v acetic acid was precipitated with SDS (Sodium dodecyl sulfate 560 mg) in a rapport of 1:100 for 2 h. The reaction was centrifuged for 10 min at 10,000 rpm in an ALC PK121 multispeed centrifuge (Concordia, Modena, Italy), the supernatant was discarded, and the precipitate was dried in a desiccator under negative pressure overnight. Simultaneously, PLGA was activated for reaction with chitosan by means of NHS-DDC technology. The covalent binding between the carboxy terminus of the polymer PLGA RG503H and the terminal amine of the peptide has been formed by standard methods, namely the activation of the carboxy group of PLGA by means of an ester with N-hydroxysuccinimide in the presence of dicyclohexylcarbodiimide, and the subsequent formation of an amidic linkage with the N-terminus of the unprotected peptide. Thus, to a solution of PLGA RG503H (1.00 g, 88 µmol) in anhydrous dioxane (5 mL), DCC (dicyclohexylcarbodiimide. 19.0 mg, 93 µmol) and N-hydroxysuccinimide (NHS, 11.0 mg, 93 µmol) were added, and the mixture was stirred for 4 h at 20 °C. After, the dicyclohexylurea was filtered away and the solution was decanted into cold anhydrous diethyl ether. The insoluble polymer was collected and purified by dissolution in DCM, followed by precipitation by the addition of anhydrous diethyl ether, then dried under reduced pressure. The content of NHS groups reacted with PLGA RG503H was determined by 1H-NMR spectroscopy (DPX 200; Bruker, Rheinstetten, Germany) in DMSO-d6, from the relative peak area of the multiplet at 2.95 ppm and of the multiplet at 1.80–1.60 ppm, corresponding to the protons of the N-succinimide and those of the methyl groups of the polymer, respectively, and resulted to be 49 µmol NHS/g of polymer. After having obtained both polymers, a fixed amount (50 mg) of the chitosan salt was then solubilized in anhydrous dimethylformamide (DMF, 10 mL) and reacted with different amounts (10, 50, or 240 mg) of activated PLGA-NHS (corresponding to ratio 1:5, 1:1, 5:1 chitosan: PLGA, respectively) and reacted for 48 h.

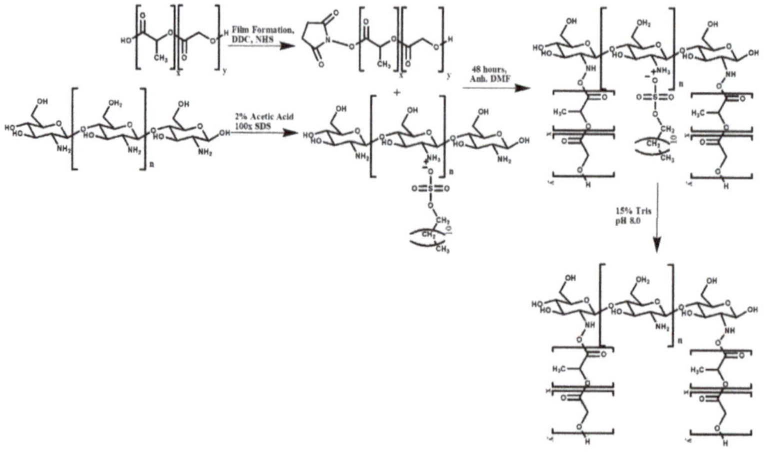

Scheme 2. Solution Phase Reaction using DMF soluble chitosan-SDS salt.

All the products were purified and isolated by means of centrifugation at 12,000 rpm for 10 min to remove any precipitated material during the reaction. The supernatant was then dried by rotoevaporation to yield a white/yellow powder containing the PLGA-chitosan/SDS salt conjugate. The SDS was then removed from the conjugate, which led to a precipitation of the final product by incubation in 50 mL 15% TRIS pH 8.0 for 48 h. The precipitate was then centrifuged at 12,000 rpm with an ALC PK121 multispeed centrifuge and the supernatant was decanted away. The final product was then dried by lyophilization and stored in a desiccator at room temperature until analysis. Solubility of the samples was tested by weighing 1 mg of each product and testing its ability to dissolve in DMSO or acetic acid (2% v/v) of concentrations of 200 ugmL^{-1}.

2.5. Characterization Protocols of PLGA-Chitosan Co-Polymer in Solution

FTIR was described as previously described. The FT-IR spectra were recorded by a Vertex 70 (Bruker Optics, Ettlingen, Germany) FT-IR spectrophotometer, equipped with a deuterium triglycine sulphate (DTGS) detector. Setting parameters are: resolution 4 cm^{-1}; apodization weak. The spectral range was 4000–600 cm^{-1} with 32 scans for each spectrum. The ATR spectra were recorded using the Golden-Gate accessory (Golden Gate™ Single Reflection Diamond ATR Series MkII).

After purification, NMR spectra were acquired at 300 K using an AVANCE III HD 600 Bruker spectrometer, equipped with a 2.5 mm H/X CPMAS probe operating at 600.13 and 150.90 MHz for 1H and 13C, respectively (Bruker, Milano, Italy). Samples were packed into 2.5 mm zirconia rotors and spun at the magic angle. 13C NMR spectra were obtained using a standard pulse sequence for cross polarization (CP), at 16 kHz magic angle spinning (MAS) rate. The relevant acquisition parameters for CP-MAS 13C NMR spectra were: 45 kHz spectral width, 10 s relaxation delay, 2.5 μs 1H 90° pulse, 62.5 kHz radio frequency field strength for Hartmann–Hahn match, 2k data points, and 2k scans. All chemical shifts were referenced by adjusting the spectrometer field to the value corresponding to 38.48 ppm chemical shift for the deshielded line of the adamantane 13C NMR spectrum.

Dynamic scanning Calorimetry of DSC was performed on a Netzsch Phox DSC 200 PC using the Netzsch Proteus analysis software (NETZSCH-Gerätebau GmbH, Selb, Germany). Samples were precisely weighed between 2–4 mg each into NETZSCH DSC-crucibles (Al; 25 uL) and sealed with their appropriate lids. An empty crucible was used as the reference sample. Samples were analyzed with the following thermometric gradient: 2 min isothermal gradient to standardize the starting point at 15 °C, 15–320 °C over 38 min increasing at 10 °C per minute, with a 2 min isothermal section.

3. Results

The most direct method of conjugating PLGA and chitosan is an amid bond formation between the amine on chitosan and the carboxylic acid of PLGA (Scheme 1). Functionally however, this reaction is complicated due to the extreme difference in solubility between the two molecules. Previous attempts reacted chitosan in solution with a PLGA film to create a positively-charged surface aiming to create nanofibers without the need for purification [24]. Another researcher produced the hybrid PLGA-chitosan polymer for the creation of nanoparticles, but it required harsh conditions (concentrated nitric acid) [26]. Therefore, to create a pure, reproducible, and controllable hybrid polymer that could be used in solution for NP formation, a reaction was performed under milder conditions on a PLGA film created by evaporating PLGA on a surface activating it with EDC and NHS and reacting it with a large excess of chitosan. After 48 h, the chitosan solution was removed and the product was purified in a biphasic solution of 0.1% acetic acid (PLGA-chitosan product) and DCM (non-reacted PLGA). Initial reactions were performed in a round bottom flask; however, to make the reaction greener by decreasing the ratio of surface area:volume (to decrease amount of solvent and reactants needed to cover the PLGA film), reactions were performed in a flat petri dish. This simple change not only decreased reaction volume (80 mL to 10 mL), but it also increased the % yield from ~25% to 50%.

Characterization of the product was performed by FTIR spectroscopy (a common technique for investigating interactions between polymers) and NMR. General FTIR points of interest for the reaction arise in the broad band between 3450 and 3200 cm^{-1} (νOH + νNH) and two weak peaks at 2940 cm^{-1} and 2890 cm^{-1} (νCH2) (Figure S1, top panel). More critical for the identification of the conjugation of PLGA to chitosan are the major characteristic absorption bands at around 1648 and 1588 cm^{-1}, corresponding to amide I (νC=O) and amide II (δNH + νCN) of the residual N-acetyl groups. Under the band centered at 1585 cm^{-1}, the contribution of δNH2 is also hidden, which overlaps the amide II peak [27]. Pure PLGA exhibits the strong characteristic adsorption peaks at 1170 and 1090 cm^{-1} (νCOC, ether), 1130 cm^{-1} (ρCH3), 1452, 1390, and 745 cm^{-1} (δCH), and peaks 3020 and 2930 cm^{-1}, which were attributed to νCH2 from glycolic acid portion, and νCH3 from the lactic acid portion. The most notable peak to discern the presence of PLGA arises at 1749 cm^{-1} (νC=O, ester) (Figure S1, bottom panel) [28]. FTIR analysis of the product showed that the mild acetic acid conditions did not result in the covalent linkage between PLGA and chitosan (Figure 1). While a shift in the amid bonds at 1648 and 1588 cm^{-1} were observed, the new peaks did not correspond to further amid bond creation, but instead showed the resemblance of the formation of a chitosan salt [29] with bands further downfield at 1627 and 1517 cm^{-1}. Also, only a small emergence of a peak indicating the presence of the C=O of PLGA at 1748 cm^{-1}, but instead a peak at 1703 cm^{-1} indicating the appearance of an acid was observed.

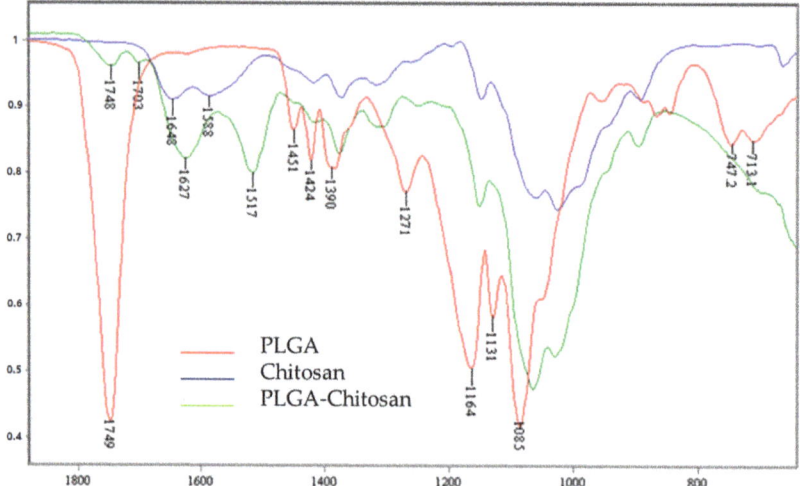

Figure 1. FT-IR/ATR spectra of unmodified poly lactic-co-glycolic acid (**red**), unmodified chitosan (**blue**), and PLGA-chitosan product (**green**).

NMR analysis confirms the poor reaction results. 1H NMR of the product and the chitosan control show little to no difference with the H2 peak at 3 ppm and the H3-6 peaks as a broad series of peaks at 3.5–4 ppm. These peaks corresponded to the literature precedence of the 13C NMR peaks of the main carbon ring at ~C1 (100 ppm), C3-5 (73–85 ppm), and C2,6 (55–60 ppm). However, the normal peaks expected for the PLGA C=O (170 ppm) or CH3 (1.5 ppm and 15 ppm 13C) or its degradation products (glycolic and lactic acid) are not present (Figure 2). Only small fragment peaks that do not cross correlate upon 2D analysis (Figure S2). Degradation of the PLGA into small fragments during the reaction would explain the FTIR results showing the formation of a small acid peak. It also could explain the amid bond shifts to that of the salt formation as small negatively-charged acidic degradation products could lead to a salt formation with the free amine of chitosan.

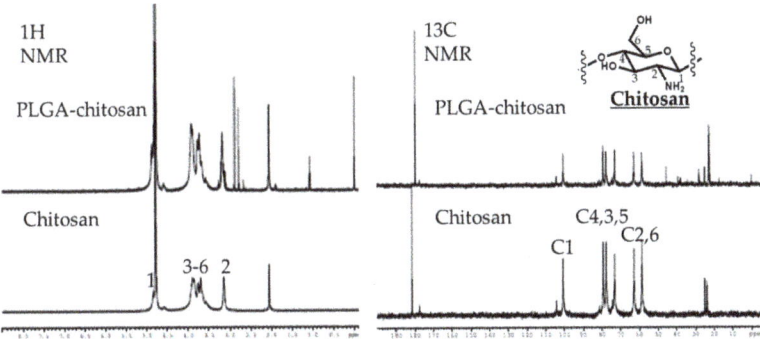

Figure 2. 600 MHz NMR Proton (**left panel**) and Carbon (**Right Panel**) analysis. Numbers indicate the Carbons (C1-6) of the chitosan ring structure or their respective hydrogens as depicted in the structure (Top right).

While the reaction of PLGA and chitosan is found in the literature, it is often performed with harsh conditions on gels or without purification. Using a solid phase reaction with mild reaction conditions did not prove successful. This ruled out this method as a viable option to create a controlled series of hybrid polymer variants for further characterization; therefore, more stable and controllable methodology was pursued.

The mild reaction conditions led to a lack of product formation. To overcome this, an alternate method was adapted in which chitosan is precipitated as an SDS salt in order to improve its solubility in organic solutions (DMSO, choloroform, and DMF) [25].

This intermediate was then conjugated to PLGA-NHS in anhydrous DMF. To test the flexibility of the reaction, and to analyze the physical characteristics of various PLGA-chitosan hybrid polymers, a series of three reactions was performed: (1) an excess of chitosan in a 5:1 molar ration, (2) 1:1 chitosan: PLGA and (3) 1 chitosan: 5 PLGA (Table 1). The latter corresponds to an average of one PLGA being available for each sugar unit of chitosan. After reacting activated PLGA with chitosan-SDS for 48 h, the salt was dissociated in Tris 15% pH 8 for 48 h. The percent yield of the reactions increased proportionally with the increasing rapport of PLGA: chitosan in the reaction as shown in Table I (55%, 75%, and 82%). The solubility of the products also suggested an increased PLGA attachment due to the decreasing solubility in 2% acetic acid (v/v).

Table 1. Solution phase reaction conditions of PLGA and Chitosan-SDS.

Reaction	Chitosan-SDS (mg Chitosan)	PLGA-NHS (mg)	Recovery (%)	PLGA:Chitosan in Initial Reaction Solution (mol)	Solubility in Acetic Acid [a]
1	50	10	55	1:5	soluble
2	50	50	75	1:1	semi soluble
3	50	240	82	5:1	Not soluble

a. 200 ug/mL 2% v/v.

The conjugation of PLGA to chitosan again was analyzed by FTIR, and 13C NMR in solid state (due to the differences in solubility between the products). The PLGA and chitosan starting materials were identical to as described previously (Figure 1, Figure S1). In a PLGA concentration dependent manner, the progressive appearance and intensification of the band at 1755 cm^{-1}, indicating the presence of PLGA (ester C=O stretching), can be observed (Figure 3). Unlike the solid-state reaction, bands indicating a chitosan salt formation were not observed (including the presence of the chitosan-SDS salt formation (1624 and 1523 cm^{-1})) (Figure S3). Instead, there was a clear and concentration dependent (based on initial PLGA amounts), shift indicating amid bond formation (Figure 3). This was confirmed by observing the amide I and II peaks (1650 and 1585 cm^{-1}) shift to 1633 cm^{-1} (amide I) and 1548 cm^{-1} (amide II) (Figure 3). As compared with pure chitosan, the δNH contribution of the primary amine for the band at 1580 cm^{-1} decreases or even disappears, because of a change of primary amine in the chitosan chain into amide groups, as already attested in the literature [30]. The displacement of the band from 1586 cm^{-1} (chitosan) to 1549 cm^{-1} (PLGA-chitosan) suggests that the grafting reaction occurred mainly by the reaction between the −NH2 chitosan groups and −COOH PLGA groups. Furthermore, the band at 3184 cm^{-1} progressively increases, associated with NH stretching of the secondary amide (Figure 3 inset).

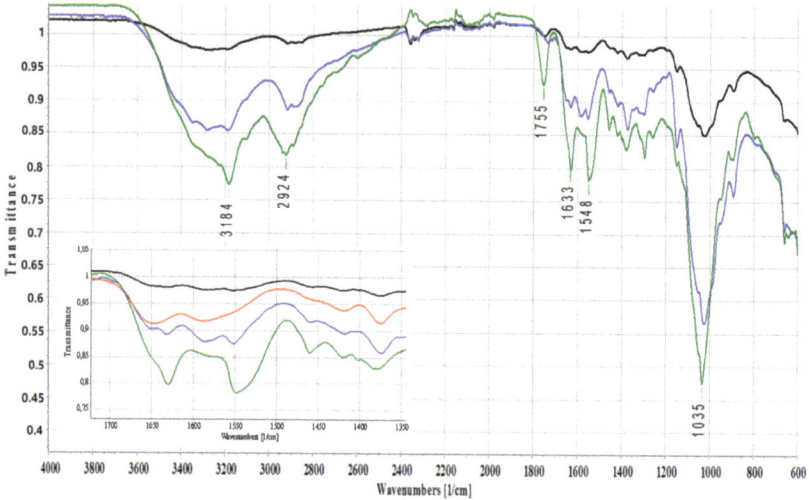

Figure 3. FTIR scan of PLGA-chitosan hybrid products of the reactions synthesized with different molar rapport of PLGA:chitosan of reaction (1) 1:5. black (2) 1:1. blue (3) 5:1. green, and (4) pure chitosan (red inset).

NMR analysis was used to support the FTIR findings (Figure 4). The solid-state NMR of pure chitosan showed the characteristic broad singlet at 100 ppm (carbon 1) along with two broad multiplet peaks between 50 pp, (carbon 2–5) and 90 ppm (carbon 6) in accordance with literature precedence (Figure 4 purple box) [31]. The pure PLGA exhibits the CH and CH2 peaks at 70 and 60 ppm, respectively, as well as the CH3 peak at 15 ppm and C=O peak at 170 ppm (Figure 4, orange boxes). In all reactions, the iconic peaks of the chitosan can be seen. In all three reactions, the PLGA peak corresponding to the CH2 is hidden under the chitosan (purple box) and residual TRIS salt (blue box) peaks from 50–75 ppm, but the emergence of the CH peak at 70 ppm is observed (Figure 4, Figure S4). More evident however; is that by increasing the initial amount of PLGA in the reaction, the peak corresponding to C=O at 170 ppm (indicated by a star) as well as that of the CH3 group (indicated by an @) are seen to directly increase in intensity (Figure 4 orange boxes). It is important to note that SDS, and Chitosan-SDS salt (peaks 20–40 ppm) are not present in any of the samples indicating full removal of the salt back to the original chitosan structure in the product (Figure S4, red box). NMR analysis showed constant and equal NMR spectra across multiple product samples indicating the homogeneity and controllability of each product.

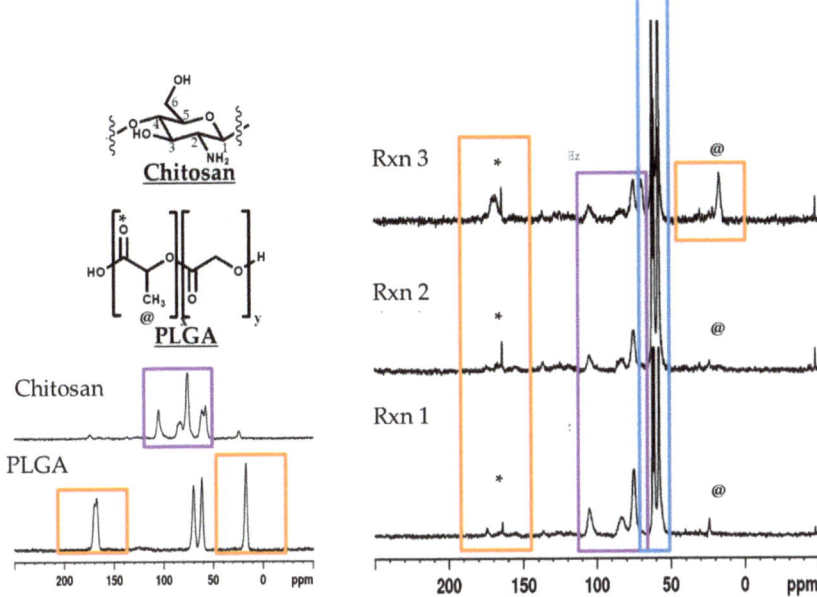

Figure 4. Solid state 13C NMR analysis with highlighted peaks of interest: chitosan (**purple**), PLGA (**orange**), TRIS salt (**blue**). Numbers indicate the Carbons (C1-6) of the chitosan ring structure or their respective hydrogens as depicted in the structure (Top left), * indicates the carbon of the PLGA carbonyl peak, and @ indicating the PLGA methyl group.

To further validate the conjugation of PLGA to chitosan, DSC analysis was performed (Figure 5). The transitional peak of PLGA was seen at 50 °C along with an endothermic transition during its degradation between 280–380 °C. The chitosan control shows the liberation of the water entrapped between the chitosan chains at 115 °C along with an exothermic transition at approximately 300 °C reasoned to be the degradation of the chitosan ring structures permitting 3-D rotation. Analysis of the polymer samples showed a shift in all transitional states dependent on the concentration of PLGA in the initial reaction solution. With increasing amounts of PLGA, the transitional phase at 50 °C disappeared due to the loss of the glass transition when bound to chitosan. In a physical mixture of PLGA and chitosan however this transition was still observed (Figure S5). Secondly, a shift to higher temperatures of the water loss from 120 °C to 150 °C in sample 3 (with the most PLGA) exhibiting numerous peaks in this range. The energy required to remove the water associated with the chitosan chains is increased by the increased encumbrance of PLGA. Finally, the disappearance of the transition peak at around 290 °C caused by the bulky PLGA sterically hindering the free rotation of the chitosan chains as well as cancellation of the endothermic (PLGA) and exothermic (Chitosan) energies were observed supporting the FTIR and NMR data of the presence of increasing amounts of PLGA chemically linked to the chitosan chain. To ensure these changes were not caused by the presence of the chitosan-sds salt, a sample was also analyzed showing none of the characteristics of the polymer products (Figure S5).

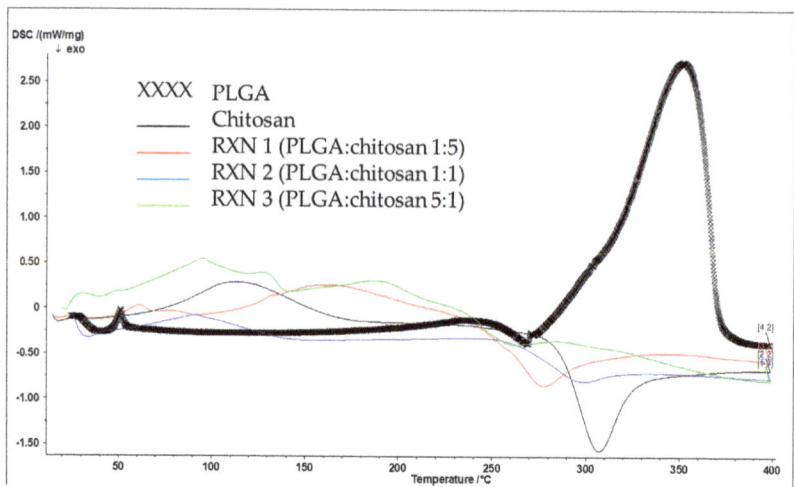

Figure 5. Dynamic scanning calorimetry analysi: PLGA (**black crosses**), chitosan (**black line**), and the three PLGA-chitosan reaction products based on PLGA:chitosan molar rapport: (1) 1:5 (**red**) (2) 1:1 (**blue**) and (3) 5:1 (**green**).

These three series of analysis not only demonstrate the formation of the hybrid polymer series using mild reaction conditions, but also show the versatility of the reaction in its ability to be stoichiometric controlled to create a uniform product unlike that seen by the solid-state reaction. The formation of the chitosan-SDS made it optimal for the reaction in organic solvents with PLGA. By varying the ration of PLGA in the reaction conditions from a 1:5 excess of chitosan, to 1:1, and finally to 5:1 excess of PLGA, it was possible to create a variation of hybrid polymers. The hybrid series was not only verified by the analytical characterizations, but also by the difference of solubility of the product. Controlling the reaction in a stoichiometric controlled manner to create such clean and reproducible product, hybrid polymers greatly increases the translatability and feasible uses of these polymers in drug delivery purposes.

4. Discussion and Conclusions

Finding new materials to stabilize molecules with poor stability, solubility, or biocompatibility properties is necessary to continue advancing new disease treatment methods with "critical" but non-compatible drugs. PLGA offers a very promising base material as it is FDA approved and has been used extensively to specifically target drugs to diseases as NPs or as site specific delivery agents inserted as a film but is limited in loading positively-charged molecules.

Creating new co-polymers in a constant and controlled manner offers an increasing utility of PLGA assemblies for a broader range of potential drug candidates in which it is currently non-compatible. To this end, two reaction methods were attempted to conjugate negatively-charged PLGA to the positively-charged chitosan to form a series of novel co-polymers. Previous works have attempted to make chitosan hybrid polymers using harsh reaction conditions (nitric acid), PLGA films, or in solution (to make polycaprolactone hybrid polymers) in a non-purified and uncontrollable manner. To truly benefit from these types of hybrid polymers, the reaction must be reproducible, controllable, create a series of pure homogenous products that can be selected dependent on the therapeutic need.

Data indicated that solid-phase synthesis using a PLGA film and mild reaction conditions was insufficient to create PLGA-chitosan hybrid polymers, but instead led to a salt formation with degradation products in solution. However, by utilizing an SDS salting out reaction to create a chitosan

SDS intermediate that is soluble in organic solvents, a series of PLGA-chitosan co-polymers with different molar ratios were produced.

Remarkably, this reaction was able to furnish a unique series of pure and reproducible PLGA-chitosan hybrids with various molar rapport and solubilities. This controlled synthesis method makes these hybrids prime candidates for protection and delivery of a wide range of previously non-combatable drugs either as NPs formed through chitosan self-assembly techniques (for those still soluble in acidic solutions) or for the encapsulation in stable and non-toxic films for long-term controlled release (for those insoluble in biological solutions).

These preliminary results could pave the way to further advances in the application of PLGA-based nanotherapeutics, expanding the tunability of the core polymer structure to be better suited for a wider range of drugs candidates to be loaded, protected, and delivered to diseased cells improving their potency and efficacy.

Supplementary Materials: The following are available online at http://www.mdpi.com/2073-4360/12/4/823/s1, Figure S1: Full FTIR Scan of PLGA and chitosan polymers, Figure S2: 2D NMR correlation analysis of the 1H and 13C PLGA-chitosan reaction product, Figure S3: FTIR analysis of chitosan control (Red), chitosan-SDS salt (green) and SDS salt (blue), Figure S4: Solid state 13C NMR analysis with highlighted peaks of interest: chitosan (purple), SDS (red), TRIS salt (blue), Figure S5: Dynamic scanning calorimetry analysis: chitosan-SDS salt (black line), chitosan: PLGA 1:1.5 physical mixture (green line).

Author Contributions: Conceptualization, J.T.D., G.T., F.F., M.A.V., and B.R.; methodology, J.T.D., C.B., M.C.G., F.F., G.T., and B.R.; formal analysis, J.T.D., C.B., I.O., and F.D.R.; investigation, J.T.D., I.O., F.D.R., and C.B.; resources, J.T.D., G.T., F.F., M.A.V., and B.R.; data curation, J.T.D., C.B., M.C.G., and F.F.; writing—original draft preparation, J.T.D., I.O., F.D.R., G.T., and B.R.; writing—review and editing, J.T.D., I.O., and F.D.R.; supervision, G.T., F.F., M.A.V., and B.R.; project administration, J.T.D., G.T., F.F., M.A.V., and B.R.; funding acquisition, J.T.D., G.T., F.F., M.A.V., and B.R. All authors have read and agreed to the published version of the manuscript.

Funding: This research was funded by the Umberto Veronesi Foundation, UNIMORE grant FAR (PI Prof. ZOLI), MAECI grant (PI Tosi, Nanomedicine for BBB-crossing in CNS oncologic pathologies), ER Project funding (POR FESR: Mat2Rep, https://mat2rep.it/).

Acknowledgments: Authors gratefully acknowledge professional technical of CIGS staff (University of Modena and Reggio Emilia) Maria Cecilia ROSSI, Cinzia Restani, and Adele Mucci (Department of Chemical and Geological Sciences) for assistance in NMR analysis.

Conflicts of Interest: The authors declare no conflicts of interest.

References

1. Tantra, R.; Oksel, C.; Puzyn, T.; Wang, J.; Robinson, K.N.; Wang, X.Z.; Ma, C.Y.; Wilkins, T. Nano(Q)SAR: Challenges, pitfalls and perspectives. *Nanotoxicology* **2015**, *9*, 636–642. [CrossRef]
2. Baig, M.H.; Khurshid, A.; Sudeep, R.; Jalaluddin, M.A.; Mohd, A.; Siddiqui, M.H.; Saif, K.; Kamal, M.A.; Ivo, P.; Inho, C. Computer Aided Drug Design: Success and Limitations. *Curr. Pharm. Des.* **2016**, *22*, 572–581. [CrossRef]
3. Li, J.; Crowley, S.T.; Duskey, J.; Khargharia, S.; Wu, M.; Rice, K.G. Miniaturization of gene transfection assays in 384- and 1536-well microplates. *Anal. Biochem.* **2015**, *470*, 14–21. [CrossRef]
4. Duskey, J.T.; Belletti, D.; Pederzoli, F.; Vandelli, M.A.; Forni, F.; Ruozi, B.; Tosi, G. Chapter One—Current Strategies for the Delivery of Therapeutic Proteins and Enzymes to Treat Brain DisordersInternational. *Rev. Neurobiol.* **2017**, *137*, 1–28. [CrossRef]
5. Tosi, G.; Duskey, J.T.; Jorg, K. Nanoparticles as carriers for drug delivery of macromolecules across the blood-brain barrier. *Expert Opin. Drug Deliv.* **2019**, *17*, 23–32. [CrossRef]
6. Hoyos-Ceballo, G.P.; Ruozi, B.; Ottonelli, I.; Da Ros, F.; Vandelli, M.A.; Forni, F.; Daini, E.; Vilella, A.; Zoli, M.; Tosi, G.; et al. PLGA-PEG-ANG-2 Nanoparticles for Blood-Brain Barrier Crossing: Proof-of-Concept Study. *Pharmaceutics* **2020**, *12*, 72. [CrossRef]
7. Oddone, N.; Pederzoli, F.; Duskey, J.T.; De Benedictis, C.A.; Grabrucker, A.M.; Forni, F.; Vandelli, M.A.; Ruozi, B.; Tosi, G. ROS-responsive "smart" polymeric conjugate: Synthesis, characterization and proof-of-concept study. *Int. J. Pharm.* **2019**, *270*, 1–11. [CrossRef]

8. Liu, J.; Postupalenko, V.; Duskey, J.T.; Palivan, C.G.; Meier, W. pH-Triggered Reversible Multiple Protein-Polymer conjugation Based on Molecular Recognition. *J. Phys. Chem. B* **2015**, *119*, 12066–12073. [CrossRef]
9. Dinu, A.I.; Duskey, J.T.; Car, A.; Palivan, C.G.; Meier, W. Engineered non-toxic cationic nanocarriers with photo-triggered slow-release properties. *Polym. Chem.* **2016**, *7*, 3451–3464. [CrossRef]
10. Khargharia, S.; Baumhover, N.J.; Crowley, S.T.; Duskey, J.T.; Rice, K.G. The uptake mechanism of PEGylated DNA polyplexes by the liver influences gene expression. *Gene Ther.* **2014**, *21*, 1021–1028. [CrossRef]
11. Najer, A.; Thamboo, S.; Duskey, J.T.; Palivan, C.G.; Beck, H.; Meier, W. Analysis of Molecular Parameters Determining the Antimalarial Activity of Polymer-Based Nanomimics. *Macromol. Rapid Commun.* **2015**, *36*, 1923–1928. [CrossRef]
12. Tosi, G.; Pederzoli, F.; Belletti, D.; Vandelli, M.A.; Forni, F.; Duskey, J.T.; Ruozi, B. Nanomedicine in Alzheimer's disease: Amyloid beta targeting strategy. *Prog. Brain Res.* **2019**, *245*, 57–88. [CrossRef]
13. Pederzoli, F.; Ruozi, B.; Duskey, J.T.; Hagmeyer, S.; Sauer, A.K.; Grabrucker, S.; Coelho, R.; Oddone, N.; Ottonelli, I.; Daini, E.; et al. Nanomedicine against Aβ aggregation by β–sheet breaker peptide delivery: In vitro evidence. *Pharmaceutics* **2019**, *11*, 572. [CrossRef]
14. Rigon, L.; Salvalaio, M.; Pederzoli, F.; Legnini, E.; Duskey, J.T.; D'Avanzo, F.; De Filippis, C.; Ruozi, B.; Marin, O.; Vandelli, M.A.; et al. Targeting brain disease in MPSII: Preclinical evaluation of IDS-loaded PLGA nanoparticles. *Int. J. Mol. Sci.* **2019**, *20*, 2014. [CrossRef]
15. Danhier, F.; Ansorena, E.; Silva, J.M.; Coco, R.; Le Breton, A.; Préat, V. PLGA-based nanoparticles: An overview of biomedical applications. *J. Control. Release* **2012**, *16*, 505–522. [CrossRef]
16. Rezvantalab, S.; Drude, N.I.; Moraveji, M.K.; Güvener, N.; Koons, E.K.; Shi, Y.; Lammers, T.; Kiessling, F. PLGA-Based Nanoparticles in Cancer Treatment. *Front. Pharmacol.* **2018**, *9*, 1–19. [CrossRef]
17. Kim, K.T.; Lee, J.; Kim, D.; Yoon, I.; Cho, H. Recent Progress in the Development of Poly(lactic-co-glycolic acid)-Based Nanostructures for Cancer Imaging and Therapy. *Pharmaceutics* **2019**, *11*, 280. [CrossRef]
18. Li, X.; Jiang, X. Microfluidics for producing poly (lactic-co-glycolic acid)-based pharmaceutical nanoparticles. *Adv. Drug Deliv. Rev.* **2018**, *128*, 101–114. [CrossRef]
19. Midoux, P.; Pichon, C.; Yaouanc, J.; Jaffrès, P. Chemical vectors for gene delivery: A current review on polymers, peptides and lipids containing histidine or imidazole as nucleic acids carriers. *Br. J. Pharmacol.* **2009**, *157*, 156–178. [CrossRef]
20. Chakravarthi, S.S.; Robinson, D.H. Enhanced cellular association of paclitaxel delivered in chitosan-PLGA particles. *Int. J. Pharm.* **2011**, *409*, 111–120. [CrossRef]
21. Simon, L.C.; Rhett, W.S.; Sabliov, C. Bioavailability of Orally Delivered Alphatocopherol by Poly(Lactic-Co-Glycolic)Acid (PLGA) Nanoparticles and Chitosan Covered PLGA Nanoparticles in F344 Rats. *NanoBiomedicine* **2016**, *3*, 1–10. [CrossRef] [PubMed]
22. Nafee, N.; Taetz, S.; Schneider, M.; Schaefer, F.; Lehr, C. Chitosan-coated PLGA nanoparticles for DNA/RNA delivery: Effect of the formulation parameters on complexation and transfection of antisense oligonucleotides. *Nanomedicine* **2007**, *3*, 173–183. [CrossRef] [PubMed]
23. Chung, Y.I.; Kim, J.C.; Kim, H.A.; Tae, G.; Lee, S.; Kim, K.; Kwon, I.C. The effect of surface functionalization of PLGA nanoparticles by heparin- or chitosan-conjugated Pluronic on tumor targeting. *J. Control. Release* **2010**, *143*, 374–382. [CrossRef]
24. Li, A.D.; Sun, Z.Z.; Zhou, M.; Xua, X.X.; Ma, J.Y.; Zheng, W.; Zhou, H.M.; Li, L.; Zheng, Y.F. Electrospun Chitosan-graft-PLGA nanofibres with significantly enhanced hydrophilicity and improved mechanical property. *Colloids Surf. B Biointerfaces* **2013**, *102*, 674–681. [CrossRef]
25. Cai, G.; Jiang, H.; Tu, K.; Wang, L.; Zhu, K. A Facile Route for Regioselective Conjugation of Organo-Soluble Polymers onto Chitosan. *Macromolecular Bioscience* **2009**, *9*, 256–261. [CrossRef]
26. Thakura, C.K.; Thotakura, N.; Kumar, R.; Kumar, P.; Singh, B.; Chitkara, D.; Raza, K. Chitosan-modified PLGA polymeric nanocarriers with better delivery potential for tamoxifen. *Int. J. Biol. Macromol.* **2016**, *93*, 381–389. [CrossRef]
27. Branca, C.; D'Angelo, G.; Crupi, C.; Khouzami, K.; Rifici, S.; Ruello, G.; Wanderlingh, U. polysaccharide-nanocomposite interactions: A FTIR-ATR study on chitosan and chitosan/clay films. *Polymer* **2016**, *99*, 614–622. [CrossRef]

28. Vey, E.; Rodger, C.; Booth, J.; Claybourn, M.; Miller, A.F.; Saiani, A. Degradation kinetics of poly(lactic-co-glycolic) acid block copolymer cast films in phosphate buffer solution as revealed by infrared and Raman spectroscopies. *Polym. Dedgradation Stab.* **2011**, *96*, 1882–1889. [CrossRef]
29. Piyamongkala, K.; Mekasut, L.; Pongstabodee, S. Cutting Fluid Effluent Removal by Adsorption on Chitosan and SDS-Modified Chitosan. *Macromol. Res.* **2008**, *16*, 492–502. [CrossRef]
30. Ma, F.K.; Li, J.; Kong, M.; Liu, Y.; An, Y.; Chen, X.G. Preparation and hydrolytic erosion of differently structured PLGA nanoparticles with chitosan modification. *Macromolecules* **2013**, *54*, 174–179. [CrossRef]
31. Heux, L.; Brugnerotto, J.; Desbrie'res, J.; Versali, M.F.; Rinaudo, M. Solid State NMR for Determination of Degree of Acetylation of Chitin and Chitosan. *Biomacromolecules* **2000**, *1*, 746–751. [CrossRef]

© 2020 by the authors. Licensee MDPI, Basel, Switzerland. This article is an open access article distributed under the terms and conditions of the Creative Commons Attribution (CC BY) license (http://creativecommons.org/licenses/by/4.0/).

Article

p47phox siRNA-Loaded PLGA Nanoparticles Suppress ROS/Oxidative Stress-Induced Chondrocyte Damage in Osteoarthritis

Hyo Jung Shin [1,2], Hyewon Park [1,2], Nara Shin [1,2], Hyeok Hee Kwon [1,2], Yuhua Yin [1,2], Jeong-Ah Hwang [1,2], Song I Kim [1,2], Sang Ryong Kim [3], Sooil Kim [2], Yongbum Joo [4], Youngmo Kim [4], Jinhyun Kim [5], Jaewon Beom [6] and Dong Woon Kim [1,2,*]

1. Department of Medical Science, Chungnam National University College of Medicine, Daejeon 35015, Korea; shinhyo1013@gmail.com (H.J.S.); phw6304@gmail.com (H.P.); s0870714@gmail.com (N.S.); kara00124@gmail.com (H.H.K.); yoonokhwa527@gmail.com (Y.Y.); ijjanghwang@gmail.com (J.-A.H.); kthddl2295@gmail.com (S.I.K.)
2. Department of Anatomy and Cell Biology, Brain Research Institute, Chungnam National University College of Medicine, Daejeon 35015, Korea; sikim@cnu.ac.kr
3. School of Life Sciences, BK21 Plus KNU Creative BioResearch Group, Institute of Life Science & Biotechnology, Brain Science and Engineering Institute, Kyungpook National University, Daegu 41566, Korea; srk75@knu.ac.kr
4. Department of Orthopedics, Chungnam National University College of Medicine, Daejeon 35015, Korea; longman76@hanmail.net (Y.J.); osdr69@cnuh.co.kr (Y.K.)
5. Division of Rheumatology, Department of Internal Medicine, Chungnam National University College of Medicine, Daejeon 35015, Korea; md228@hanmail.net
6. Department of Physical Medicine and Rehabilitation, Chung-Ang University Hospital, Chung-Ang University College of Medicine, Seoul 06973, Korea; powe5@cau.ac.kr
* Correspondence: visnu528@cnu.ac.kr

Received: 28 November 2019; Accepted: 10 February 2020; Published: 13 February 2020

Abstract: Osteoarthritis (OA) is the most common joint disorder that has had an increasing prevalence due to the aging of the population. Recent studies have concluded that OA progression is related to oxidative stress and reactive oxygen species (ROS). ROS are produced at low levels in articular chondrocytes, mainly by the nicotinamide adenine dinucleotide phosphate (NADPH) oxidase, and ROS production and oxidative stress have been found to be elevated in patients with OA. The cartilage of OA-affected rat exhibits a significant induction of p47phox, a cytosolic subunit of the NADPH oxidase, similarly to human osteoarthritis cartilage. Therefore, this study tested whether siRNA p47phox that is introduced with poly (D,L-lactic-co-glycolic acid) (PLGA) nanoparticles (p47phox si_NPs) can alleviate chondrocyte cell death by reducing ROS production. Here, we confirm that p47phox si_NPs significantly attenuated oxidative stress and decreased cartilage damage in mono-iodoacetate (MIA)-induced OA. In conclusion, these data suggest that p47phox si_NPs may be of therapeutic value in the treatment of osteoarthritis.

Keywords: osteoarthritis; monosodium iodoacetate; p47phox; PLGA nanoparticles; reactive oxygen species

1. Introduction

Osteoarthritis (OA) is the most prevalent form of joint disease, a hallmark of which is cartilage loss [1]. However, OA also affects various tissues and ultimately triggers articular chondrocyte degeneration, chondrocyte clustering, synovial inflammation, osteophyte formation, and subchondral bone remodeling in affected articular cartilage [2–4]. Patients with degenerative arthritis exhibit varying degrees of oxidative stress and cartilage destruction [5]. Additionally, reactive oxygen species (ROS) (in particular chondrocytes) are produced at low levels, mainly by the nicotinamide adenine dinucleotide phosphate (NADPH) oxidase, where they act as integral mediators of intracellular signaling mechanisms that contribute to the maintenance of cartilage homeostasis [6]. ROS production and oxidative stress are elevated in patients with OA; compared to normal cartilage, OA cartilage has a significantly greater degree of ROS-induced DNA damage, which is mediated by interleukin-1 [7]. In contrast, the levels of antioxidant enzymes (e.g., superoxide dismutase, catalase, and glutathione peroxidase) are lower in patients with OA, which indicates that oxidative stress plays a role in OA pathogenesis [8,9].

The NOX family of NADPH oxidases is a major source of ROS production under various pathological conditions. Currently, this family is known to include seven members that utilize various combinations of subunits to form active enzyme complexes: NOX1, NOX2, NOX3, NOX4, NOX5, dual oxidase 1 (DUOX1), and DUOX2 [10–12]. Of these combinations, the NOX2 complex is composed of two membrane-bound subunits (p22phox and gp91phox), various cytosolic proteins (p40phox, p47phox, and p67phox), and an Rac1 GTPase; these components assemble at membrane sites upon cell activation. Recently, phagocyte-type NADPH oxidases have been identified in other cell types (e.g., adventitial fibroblasts, vascular smooth muscle cells, endothelial cells, and renal mesangial cells), where this enzyme is thought to serve a signaling function [13].

Upon exposure to cytokines, synovial NOX is known to produce superoxide anions that activate multiple signaling pathways and lead to the expression of inflammatory genes in rheumatoid arthritis synovial fibroblasts [14–16]. Many studies have suggested that excessive levels of NOX-mediated ROS generation appear to be a major mediator of the pathogenesis of OA. For example, NOX4/p22phox-induced ROS production plays an essential role in the chondrocytes of patients with OA [17–20]. Moreover, a recent study showed that, compared to a control group, patients with knee OA exhibited 4.8- and 8.4-fold increases in the protein contents of prolidase and NOX2, respectively, while xanthine oxidase levels tended to increase, and a 5.4-fold increase was observed in NALP3 inflammasomes [21]. The van Dan Bosch group reported that NOX2-derived ROS enhance joint destruction during collagenase-induced OA [22], while another study showed that NOX2 is involved in the production of superoxides in synovial cells that were obtained from patients with rheumatoid arthritis and patients with OA [23]. Furthermore, NOX2-generated ROS are involved in chondrocyte death that is induced by interleukin-1β and the loss of the extracellular matrix due to the activation of hyaluronidase via acidification [24–26]. However, to the best of our knowledge, the role of p47phox, an NADPH oxidase subunit that is the major NOX that is involved in OA pathogenesis, has not yet been fully elucidated. It is known that p47phox plays a pivotal role in neutrophil NOX2 activation by providing domains for physical binding to cytochrome b558 (a complex that is associated with gp91phox and p22phox) and p67phox [27,28].

Local treatment via intra-articular injection is an appropriate strategy due to the fact that OA only affects the joints. When small molecular drugs have been introduced into intra-articular space, they have been easily and quickly removed by blood vessels and lymphatics [29]. Therefore, a suitable drug delivery system, such as biodegradable and bioeliminable material nanoparticles, is required for these drugs to increase their solubility and prolong their retention time in the articular cavity. In this study, we adopted poly (D,L-lactic-co-glycolic acid) (PLGA) polymers as a carrier system, because they have shown the ability to protect the loaded drug or siRNA from inactivation, reduce unwanted side effects, and enhance the efficacy of the active pharmaceutical ingredient due to improved solubility and bioavailability. Notably, only nanoparticles (NPs) with a size of less than 200 nm have the ability to easily permeate through mucus without being immobilized by the natural size-filtering mechanism [30].

Thus, the present study investigated the roles of p47phox in ROS production and cartilage damage in an OA model by using NPs.

2. Materials and Methods

2.1. Animals and Osteoarthritis Model

Male Sprague–Dawley rats weighing 120–140 g at the time of OA induction were used. Under brief isoflurane anesthesia, the animals were intra-articularly injected in the left knee with 2 mg of mono-iodoacetate (MIA; #I2512; Sigma-Aldrich, St. Louis, MI, USA) that was dissolved in 20 µL saline; controls received saline only (day 0). MIA was delivered through the left patellar tendon by using a 30-G needle.

2.2. Human Chondrocyte

We used a scalpel to excise cartilage from the femoral condyles and posterior patellar surfaces of OA patients that were treated at Chungnam National University Hospital (CNUH) (approval no. IRB-2016-06-007). The cartilage was cut into 2 mm thick pieces, and the digested suspension was passed through a 40 µM pore-size cell strainer to isolate individual chondrocytes. Cells were counted by using a cell counter. Cells (5×10^6) were seeded into 10 mm diameter dishes and cultured for 10 days in Dulbecco's minimal essential medium (DMEM) supplemented with 10% (v/v) fetal bovine serum (FBS). The medium was changed every 2 days [31].

2.3. Hematoxylin and Eosin Staining

Cartilage from humans with OA was frozen, sectioned to a 4-µm thickness, and fixed in 4% paraformaldehyde (PFA). MIA-injected knees were fixed in 4% (v/v) paraformaldehyde for 2 days, decalcified in a Calci-Clear solution (catalog no. HS-105; National Diagnostics, Atlanta, GA, USA) for 2 days, sectioned in the coronal plane (4 µm thickness), embedded in paraffin wax, and used to prepare slides after staining with hematoxylin. Then, the slides were sequentially dehydrated in 70%, 80%, 90%, and 100% ethanol. Finally, sections were cleared in xylene. A light microscope and a digital camera were used to capture and evaluate the histopathological features of the articular cartilage.

2.4. Behavior Test

Mechanical paw withdrawal thresholds were measured via up–down von Frey testing [32]. Rats were placed on an elevated metal grid. Fifty percent withdraw threshold values were determined by using the up–down method. Briefly, mechanical allodynia was assessed by measuring foot withdrawal thresholds in response to mechanical stimuli to the hind paw. The withdrawal threshold was determined by using the up–down method with a set of von Frey filaments from 0.008 to 1.4 g (0.008, 0.02, 0.04, 0.07, 0.16, 0.4, 0.6, 1, and 1.4 g).

2.5. Immunohistochemistry

After incubating the tissues with a blocking buffer (5% normal serum/0.3% Triton X-100, Bio-Rad, Irvine, CA, USA) for 1 h to prevent nonspecific binding, the sections were incubated with primary antibodies (p47phox, #sc-17844, 1:200, SantaCruz, Dallas, TX, USA) and diluted in a blocking buffer. Immunostaining was performed by using the avidin–biotin peroxidase complex (ABC) method, as reported previously [33,34]. The sections were mounted with Vectashield (Vector Laboratories, Burlingame, CA, USA), and images were obtained with a confocal microscope. The immunodensities in the graphs were quantified by the Image J program software.

2.6. DHE Staining

Superoxide anion levels in the spinal cord were determined by using dihydroethidium (DHE; Thermo Fisher Scientific), as described previously [34]. Cartilage sections were incubated with DHE (1 µM) at room temperature for 5 min and mounted on slides.

2.7. Cytotoxicity Assay

An MTT assay was performed in order to test the cell viability caused by siRNA-encapsulated PLGA nanoparticles as per the manufacturer's instructions. Detailed procedures can also be found in previous reports [35].

2.8. Preparation of siRNA-Encapsulated PLGA Nanoparticles

p47phox siRNA-encapsulated PLGA NPs were prepared from a PLGA nanoparticle synthesis service from the Nanoglia company (Daejeon, Republic of Korea) with minor modifications, as reported previously [33,35]. A copolymer of DL-lactic and glycolide in a 50/50 molar ratio and with an inherent viscosity midpoint of 0.2 dL/g was used. To produce p47phox siRNA-encapsulated PLGA NPs, 200 µL of 200 µM siRNA of a TE 8.0 buffer was added in drops to 800 µL of dichloromethane (DCM) that contained 25 mg of PLGA (Corbion, Amsterdam, the Netherlands) and then emulsified by sonication (50 W, 1 min; Vibra-Cell™ VCX 130; Sonics, Newtown, CT, USA) into a primary W1/O emulsion. Later, 2 mL of 1% PVA1500 (w/v; Thermo Fisher Scientific, Waltham, MA, USA) was directly added into the primary emulsion and further emulsified by sonication for 1 min to form a W1/O/W2 double emulsion. The resulting product was then diluted with 6 mL of 1% PVA1500 (w/v) and stirred magnetically for 3 h at room temperature to evaporate the DCM in a fume hood. Finally, the PLGA NPs were collected by centrifugation at 15,000× g for 15 min at 4 °C, washed twice with deionized water, and freeze-dried with 10 vials.

2.9. Colloidal Characterization of Nanoparticles

Two milligrams of lyophilized particles were dispersed in 1 mL of deionized water to determine the size distribution, zeta potential and polydispersity index (PDI) with the Zetasizer Nano ZS (Malvern Instruments, Malvern, UK), and the diameter and shape were determined with scanning electron microscopy (SEM; SNE-4500 M; SEC Co., Ltd., Suwon, Korea). The each 20 µM of siRNA-encapsulated PLGA nanoparticles, were collected and incubated in a Eppendorf tube with 250 µL of PBS, incubated at 37 °C for 48 h. At the designated time, 200 µL of the released medium was taken and replaced by the same amount of a fresh buffer. The amount of the p47phox siRNA was measured in the released buffer by using NanoDrop (Thermo Fisher Scientific). The accumulated release percentage of the p47phox siRNA and the entrapment efficiency were evaluated according to a previous report [36].

2.10. Statistical Analysis

The data are expressed as mean ± standard error of mean (SEM). The statistical significance between multiple groups was compared by a one-way analysis of variance (ANOVA) followed by an appropriate multiple comparison test. p-values of less than 0.05 were considered statistically significant. All statistical analyses were performed by using GraphPad Prism 6 (GraphPad Software Inc., San Diego, CA, USA).

3. Results

3.1. p47phox and ROS Were Highly Expressed in Chondrocyte Clusters in Human OA Tissues

Chondrocyte clustering is a histological sign of late-stage OA [1,37]; this type of clustering and its associated morphological abnormalities are particularly evident in the superficial zones of OA tissues. To explore whether p47phox was associated with ROS production and cartilage damage, samples of degenerating articular cartilage from patients with OA were assessed. In osteoarthritic joint cartilage, cluster cell numbers exceeded eight in affected lacunae, but this did not occur in non-affected lacunae (Figure 1A,B); DHE staining revealed an increased ROS production at injured sites, relative to the less involved areas (Figure 1A,B). Immunohistochemistry analyses of p47phox also revealed high levels of p47phox expression in chondrocytes in joint cartilage samples with clusters (Figure 1C,D). Taken together, these results indicated that increased levels of ROS production and p47phox were associated with cartilage degeneration in patients with OA.

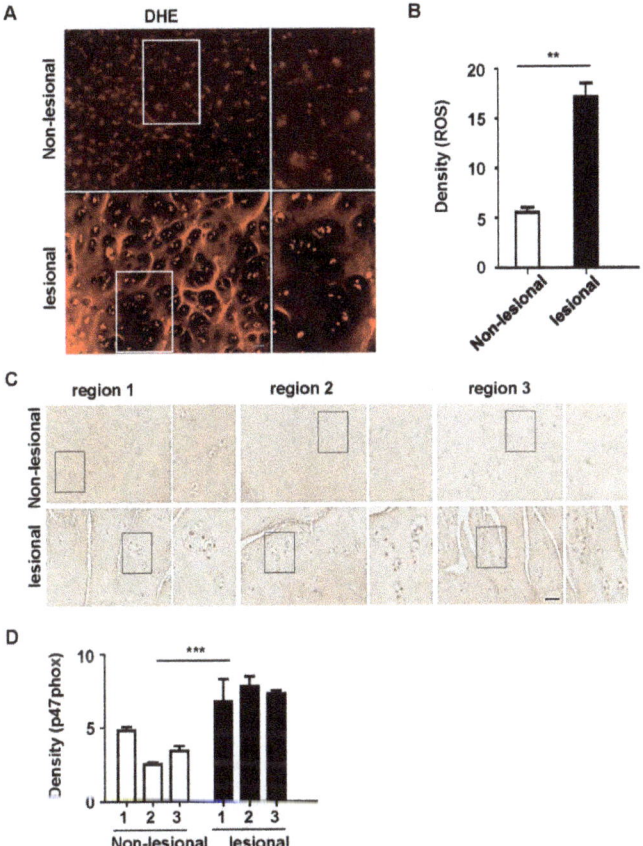

Figure 1. Reactive oxygen species (ROS) production and p47phox expression increased in the articular cartilage of human osteoarthritis (OA) knee joints. (**A**) Representative images of ROS-dependent dihydroethidium (DHE) fluorescence in human OA knee joints. (**B**) DHE staining density measured by using ImageJ software. (**C**) p47phox exhibited prominent expression at lesional sites, compared to non-lesional sites; immunohistochemical analyses show the expression levels of p47phox in knee cartilage at different sites. (**D**) p47phox immunostaining density quantified by ImageJ software; scale bar = 50 µm.

3.2. p47phox Was Highly Expressed in the Chondrocytes of MIA-Induced OA Rats

The progression of OA is significantly related to oxidative stress and ROS [7]. The major sites of ROS generation include the mitochondria (via oxidative phosphorylation), the non-mitochondrial membrane-bound NADPH oxidase, and the xanthine oxidase [38]. To verify the effects of the NADPH oxidase on OA progression, the present study examined the expression levels of five components of the NADPH oxidase: three cytosolic fractions (p40phox, p47phox, and p67phox) and two membrane fractions (p22phox and gp91phox). As proof-of-concept, monosodium iodoacetate (MIA) was injected into the knees of rats to assess changes in the levels of NADPH oxidase components in an animal model of toxin-induced OA.

The most common experimental OA model features intra-articular injections of MIA, which is a metabolic inhibitor, into the knee joints of rats [39]. This process results in the progressive loss of articular cartilage and the development of subchondral bone lesions; these closely resemble clinical findings in patients with OA. In the present study, a preliminary experiment was conducted by using a 2 mg/20 µL dose of MIA, which revealed that articular cartilage loss from the extracellular matrix was more pronounced on day 3 (data not shown). These results showed that MIA induced significant cartilage damage in the medial tibial plateau and femoral condyle at day 3 after intra-articular injection. Additionally, mechanical hypersensitivity was observed after approximately seven days and persisted at 14 days in the MIA-treated group (Figure 2B).

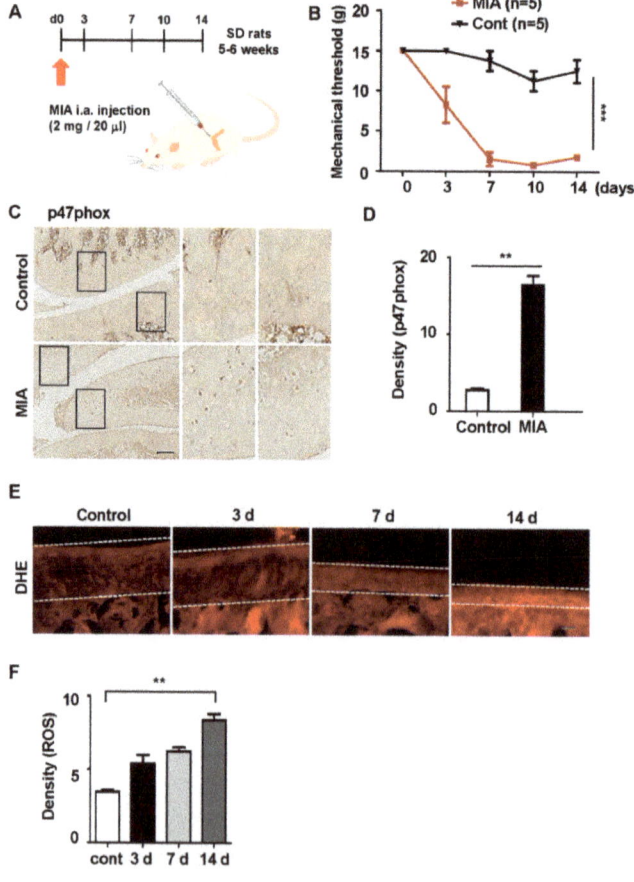

Figure 2. Monosodium iodoacetate (MIA)-induced expression of p47phox and the production of cellular ROS. (**A**) Prior to MIA injection, the rats were subjected to a von Frey filament test; only those that met a predefined threshold were selected for MIA injections. (**B**) The von Frey test was repeated on days 3, 7, 10, and 14 after injection; all data are presented as the mean ± standard error of the mean. (**C**) Rat knee tissues were immunostained with an anti-p47phox antibody at 3 days after MIA injection; scale bar = 50 µm. (**D**) The density of p47phox expression in the knee was measured by using ImageJ software; all data are presented as the mean ± standard error of the mean. (**E**) DHE fluorescence imaging of the knee in the OA rat model; white lines indicate cartilage in the tissue. (**F**) Quantification of DHE fluorescence; all data represent the mean ± standard error of the mean (error bars) of three experiments.

Because the cartilage damage that was caused by MIA was associated with ROS production, the expression levels of p47phox were measured. Animals injected with 2 mg of MIA exhibited an increased expression of p47phox by day 3 (Figure 2C,D). Next, the effects of MIA on ROS levels in chondrocytes were investigated; ROS production significantly increased in the cartilage of the MIA-treated group (Figure 2E,F). Taken together, these data suggested that the upregulation of p47phox in cartilage was associated with articular cartilage loss and ROS production.

3.3. Colloidal Characterization of p47phox siRNA-Encapsulated PLGA NPs

The present study also assessed whether the inhibition of p47phox would affect cartilage damage in the rat model of MIA-induced OA. The efficient delivery of p47phox siRNA into the joints by using a gene delivery system requires the consideration of several factors. Thus, the present study employed poly(D,L-lactic-co-glycolic acid) (PLGA) copolymers; these are biodegradable in and biocompatible with humans, and some products from PLGA are widely used in many Food and Drug Administration (FDA)-approved drugs. There are currently 15 FDA-approved PLA/PLGA-based drug products that are available on the US market [40].

Our research group previously demonstrated that p38 MAPK siRNA that is encapsulated in PLGA NPs attenuates spinal nerve ligation-induced neuropathic pain [35]. Moreover, our group reported that Foxp3 plasmid-loaded PLGA NPs can effectively relieve neuropathic pain in animals by reducing microglia activity and subsequently modulating neuroinflammation [33]. For the present study, p47phox siRNA-loaded PLGA NPs (p47phox si_PLGA NPs) were prepared via sonication by using the double emulsion (W/O/W) method. To prepare PLGA nanoparticles, 200 µL of 20 µM siRNA or scrambled siRNA in a TE7.5 buffer was added to 800 µL of dichloromethane (DCM) that contained 25 mg of PLGA. The average size and zeta potential of siRNA p47phox NPs were 126 ± 55 nm and −23 ± 2 mV, respectively (Figure 3A,B), and those of scrambled siRNA-encapsulated NPs were 117 ± 52 nm and −20 ± 2 mV, respectively, when measured with a Zetasizer ZS90 (Figure 3C,D). In addition, Zetasizer measurements revealed the formation of monodisperse particles (PDI ≤ 0.2) within the desired size range on scrambled or siRNA encapsulation. It should be noted that although PDI values smaller than 0.3 are considered acceptable for drug delivery applications, more specific standards and guidelines have yet to be established by regulatory authorities [30,41]. Furthermore, the uniformity and morphology of the NPs were confirmed by using scanning electron microscopy (Figure 3E). Prior to the in vivo administration of p47phox NPs to the MIA-induced OA rats, the present study assessed whether PLGA NPs would preferentially localize to articular cartilage. To accomplish this, PLGA NPs that encapsulated plasmid-expressing mCherry (pAAV-EF1a-MCS-T2A-mCherry) were administered into the knee joints of rats [33]; on post-injection day 3, mCherry was observed in the cartilage (Figure 3F).

Next, the p47phox siRNA release profiles from p47phox si_NPs were analyzed by using a cumulative percentage approach. The results showed that p47phox si_NPs gradually released siRNA; this cumulative release peaked at 48 h after injection. After an initial burst of release (53.2%) at 24 h, the release of siRNA was consequently observed in a sustained manner at different times. The percentage of encapsulation efficiency was calculated as the amount of siRNA released from the lyophilized PLGA NPs/the amount of siRNA initially taken to prepare the NPs × 100. Encapsulation efficiency was 36.4 ± 0.17% (Figure 3G). We set up the in vivo administration of p47phox si_NPs in the OA model via intra-articular injections at three days after MIA treatment (Figure 3H).

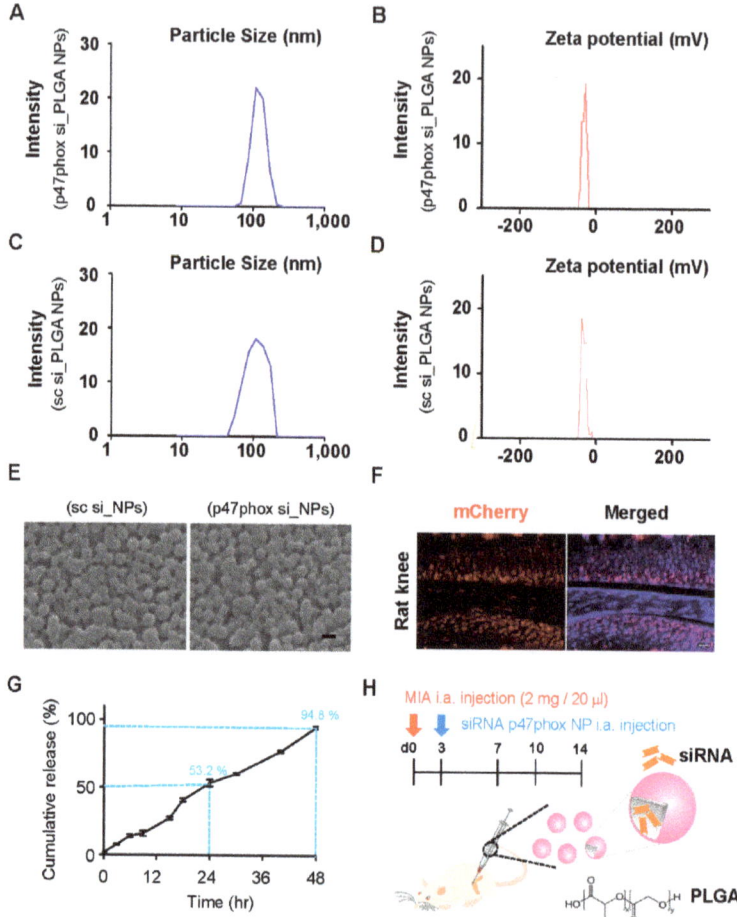

Figure 3. Characterization of p47phox siRNA-loaded poly (D,L-lactic-co-glycolic acid) (PLGA) nanoparticles (NPs). siRNA p47phox-encapsulated PLGA nanoparticles were dissolved in water and measured in terms of (**A,C**) size and (**B,D**) zeta potential by using a Zetasizer ZS90. (**E**) Suspended NPs were also assessed by using scanning electron microscopy; scale bar = 200 nm. (**F**) After 3 days, normal rat knees received intra-articular injections of AAV-mCherry expression vector-loaded PLGA NPs, and they were then were examined under a fluorescent microscope to assess uptake; scale bar = 50 μm. (**G**) In vitro cumulative siRNA release of PLGA NPs over 48 h. (**H**) Experimental schematic of the present study that used an MIA-induced OA animal model.

3.4. Inhibition of p47phox by NP-Delivered siRNA Attenuated Pain Behaviors, Cartilage Damage, and ROS Production in Knee Joints with MIA-Induced OA

To address the question whether siRNA-encapsulated nanoparticles could work in cartilage, the determination of the effective dose was firstly performed. Different doses of siRNA p47phox NPs—0.2, 0.4, and 0.8 μM—were applied to the MIA-induced OA (Supplementary Figure S1A). The mechanical thresholds were reduced in all dose dependent groups. As the lowest dose of 0.2 μM could reduce pain hypersensitivity, we used it for next investigations. Next, PLGA NPs that contained scrambled siRNA or p47phox siRNA were delivered to the cartilage via intra-articular injections. Because subchondral changes are closely associated with pain and are predictive of the severity of cartilage damage in

OA [42], the present study assessed whether p47phox si_NPs could attenuate OA-related pain behavior by using the von Frey filament test. Compared to saline controls, MIA injections (2 mg/20 μL) induced mechanical allodynia in the ipsilateral paw. However, injections of p47phox si_PLGA NPs alleviated mechanical allodynia in MIA rats for up to 14 days after injection (Figure 4A).

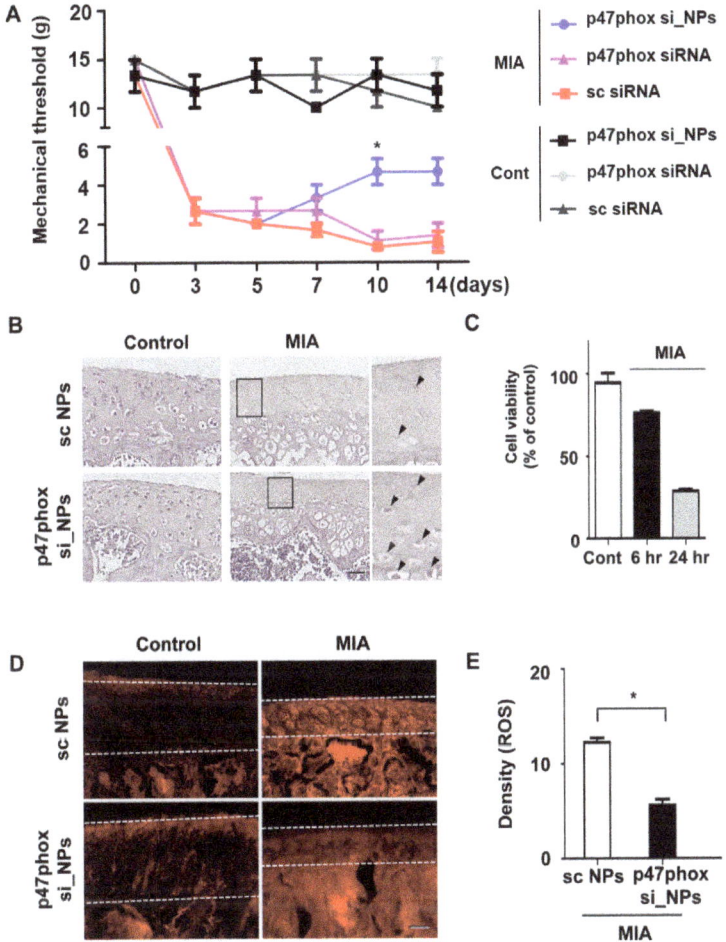

Figure 4. p47phox siRNA-encapsulated PLGA NPs reduced chondrocyte cell death by decreasing ROS production in OA rats. (**A**) On day 3 after MIA-induced OA, p47phox siRNA-loaded NPs were directly administered to the knee via intra-articular injections, and von Frey filament tests were performed on days 7, 10, and 14 after injection; all data are presented as the mean ± standard error of the mean. (**B**) Representative hematoxylin-stained sections of knee joints from rats with MIA-induced OA after 14 days, scale bar = 50 μm. (**C**) Cell viability over time in human primary chondrocytes following MIA injection. (**D**) DHE fluorescence imaging of knees in the OA rat model at 1 week after injection of PLGA NPs; white lines indicate cartilage in the tissue, scale bar = 50 μm. (**E**) Density of fluorescence intensity showing ROS production; all data represent the mean ± standard error of the mean (error bars) of three experiments.

Next, the present study investigated whether p47phox si_NPs could attenuate the loss of proteoglycan and calcification of articular cartilage on day 3 after MIA injection. Compared to scrambled siRNA-loaded NPs, treatment with p47phox si_NPs reduced the thickness of the subchondral bone plate and attenuated the loss of cartilage lacunae (Figure 4B). To observe whether MIA treatment affected cell viability, the human primary chondrocyte cells were incubated alone or in the presence of 5 μM MIA for 6 or 24 h. Following treatment, the viability of the cells was decreased to approximately 30% of their pre-treatment levels (Figure 4C). These results further support the idea that MIA-mediated OA could increase chondrocytic cell death in OA, and ROS production was examined in the MIA-injected joints of rats that received p47phox si_NPs. DHE staining revealed that administration of p47phox si_NPs attenuated ROS production in cartilage (Figure 4D,E). Taken together, these findings suggested that the inhibition of p47phox attenuated pain behaviors, cartilage damage, and ROS production in knee joints with MIA-induced OA.

4. Discussion

The signaling pathways by which ROS contribute to the pathophysiology of OA are complex and require further investigation. In general, antioxidant therapies are inefficient treatments for the relief of OA symptoms, whereas antioxidant drugs have shown promising in vitro results; thus, further human studies are required.

In the present study, MIA-induced OA chondrocytes produced ROS, a result that is consistent with findings from MIA-treated human chondrocytes. Excessive ROS production causes apoptotic cell death in OA chondrocytes; once induced, ROS are synthesized at a constant rate for a substantial period of time. The present results demonstrated a marked increase in the release of ROS in MIA-induced OA chondrocytes. Excessive levels of intracellular ROS are due to oxidative stress and enhanced levels of oxidative markers, which are the primary causes of cell damage and death. Thus, the inhibition of ROS-mediated injuries can reduce oxidative stress, protect chondrocytes, and treat OA. The present results demonstrated that the suppression of MIA-mediated injuries attenuated oxidative stress and protected chondrocytes. Therefore, the present findings contribute to the expanding understanding of proteins that are involved in the regulation of p47phox activation, and they also provide further insights regarding possible mechanisms that are involved in the regulation of cellular fates following OA.

The permeability transition pore is generally regarded as the major ROS target inside mitochondria [43]. The oxidative modification of mitochondrial permeability transition pore proteins has a significant impact on mitochondrial anion fluxes. For example, in response to pro-apoptotic stimuli, including ROS overload, mitochondrial permeability transition pores assume a high-conductance state that allows for the deregulated entry of small solutes into the mitochondrial matrix along their electrochemical gradients. The redox regulation of proteins by moderate levels of ROS has been observed in various signaling pathways, including the autophagy pathway, which is a catabolic pathway for the degradation of intracellular proteins and organelles via lysosomes. Some studies have shown that autophagy and cartilage damage are increased in MIA-induced OA models [44], while others have shown that OA chondrocytes exhibit reduced mitochondrial membrane potentials. Thus, the present study aimed to further clarify that these findings were not due to the blockage of autophagy.

The activation and regulation of the NADPH oxidase are controlled by the phosphorylation of its cytosolic component (p47phox) on serine subunits that are located between Ser303 and Ser379 [45]. Notably, the stimulation of neutrophils by high concentrations of the chemotactic peptide n-formyl methionyl-leucyl-phenylalanine, due to the protein kinase C agonist (phorbol 12-myristate 13-acetate), induces the complete phosphorylation of p47phox, which is required for the activation of the NADPH oxidase [46]. The present study showed that p47phox phosphorylation was required for MIA-induced OA. During the activation of the NADPH oxidase, approximately 10%–20% of p47phox proteins migrate to the plasma membrane, whereas 80%–90% remain in the cytosol. During activation, p47phox is presumed to bind to gp91phox/NOX2 and p22phox, because the translocation of p47phox to the plasma membranes impairs neutrophils from gp91phox- or p22phox-deficient patients. Though gp91phox is the central docking site for cytosolic components that translocate to the plasma membrane, p47phox is the subunit that is responsible for transporting the whole cytosolic complex to the docking site during the activation of the NADPH oxidase. Thus, p47phox is regarded as the organizer subunit because it coordinates the interactions of different NADPH oxidase subunits and allows for the formation of an active complex.

OA is a systemic inflammatory disorder that most commonly targets the joints. The synovial fluid in patients with OA contains large numbers of neutrophils and macrophages, which suggests that these factors may contribute to tissue injury. NADPH oxidase activity and the phosphorylation of p47phox are markedly increased in neutrophils from patients with OA. This upregulation could be due to the actions of pro-inflammatory cytokines, such as TNF-α, which is found in high concentrations in the synovial fluids of these patients [45].

Our study presented, for the first time, that p47phox siRNA-encapsulated PLGA NPs could have therapeutic effects on OA patients, leading to reduced chondrocytic cell death and cartilage damages. Additionally, the administration of NPs to the knees of patients with OA may influence their in vivo efficacy. However, these considerations are beyond the scope of this study and will be investigated in future work from our research group.

5. Conclusions

In the present study, the successful encapsulation of siRNA into PLGA NPs resulted in near-monodispersed and spherical particles with a low polydispersity index, and these particles maintained their effects for an extended duration. Furthermore, the present study demonstrated that the p47phox siRNA-encapsulated PLGA NPs acted synergistically to reduce ROS production in an animal model of OA; the formulated PLGA NPs provided a sustained release of siRNA. Additionally, the capability of the manufactured NPs to permit sustained release suggests the potential for a delivery system that could reduce dosing frequency to a weekly regimen. Therefore, p47phox siRNA PLGA NPs may represent a promising novel therapeutic avenue for the treatment of OA.

Supplementary Materials: The following are available online at http://www.mdpi.com/2073-4360/12/2/443/s1, Figure S1: Intra-articular injection of siRNA p47phox-encapsulated PLGA nanoparticles into the cartilage alleviates mechanical allodynia following MIA-induced OA pain in a dose-dependent manner.

Author Contributions: H.J.S., designed, performed and analyzed the experiments. H.P., N.S., H.H.K., Y.Y., J.-A.H., S.I.K., and S.R.K., formal analysis and data curation. S.K., J.K., Y.J., and Y.K., evaluated human samples. J.B. investigated. D.W.K. wrote the paper. All authors have read and agreed to the published version of the manuscript.

Funding: The authors gratefully acknowledge the financial supports by research fund of Chungnam National University and by the National Research Foundation of Korea (NRF) funded by the Ministry of Science, ICT & Future Planning (NRF-2019R1A2C2004884) and the Brain Korea 21 PLUS Project for Medical Science, Chungnam National University school of medicine.

Conflicts of Interest: The authors declare that they have no conflict of interest.

References

1. Hoshiyama, Y.; Otsuki, S.; Oda, S.; Kurokawa, Y.; Nakajima, M.; Jotoku, T.; Tamura, R.; Okamoto, Y.; Lotz, M.K.; Neo, M. Chondrocyte clusters adjacent to sites of cartilage degeneration have characteristics of progenitor cells. *J. Orthop. Res.* **2015**, *33*, 548–555. [CrossRef] [PubMed]
2. Lee, C.M.; Kisiday, J.D.; McIlwraith, C.W.; Grodzinsky, A.J.; Frisbie, D.D. Synoviocytes protect cartilage from the effects of injury in vitro. *BMC Musculoskelet. Disord.* **2013**, *14*, 54. [CrossRef] [PubMed]
3. Sandell, L.J.; Aigner, T. Articular cartilage and changes in arthritis. An introduction: Cell biology of osteoarthritis. *Arthritis Res.* **2001**, *3*, 107–113. [CrossRef] [PubMed]
4. Khan, I.M.; Palmer, E.A.; Archer, C.W. Fibroblast growth factor-2 induced chondrocyte cluster formation in experimentally wounded articular cartilage is blocked by soluble Jagged-1. *Osteoarthr. Cartil.* **2010**, *18*, 208–219. [CrossRef]
5. Soto-Hermida, A.; Fernandez-Moreno, M.; Pertega-Diaz, S.; Oreiro, N.; Fernandez-Lopez, C.; Blanco, F.J.; Rego-Perez, I. Mitochondrial DNA haplogroups modulate the radiographic progression of Spanish patients with osteoarthritis. *Rheumatol. Int.* **2015**, *35*, 337–344. [CrossRef]

6. Henrotin, Y.; Kurz, B.; Aigner, T. Oxygen and reactive oxygen species in cartilage degradation: Friends or foes? *Osteoarthr. Cartil.* **2005**, *13*, 643–654. [CrossRef]
7. Lepetsos, P.; Papavassiliou, A.G. ROS/oxidative stress signaling in osteoarthritis. *Biochim. Biophys. Acta* **2016**, *1862*, 576–591. [CrossRef]
8. Altindag, O.; Erel, O.; Aksoy, N.; Selek, S.; Celik, H.; Karaoglanoglu, M. Increased oxidative stress and its relation with collagen metabolism in knee osteoarthritis. *Rheumatol. Int.* **2007**, *27*, 339–344. [CrossRef]
9. Davies, C.M.; Guilak, F.; Weinberg, J.B.; Fermor, B. Reactive nitrogen and oxygen species in interleukin-1-mediated DNA damage associated with osteoarthritis. *Osteoarthr. Cartil.* **2008**, *16*, 624–630. [CrossRef]
10. Dupuy, C.; Ohayon, R.; Valent, A.; Noel-Hudson, M.S.; Deme, D.; Virion, A. Purification of a novel flavoprotein involved in the thyroid NADPH oxidase. Cloning of the porcine and human cdnas. *J. Biol. Chem.* **1999**, *274*, 37265–37269. [CrossRef]
11. Geiszt, M.; Leto, T.L. The Nox family of NAD(P)H oxidases: Host defense and beyond. *J. Biol. Chem.* **2004**, *279*, 51715–51718. [CrossRef] [PubMed]
12. Ma, M.W.; Wang, J.; Zhang, Q.; Wang, R.; Dhandapani, K.M.; Vadlamudi, R.K.; Brann, D.W. NADPH oxidase in brain injury and neurodegenerative disorders. *Mol. Neurodegener.* **2017**, *12*, 7. [CrossRef] [PubMed]
13. Li, J.M.; Mullen, A.M.; Yun, S.; Wientjes, F.; Brouns, G.Y.; Thrasher, A.J.; Shah, A.M. Essential role of the NADPH oxidase subunit p47(phox) in endothelial cell superoxide production in response to phorbol ester and tumor necrosis factor-alpha. *Circ. Res.* **2002**, *90*, 143–150. [CrossRef]
14. Chenevier-Gobeaux, C.; Lemarechal, H.; Bonnefont-Rousselot, D.; Poiraudeau, S.; Ekindjian, O.G.; Borderie, D. Superoxide production and NADPH oxidase expression in human rheumatoid synovial cells: Regulation by interleukin-1beta and tumour necrosis factor-alpha. *Inflamm. Res.* **2006**, *55*, 483–490. [CrossRef] [PubMed]
15. Sung, J.Y.; Hong, J.H.; Kang, H.S.; Choi, I.; Lim, S.D.; Lee, J.K.; Seok, J.H.; Lee, J.H.; Hur, G.M. Methotrexate suppresses the interleukin-6 induced generation of reactive oxygen species in the synoviocytes of rheumatoid arthritis. *Immunopharmacology* **2000**, *47*, 35–44. [CrossRef]
16. Chi, P.L.; Chen, Y.W.; Hsiao, L.D.; Chen, Y.L.; Yang, C.M. Heme oxygenase 1 attenuates interleukin-1beta-induced cytosolic phospholipase A2 expression via a decrease in NADPH oxidase/reactive oxygen species/activator protein 1 activation in rheumatoid arthritis synovial fibroblasts. *Arthritis Rheum.* **2012**, *64*, 2114–2125. [CrossRef]
17. Drevet, S.; Gavazzi, G.; Grange, L.; Dupuy, C.; Lardy, B. Reactive oxygen species and NADPH oxidase 4 involvement in osteoarthritis. *Exp. Gerontol.* **2018**, *111*, 107–117. [CrossRef]
18. Rousset, F.; Hazane-Puch, F.; Pinosa, C.; Nguyen, M.V.; Grange, L.; Soldini, A.; Rubens-Duval, B.; Dupuy, C.; Morel, F.; Lardy, B. IL-1beta mediates MMP secretion and IL-1beta neosynthesis via upregulation of p22(phox) and NOX4 activity in human articular chondrocytes. *Osteoarthr. Cartil.* **2015**, *23*, 1972–1980. [CrossRef]
19. Lepetsos, P.; Pampanos, A.; Lallos, S.; Kanavakis, E.; Korres, D.; Papavassiliou, A.G.; Efstathopoulos, N. Association of NADPH oxidase p22phox gene C242T, A640G and -930A/G polymorphisms with primary knee osteoarthritis in the Greek population. *Mol. Biol. Rep.* **2013**, *40*, 5491–5499. [CrossRef]
20. Grange, L.; Nguyen, M.V.; Lardy, B.; Derouazi, M.; Campion, Y.; Trocme, C.; Paclet, M.H.; Gaudin, P.; Morel, F. NAD(P)H oxidase activity of Nox4 in chondrocytes is both inducible and involved in collagenase expression. *Antioxid. Redox Signal.* **2006**, *8*, 1485–1496. [CrossRef]
21. Clavijo-Cornejo, D.; Martinez-Flores, K.; Silva-Luna, K.; Martinez-Nava, G.A.; Fernandez-Torres, J.; Zamudio-Cuevas, Y.; Guadalupe Santamaria-Olmedo, M.; Granados-Montiel, J.; Pineda, C.; Lopez-Reyes, A. The Overexpression of NALP3 Inflammasome in Knee Osteoarthritis Is Associated with Synovial Membrane Prolidase and NADPH Oxidase 2. *Oxid. Med. Cell. Longev.* **2016**, *2016*, 1472567. [CrossRef] [PubMed]
22. Van Dalen, S.C.M.; Kruisbergen, N.N.L.; Walgreen, B.; Helsen, M.M.A.; Sloetjes, A.W.; Cremers, N.A.J.; Koenders, M.I.; van de Loo, F.A.J.; Roth, J.; Vogl, T.; et al. The role of NOX2-derived reactive oxygen species in collagenase-induced osteoarthritis. *Osteoarthr. Cartil.* **2018**, *26*, 1722–1732. [CrossRef] [PubMed]
23. Chenevier-Gobeaux, C.; Simonneau, C.; Therond, P.; Bonnefont-Rousselot, D.; Poiraudeau, S.; Ekindjian, O.G.; Borderie, D. Implication of cytosolic phospholipase A2 (cPLA2) in the regulation of human synoviocyte NADPH oxidase (Nox2) activity. *Life Sci.* **2007**, *81*, 1050–1058. [CrossRef] [PubMed]

24. Yasuhara, R.; Miyamoto, Y.; Akaike, T.; Akuta, T.; Nakamura, M.; Takami, M.; Morimura, N.; Yasu, K.; Kamijo, R. Interleukin-1beta induces death in chondrocyte-like ATDC5 cells through mitochondrial dysfunction and energy depletion in a reactive nitrogen and oxygen species-dependent manner. *Biochem. J.* **2005**, *389*, 315–323. [CrossRef]
25. Funato, S.; Yasuhara, R.; Yoshimura, K.; Miyamoto, Y.; Kaneko, K.; Suzawa, T.; Chikazu, D.; Mishima, K.; Baba, K.; Kamijo, R. Extracellular matrix loss in chondrocytes after exposure to interleukin-1beta in NADPH oxidase-dependent manner. *Cell Tissue Res.* **2017**, *368*, 135–144. [CrossRef]
26. Yoshimura, K.; Miyamoto, Y.; Yasuhara, R.; Maruyama, T.; Akiyama, T.; Yamada, A.; Takami, M.; Suzawa, T.; Tsunawaki, S.; Tachikawa, T.; et al. Monocarboxylate transporter-1 is required for cell death in mouse chondrocytic ATDC5 cells exposed to interleukin-1beta via late phase activation of nuclear factor kappaB and expression of phagocyte-type NADPH oxidase. *J. Biol. Chem.* **2011**, *286*, 14744–14752. [CrossRef]
27. Panday, A.; Sahoo, M.K.; Osorio, D.; Batra, S. NADPH oxidases: An overview from structure to innate immunity-associated pathologies. *Cell. Mol. Immunol.* **2015**, *12*, 5–23. [CrossRef]
28. Morozov, I.; Lotan, O.; Joseph, G.; Gorzalczany, Y.; Pick, E. Mapping of functional domains in p47(phox) involved in the activation of NADPH oxidase by "peptide walking". *J. Biol. Chem.* **1998**, *273*, 15435–15444. [CrossRef]
29. Kou, L.; Xiao, S.; Sun, R.; Bao, S.; Yao, Q.; Chen, R. Biomaterial-engineered intra-articular drug delivery systems for osteoarthritis therapy. *Drug Deliv.* **2019**, *26*, 870–885. [CrossRef]
30. Operti, M.C.; Dolen, Y.; Keulen, J.; van Dinther, E.A.W.; Figdor, C.G.; Tagit, O. Microfluidics-Assisted Size Tuning and Biological Evaluation of PLGA Particles. *Pharmaceutics* **2019**, *11*, 590. [CrossRef]
31. Bakker, B.; Eijkel, G.B.; Heeren, R.M.; Karperien, M.; Post, J.N.; Cillero-Pastor, B. Oxygen Regulates Lipid Profiles in Human Primary Chondrocyte Cultures. *Osteoarthr. Cartil.* **2016**, *24*, S456–S457. [CrossRef]
32. Roh, D.H.; Kim, H.W.; Yoon, S.Y.; Seo, H.S.; Kwon, Y.B.; Kim, K.W.; Han, H.J.; Beitz, A.J.; Na, H.S.; Lee, J.H. Intrathecal injection of the sigma(1) receptor antagonist BD1047 blocks both mechanical allodynia and increases in spinal NR1 expression during the induction phase of rodent neuropathic pain. *Anesthesiology* **2008**, *109*, 879–889. [CrossRef] [PubMed]
33. Shin, J.; Yin, Y.; Kim, D.K.; Lee, S.Y.; Lee, W.; Kang, J.W.; Kim, D.W.; Hong, J. Foxp3 plasmid-encapsulated PLGA nanoparticles attenuate pain behavior in rats with spinal nerve ligation. *Nanomedicine* **2019**, *18*, 90–100. [CrossRef] [PubMed]
34. Shin, N.; Kim, H.G.; Shin, H.J.; Kim, S.; Kwon, H.H.; Baek, H.; Yi, M.H.; Zhang, E.; Kim, J.J.; Hong, J.; et al. Uncoupled Endothelial Nitric Oxide Synthase Enhances p-Tau in Chronic Traumatic Encephalopathy Mouse Model. *Antioxid. Redox Signal.* **2019**, *30*, 1601–1620. [CrossRef]
35. Shin, J.; Yin, Y.; Park, H.; Park, S.; Triantafillu, U.L.; Kim, Y.; Kim, S.R.; Lee, S.Y.; Kim, D.K.; Hong, J.; et al. p38 siRNA-encapsulated PLGA nanoparticles alleviate neuropathic pain behavior in rats by inhibiting microglia activation. *Nanomedicine* **2018**, *13*, 1607–1621. [CrossRef]
36. Peltonen, L.; Aitta, J.; Hyvonen, S.; Karjalainen, M.; Hirvonen, J. Improved entrapment efficiency of hydrophilic drug substance during nanoprecipitation of poly(l)lactide nanoparticles. *AAPS PharmSciTech* **2004**, *5*, 115. [CrossRef]
37. Karim, A.; Amin, A.K.; Hall, A.C. The clustering and morphology of chondrocytes in normal and mildly degenerate human femoral head cartilage studied by confocal laser scanning microscopy. *J. Anat.* **2018**, *232*, 686–698. [CrossRef]
38. Turrens, J.F. Mitochondrial formation of reactive oxygen species. *J. Physiol.* **2003**, *552*, 335–344. [CrossRef]
39. Pitcher, T.; Sousa-Valente, J.; Malcangio, M. The Monoiodoacetate Model of Osteoarthritis Pain in the Mouse. *J. Vis. Exp.* **2016**, e53746. [CrossRef]
40. Jain, A.; Kunduru, K.R.; Basu, A.; Mizrahi, B.; Domb, A.J.; Khan, W. Injectable formulations of poly(lactic acid) and its copolymers in clinical use. *Adv. Drug Deliv. Rev.* **2016**, *107*, 213–227. [CrossRef]
41. Lababidi, N.; Sigal, V.; Koenneke, A.; Schwarzkopf, K.; Manz, A.; Schneider, M. Microfluidics as tool to prepare size-tunable PLGA nanoparticles with high curcumin encapsulation for efficient mucus penetration. *Beilstein J. Nanotechnol.* **2019**, *10*, 2280–2293. [CrossRef] [PubMed]
42. Hunter, D.J.; Zhang, Y.; Niu, J.; Goggins, J.; Amin, S.; LaValley, M.P.; Guermazi, A.; Genant, H.; Gale, D.; Felson, D.T. Increase in bone marrow lesions associated with cartilage loss: A longitudinal magnetic resonance imaging study of knee osteoarthritis. *Arthritis Rheum.* **2006**, *54*, 1529–1535. [CrossRef] [PubMed]

43. Marchi, S.; Giorgi, C.; Suski, J.M.; Agnoletto, C.; Bononi, A.; Bonora, M.; De Marchi, E.; Missiroli, S.; Patergnani, S.; Poletti, F.; et al. Mitochondria-ros crosstalk in the control of cell death and aging. *J. Signal Transduct.* **2012**, *2012*, 329635. [CrossRef]
44. Zhang, X.; Yang, Y.; Li, X.; Zhang, H.; Gang, Y.; Bai, L. Alterations of autophagy in knee cartilage by treatment with treadmill exercise in a rat osteoarthritis model. *Int. J. Mol. Med.* **2019**, *43*, 336–344. [CrossRef] [PubMed]
45. El-Benna, J.; Dang, P.M.; Gougerot-Pocidalo, M.A.; Marie, J.C.; Braut-Boucher, F. p47phox, the phagocyte NADPH oxidase/NOX2 organizer: Structure, phosphorylation and implication in diseases. *Exp. Mol. Med.* **2009**, *41*, 217–225. [CrossRef] [PubMed]
46. Faust, L.R.; El Benna, J.; Babior, B.M.; Chanock, S.J. The phosphorylation targets of p47phox, a subunit of the respiratory burst oxidase. Functions of the individual target serines as evaluated by site-directed mutagenesis. *J. Clin. Investig.* **1995**, *96*, 1499–1505. [CrossRef]

© 2020 by the authors. Licensee MDPI, Basel, Switzerland. This article is an open access article distributed under the terms and conditions of the Creative Commons Attribution (CC BY) license (http://creativecommons.org/licenses/by/4.0/).

Article

Poly(D,L-lactide-*co*-glycolide) (PLGA) Nanoparticles Loaded with Proteolipid Protein (PLP)—Exploring a New Administration Route

Alexandre Ferreira Lima, Isabel R. Amado * and Liliana R. Pires *,†

INL–International Nanotechnology Laboratory, Avenida Mestre José Veiga s/n, 4715-330 Braga, Portugal; alexandre98.lima@gmail.com
* Correspondence: isabel.rodriguez@inl.int (I.R.A.); liliana.pires@inl.int (L.R.P.)
† Current affiliation: RUBYnanomed, Unipessoal Lda, Av. Mestre José Veiga s/n, 4715-330 Braga, Portugal.

Received: 20 November 2020; Accepted: 15 December 2020; Published: 21 December 2020

Abstract: The administration of specific antigens is being explored as a mean to re-establish immunological tolerance, namely in the context of multiple sclerosis (MS). PLP139-151 is a peptide of the myelin's most abundant protein, proteolipid protein (PLP), which has been identified as a potent tolerogenic molecule in MS. This work explored the encapsulation of the peptide into poly(lactide-*co*-glycolide) nanoparticles and its subsequent incorporation into polymeric microneedle patches to achieve efficient delivery of the nanoparticles and the peptide into the skin, a highly immune-active organ. Different poly(D,L-lactide-*co*-glycolide) (PLGA) formulations were tested and found to be stable and to sustain a freeze-drying process. The presence of trehalose in the nanoparticle suspension limited the increase in nanoparticle size after freeze-drying. It was shown that rhodamine can be loaded in PLGA nanoparticles and these into poly(vinyl alcohol)–poly(vinyl pyrrolidone) microneedles, yielding fluorescently labelled structures. The incorporation of PLP into the PLGA nanoparticles resulted in nanoparticles in a size range of 200 µm and an encapsulation efficiency above 20%. The release of PLP from the nanoparticles occurred in the first hours after incubation in physiological media. When loading the nanoparticles into microneedle patches, structures were obtained with 550 µm height and 180 µm diameter. The release of PLP was detected in PLP–PLGA.H20 nanoparticles when in physiological media. Overall, the results show that this strategy can be explored to integrate a new antigen-specific therapy in the context of multiple sclerosis, providing minimally invasive administration of PLP-loaded nanoparticles into the skin.

Keywords: dissolving microneedles; multiple sclerosis; PLP; transdermal delivery; PLGA

1. Introduction

The use of antigen-specific therapies has been explored for the treatment of autoimmune diseases. The premise is to selectively disarm autoimmune responses without suppressing global immunity [1]. Multiple sclerosis (MS) is an autoimmune and demyelinating disease characterized by the presence of inflammatory infiltrates (T cells, B cells, macrophages) within the central nervous system (CNS) that leads to immune-mediated myelin and axonal damage. In the case of MS, peptides from the three major myelin proteins—myelin basic protein (MBP), myelin oligodendrocyte glycoprotein (MOG), and proteolipid protein (PLP)—have been identified to be related to autoimmunity. The use of peptides from myelin proteins to restore immunological tolerance has been extensively investigated, showing positive results in experimental autoimmune encephalomyelitis (EAE) animal models and also in clinical trials.

In the early studies, peptides such as PLP139-151 were chemically crosslinked at the surface of syngeneic splenic leukocytes using ethylene carbodiimide. The infusion of these modified cells induced antigen-specific immune tolerance, overcoming drawbacks in trials directly related to the delivery of soluble peptides or antibodies [2]. However, the direct administration of engineered cells involves limitations and relevant costs for cell isolation under good manufacturing practice (GMP) conditions. In this context, and to translate the technology into the clinical practice, the delivery of antigen crosslinked micro- and nanoparticles was explored. Additionally, the use of nano- and microparticles takes advantage of the intrinsic properties of biomaterials and nanoparticulate systems, which can enhance delivery and cell targeting, namely toward dendritic cells, which play a critical role in the immune response [3].

The intravenous administration of poly(D,L-lactide-co-glycolide) (PLGA) microparticles crosslinked with a PLP139-151 peptide (the immunodominant T cell myelin epitope in SJL mice from the myelin's most abundant protein, PLP) demonstrated an ability not only to reduce the clinical score when administrated prophylactically, but also to treat the disease [4]. Interestingly, the same peptide administered in the form of colloidal hydrogel was effective only if administered before disease onset [5]. The incorporation of poly(ethylene-co-maleic acid) (PEMA) as a surfactant in a PLGA formulation allowed the preparation of smaller nanoparticles, providing a reliable platform for different antigen crosslinking, as demonstrated by the relevant results in the induction of immunological tolerance both in the context of EAE [6] and in a transplantation model [7]. Alternatively to antigen crosslinking, nanoparticles can be loaded with an antigen of interest. This concept can be extended to the development of multifunctional systems that combine the delivery of antigens with the encapsulation of other molecules/drugs as a means to make the immune response more specific and/or more effective. Relevant results were obtained by Yeste et al., loading gold nanoparticles with the antigen (MOG35-55) along with a tolerogenic molecule, ITE (2-(1'H-indole-3'-carbonyl)-thiazole-4-carboxylic acid methyl ester). The authors showed that the combination could achieve functional regulatory T cells in an EAE animal model more efficiently than MOG-loaded particles [8]. A different study reported treatment with PLGA nanoparticles containing MOG35-55 and interleukin-10 (IL-10). Although the results in terms of regulatory T cell expansion were not as impressive as those obtained for crosslinked nanoparticles, the nanoparticles caused a reduction in the severity of the disease via subcutaneous administration [9].

Microneedles have been investigated for the minimally invasive delivery of drugs through the skin, overcoming the *stratum corneum* barrier [10–12]. Early reports showed improved immunogenicity of molecules when administered via microneedle devices [13]. This promising result was considered to be related to delivery at the epidermal and intradermal layers of the skin, which is highly rich in immunologically active antigen-presenting cells (APCs). These cells deliver antigens to the proximal lymph nodes where T and B cells are activated, triggering an immune response [14]. Additionally, in the skin (particularly in the epidermis and the epithelium from the hair follicles), monocytes and Langerhans cells are abundant Langerhans cells display intrinsic tolerogenic properties in vivo [15].

To take advantage of skin immunogenicity, the use of microneedle patches is presented as an interesting minimally invasive means to deliver molecules that target the immune system [11,13]. In previous work, we designed and prepared dissolvable microneedles for the transdermal administration of molecules [16,17]. We showed the incorporation of the PLP139-151 peptide and its release at therapeutic doses after 3 days in physiological media [16]. As means to improve peptide stability and add a second layer of control to the release of the molecules, in the present work, we explore the encapsulation of the PLP peptide in PLGA nanoparticles and subsequent incorporation into microneedles.

2. Materials and Methods

2.1. Materials

Poly(D,L-lactide-co-glycolide) (PLGA) and acid terminated PLGA (PLGA.H) (50:50) were kindly offered from PURASORB (Corbion, Amsterdam, The Netherlands). Ethyl acetate 99.8%, poly(vinyl alcohol) (PVA, Mowiol 18–88, Mw ~130000), poly(vinyl pyrrolidone) (PVP, Mw ~10000), rhodamine 6G, and trehalose were purchased from Sigma-Aldrich (Saint Louis, MO, USA). The PLP139-151 peptide with the sequence HSLGKWLGHPDKF (purity >95%) was purchased from Primm Biotech (Cambridge, MA, USA) and stored at −80 °C in a 50 mg.ml^{-1} stock solution in water.

2.2. PLGA Nanoparticle Preparation

PLGA nanoparticles were prepared by double emulsion based on a previously described procedure [18]. Briefly, PLGA (PLGA20: 20 mg; PLGA60: 60 mg) was dissolved in 2 mL of ethyl acetate and sonicated for 15 s (70% amplitude) using an ultrasonic homogenizer (Branson Digital Sonicator, Saint Louis, MO, USA), resulting in a w/o emulsion. An equal volume of PVA solution (7% (w/v) in water) was added and sonicated for an additional 30 s (70% amplitude), resulting in a w/o/w double emulsion. The organic solvent was evaporated in a Savant SPD121P vacuum centrifuge (Thermo Fisher Scientific, Waltham, MA, USA) at 10,000 rpm for approximately 1 h at 40 °C. Fluorescent nanoparticles were prepared, adding rhodamine (0.5 mg·mL^{-1}, 1 mg) to the organic solution. To prepare peptide-loaded nanoparticles, 400 µg of PLP139-151 was added to the PLGA organic solution. The prepared nanoparticle suspensions were stored at 4 °C until further use.

2.3. Nanoparticle Freeze-Drying, Stability, and Storage

The nanoparticle suspensions were freeze-dried as follows. To remove debris and non-encapsulated materials, the nanoparticle suspensions were diluted and subsequently centrifuged at 15,000 rpm at 4 °C for 1 h. The precipitated nanoparticles were resuspended in the same volume of water or trehalose 10% (w/v) solution before freeze-drying for 24 h (LyoQuest, Telstar, Terrassa, Spain).

The nanoparticle suspensions prepared to load in the microneedle patches were frozen without dilution. In brief, 1 mL of nanoparticle suspension was added to 0.5 mL of trehalose 10% (w/v) and subsequently freeze-dried.

2.4. Characterization of Nanoparticles by Dynamic Light Scattering (DLS)

PLGA nanoparticles were characterized in terms of size and zeta potential by dynamic light scattering (DLS) using an SZ–100 nanoparticle analyzer series from Horiba Scientific. Measurements were performed using diluted nanoparticle suspensions (20×) at 25 °C. Size measurements were performed at 90° and assessed in triplicate. Zeta potential measurements were also performed in triplicate.

2.5. Polymeric Microneedle Preparation

Polydimethylsiloxane (PDMS) molds were obtained from silicon masters, as previously described [16]. The 3 cm × 3 cm masters contained 33 × 33 needles with a 600 µm tip-to-tip distance, a height of approximately 600 µm, and a diameter of 200 µm.

To prepare the polymeric microneedles, mixtures of PVA and PVP were prepared. The PVA (10% (w/v) in water) and PVP (15% (w/v) in phosphate buffer 0.1 M, pH 7.4) solutions were mixed at a 3:2 (v:v) ratio [16]. The mixture was then poured in the central part of the mold. After applying a vacuum for 20 min and drying about 1 h, more PVA:PVP mix was added, covering all the mold (including the borders to facilitate handling). After eliminating all the air bubbles, the polymeric patches were allowed to dry at room temperature for 24 h. The MN patches were peeled off and observed under optical microscopy (Nikon Eclipse Ni-E, Isaza, Portugal). Polymeric microneedle size was assessed from the optical microscopy images of at least 15 needles from three independent patches.

For the preparation of patches loaded with nanoparticles, freeze-dried nanoparticles were resuspended in 0.7 mL of a PVA:PVP mixture and, after overnight under-rotation at 4 °C, were then added to the PDMS mold, as described above.

2.6. Rhodamine Loading and Release

The association efficiency of rhodamine 6G in PLGA nanoparticles was determined by quantifying the free molecule in the supernatant after centrifugation at 15,000 rpm for 1 h at 4 °C (MIKRO 200R, Hettich, Kirchlengern, Germany). The fluorescence was assessed at $\lambda ex = 570$ nm and $\lambda em = 595$ nm using a microplate reader (Synergy H1MFD box, Biotek, Shoreline, WA, USA). The nanoparticles were incubated for 24 h in a phosphate buffer (0.1 M, pH 7.4) at 37 °C, and the released rhodamine was again determined after a centrifugation step.

2.7. PLP Quantification–Loading and Release

To quantify the PLP139-151 peptide, a high-performance liquid chromatography (HPLC) analysis was conducted using a 1260 Infinity II LC System (Agilent Technologies, Madrid, Spain) with diode-array detection (DAD) detection. The separation was performed in a PEPTIDE XB-C18 column (3.6 µm, 150 mm × 21 mm, Aeris, Phenomenex, Alcobendas, Spain). The mobile phase consisted of a mixture of acetonitrile with 0.1% trifluoroacetic acid (TFA; A) and water containing 0.1% TFA (B). Gradient elution was performed at a constant 0.3 mL·min^{-1} flow rate from 100% A to 35% during 20 min, followed by isocratic elution at 75% B for 3 min. A post-run of 5 min was conducted to equilibrate the column at the initial mobile phase composition. PLP was detected at 220 nm after a 50 µL sample injection.

PLP loading in the nanoparticles was calculated after the quantification of the PLP detected in the supernatant of centrifuged nanoparticles (15,000 rpm, 1 h, 4 °C).

To assess the release of PLP from the microneedle patches under physiological conditions, the patches were dialyzed (3 kDa Membrane, Orange Scientific, Braine-l'Alleud, Belgium) against a phosphate buffer (0.1 M, pH 7.4) at 37 °C. The medium was collected and refreshed at different time points (4, 24, and 48 h). The collected samples were filtered (0.45 µm) and stored at −20 °C until further analysis. The amount of PLP in the polymeric patches and PLGA nanoparticles was interpolated from a calibration curve of the purified peptide, ranging from 50 to 0.1 µg·mL^{-1}.

2.8. Statistical Analysis

Statistical analysis was performed using the software Graphpad Prism version 8.0 (GraphPad, San Diego, CA, USA). Statistical differences were calculated using one-way ANOVA followed by Dunnett's tests for multiple comparisons. A $p < 0.05$ was considered statistically significant and is denoted by "*", whereas $p < 0.001$ is denoted by "**".

3. Results and Discussion

3.1. PLGA Nanoparticles

In the development of new drug delivery systems for antigen-specific therapies, different strategies using nanoparticles have shown promising results due to the properties of the systems. However, incorporating these strategies into macro drug delivery systems that can assure a painless administration of these nanoparticles and antigens brings a novelty to the real-world drug administration scene that surpasses the novelty of the nano. To achieve this goal, we herein prepared and characterized PLGA nanoparticles, subsequently loaded with the PLP139–151 peptide, to an immunodominant T cell myelin epitope found in multiple sclerosis.

In the literature, one can find different procedures for the preparation of PLGA nanoparticles and also different outcomes of the synthesis [19]. Particles can be prepared with a large diameter range from 50 nm to 1000 nm, depending on the fabrication process and the publication [19,20]. In this work, we opted for a double emulsion technique, based on previous reports [18,21]. Under the experimental conditions set, the PLGA nanoparticles showed an average diameter of around 200 nm (PLGA20: 201.6 ± 19.0 nm; PLGA.H20: 210.4 ± 7.0 nm; PLGA60: 225.6 ± 14.8 nm) (Figure 1). When increasing the amount of PLGA in the formulation (PLGA60), the particles tended to increase their average size

(~220 nm) over what is seen in the literature [18,21]. The average polydispersity index was found to be below 0.17, showing the homogeneity of the particles (Figure 1). The prepared nanoparticles held a negative surface charge in the range of −20 mV for all formulations, as measured by assessing the zeta potential. The results are in accordance with the nature of the used materials, as PLGA holds carboxylic acid terminal groups, which are expected not to be protonated in the media, and a residual charge can also be provided by PVA acetate groups.

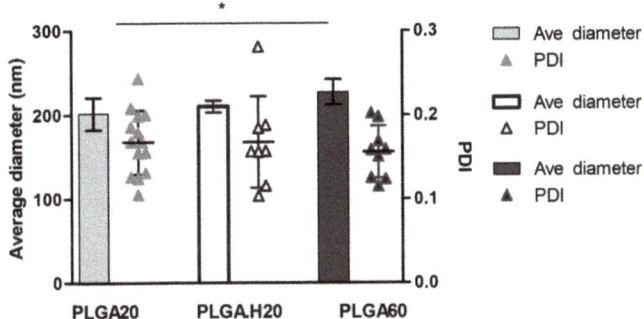

Figure 1. Characterization of poly(lactide-co-glycolide) (PLGA) nanoparticles by average diameter and polydispersity index (PDI) ($n = 3$, measured in triplicate).

Foreseeing incorporation of the particles into polymeric microneedle patches, it became interesting to study the stability of the particles to a freeze–thaw cycle. Freeze-drying is a technique that allows nanoparticle formulations to be stored in solid form and, consequently, to increase their storage stability [22]. In this process, the presence of cryoprotectants, such as sucrose, trehalose, mannitol, among others, has been explored in diversified types of samples, including polymer nanoparticles, liposomes, and drugs [22–25]. We tested two formulations: PLGA20 and the acid form of the polymer, PLGA.H20, also exploring the presence of a cryoprotectant, trehalose 10% (w/v) [25]. Trehalose was recently shown to be more efficient in preserving protein structure for long-term storage [26]. In general, after freeze-drying, the nanoparticles presented an increased average diameter. This increase was reversed when the process occurred in the presence of trehalose for the PLGA.H nanoparticles (Figure 2A), whereas the same effect was not detected in the PLGA20 formulation (Figure 2B). The particles presented good stability after resuspension, making it possible to consider its inclusion in a polymeric microneedle matrix.

3.2. Fluorescent PLGA Nanoparticles into Microneedles

To assess the feasibility of nanoparticle incorporation into polymeric, dissolvable microneedle patches, tests were first performed using fluorescent nanoparticles. Initial attempts were performed with fluorescein, but interestingly, we found that the microneedle composition, namely PVP, reduced fluorescein fluorescence, forming nonfluorescent complexes, as reported in [27]. Alternatively, we explored the incorporation of rhodamine into PLGA nanoparticles.

The characterization of the nanoparticles in terms of size and PDI yielded similar results to the empty nanoparticles. Average nanoparticle diameter was around 200 μm (Figure 3A). It was observed that there is a tendency to increase the polydispersity index when loading rhodamine in the nanoparticles; this difference was found to be nonstatistical, however.

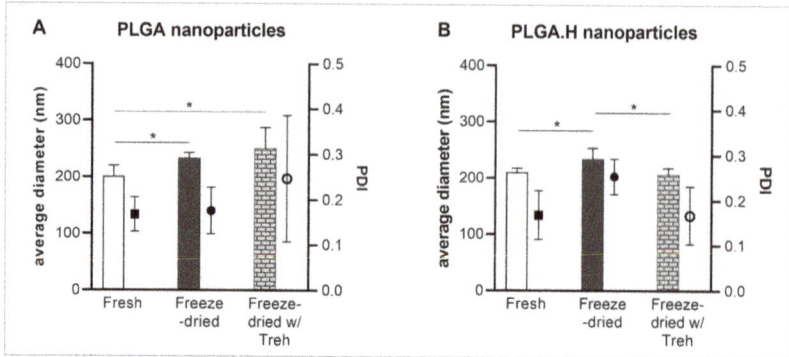

Figure 2. Characterization of the (**A**) PLGA20 and (**B**) PLGA.H20 nanoparticles after freeze-drying with or without trehalose ($n = 3$, assessed in triplicate).

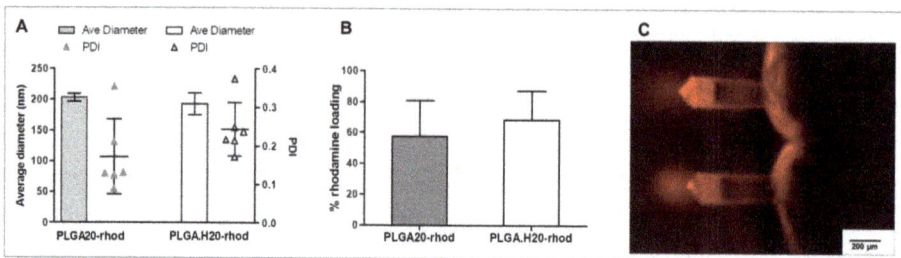

Figure 3. Characterization of PLGA–rhodamine nanoparticles. (**A**) Average diameter of PLGA20 and PLGA.H20 nanoparticles loaded with rhodamine. (**B**) Rhodamine-loading capacity of the PLGA and PLGA.H20 nanoparticles. (**C**) Fluorescence microscopy image of polymer microneedles loaded with PLGA–rhodamine nanoparticles (exposure = 10 ms).

When analyzing the loading capacity of the nanoparticles, it was found to be 57.8 ± 23.1% in the PLGA20 formulation and slightly higher, 68.4 ± 18.8%, in the case of PLGA.H20 nanoparticles (Figure 3B). Others report rhodamine loading capacity in PLGA nanoparticles to be around 40–50% [28,29].

The rhodamine-loaded nanoparticles were incorporated into microneedle patches. To do so, the nanoparticles were first freeze-dried and subsequently resuspended into the polymeric matrix. This solution was then applied into the microneedle molds and, after drying, peeled off. As the nanoparticles were fluorescently labeled, it was possible to localize them inside the needle patch structure. Interestingly, the results suggest that nanoparticles might be preferentially located closer to the needle tip, as there was more fluorescence found in that part of the structure (Figure 3C). Particles cannot be distinguished in the structure, and, in fact, it cannot be excluded that free rhodamine was present in the needles. Fluorescence signal in other fluorescence channels can only be obtained with higher exposures, pointing out the specificity of rhodamine detection.

3.3. Microneedle Loaded with PLP–PLGA Nanoparticles

The analysis of the physical properties of the PLP–PLGA nanoparticles showed that loading PLP does not significantly change the size of the nanoparticles in comparison with the empty nanoparticle formulation (Figure 4A). A slight increase of about 20 μm in the average size could be pinpointed, but the increase was not statistically significant. Results in Figure 4A also denote that PLGA.H20 (195.1 ± 10.1 nm) showed a significantly smaller size as compared with PLGA60 (238.7 ± 13.6 nm). This same tendency was detected in unloaded PLGA nanoparticles, as discussed above.

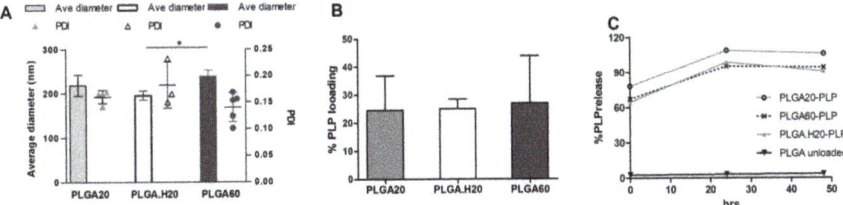

Figure 4. Characterization of PLGA nanoparticles loaded with proteolipid protein (PLP). (**A**) Average diameter and polydispersity index (PDI) of nanoparticles loaded with PLP. ($n = 3$, in triplicate) (**B**) loading efficiency of the different nanoparticle formulations. ($n = 3$) (**C**) Release of the PLP peptide into physiological media, as determined by HPLC (representative experiment out of 2).

The loading of the peptide was found to be on average above 20% for all the formulations tested (Figure 4B). Interestingly, rhodamine loading was significantly higher than for PLP, probably due to the larger size and hydrophobicity of rhodamine that may improve interactions with PLGA. The percentage of peptide PLP encapsulated on PLGA nano- and microparticles was not disclosed in some of the reference publications [6,30]. A higher association efficiency in PLGA particles was described for larger peptides, such as glucagon-like peptide-1 (GLP-1) (67%) [21] or LL37, in which 70% efficiency was achieved in particles above 300 nm [31].

Figure 4C shows the release of PLP from the nanoparticles. The release is very fast, occurring in the first hours/minutes of incubation at 37 °C. Figure 4C represents the initial point at which over 60% of the peptide was not incorporated in the nanoparticles, and, from this point on, the complete release was detected. Release in the first hours of incubation at physiological conditions is also reported in the literature [21,32].

With the final goal to develop a system that allows minimally invasive administration, we incorporated the PLP-loaded nanoparticles into polymeric microneedles. Other systems are reported in the literature, such as exploring the use of microneedles coated with nanoparticles [29,32,33] or incorporating them in a dissolvable structure for the delivery of DNA [34].

The images (Figure 5A,B) show that the incorporation of nanoparticles does not affect the morphology of the microneedles. The prepared patches contained microneedles similar to the ones reported in our previous work [16], being an average height of below 550 μm and a diameter of around 180 μm (Figure 5C). Interestingly, when placed in physiological media, the release of the peptide was likely to occur in a more sustained way (Figure 5D) when compared with the release from the nanoparticles. A burst release was observed in the first period of incubation due to the peptide not being encapsulated or being loosely bound at the surface. The detection of the peptide was challenging due to the presence of large amounts of polymer from the microneedles' dissolution. To assess the release of the peptide, the patches were maintained in a dialysis bag in physiological media. This procedure aimed to reduce the amount of polymer from the patch that was also released to the media. However, smaller polymer chains can cross the dialysis membrane making it more difficult to detect small amounts of peptide. Thus, in these experiments, the PLP signal in the HPLC chromatogram was possible to isolate only for the microneedle patches containing PLGA.H20 nanoparticles. The amount of PLP quantified was below 1 μg, representing about 2% of the theoretical amount of PLP expected in the sample. If on one side, the graph (Figure 5D) suggests that the release is in its increasing phase, further experiments using higher amounts of loaded nanoparticles and, consequently, of the peptide could help to isolate and quantify the PLP within the polymer mixture. Still, the presented results show that the incorporation of PLGA nanoparticles loaded with PLP onto dissolvable microneedles could bring a new approach to the delivery of antigen-specific therapies.

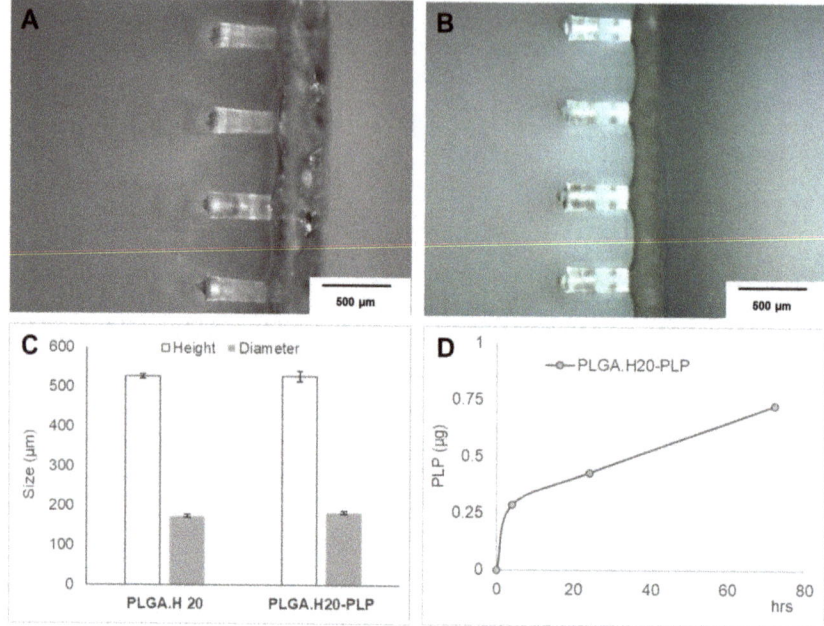

Figure 5. Polymer microneedles loaded with PLGA.H20 nanoparticles. (**A**,**B**) Optical microscopy images of microneedles loaded with (**A**) plain or (**B**) PLP-loaded PLGA.H20 nanoparticles. (**C**) Characterization of the prepared polymer microneedles in terms of diameter and height. (**D**) Quantification of the PLP released from the polymeric system when immersed in PBS a 37 °C, as quantified by HPLC.

4. Conclusions

This study presented the development of a dissolvable microneedle patch in which PLGA nanoparticles containing PLP were loaded. The nanoparticles showed good stability and the PLP peptide was successfully incorporated.

Although the dose of PLP detected in the microneedle patches was still beyond the therapeutic doses described in the literature, this strategy brings novelty to the administration of antigen-specific therapies, namely in the context of multiple sclerosis.

Author Contributions: A.F.L.: Investigation, writing—original draft. I.R.A.: Investigation, formal analysis, methodology, writing—review and editing. L.R.P.: Conceptualization, methodology, investigation, writing, funding acquisition. All authors have read and agreed to the published version of the manuscript.

Funding: This project received funding from the European Union's Seventh Framework Program for research, technological development, and demonstration under grant agreement nos. 600375 and (MSCA-2015-COFUND-FP) 713640. The authors acknowledge financial support from the Norte2020 Funding Program, N2020-PE—Nanothechnology-Based Functional Solutions (NBFS) project, under contract no. NORTE-01-0145-FEDER-000019.

Conflicts of Interest: The authors declare that they have no known competing financial interests or personal relationships that could have appeared to influence the work reported in this paper.

References

1. Lutterotti, A.; Martin, R. Antigen-specific tolerization approaches in multiple sclerosis. *Expert Opin. Investig. Drugs* **2014**, *23*, 9–20. [CrossRef] [PubMed]

2. Getts, D.R.; Turley, D.M.; Smith, C.E.; Harp, C.T.; McCarthy, D.; Feeney, E.M.; Getts, M.T.; Martin, A.J.; Luo, X.; Terry, R.L.; et al. Tolerance Induced by Apoptotic Antigen-Coupled Leukocytes Is Induced by PD-L1(+) and IL-10-Producing Splenic Macrophages and Maintained by T Regulatory Cells. *J. Immunol.* **2011**, *187*, 2405–2417. [CrossRef] [PubMed]
3. Northrup, L.; Christopher, M.A.; Sullivan, B.P.; Berkland, C. Combining antigen and immunomodulators: Emerging trends in antigen-specific immunotherapy for autoimmunity. *Adv. Drug Deliv. Rev.* **2016**, *98*, 86–98. [CrossRef]
4. Getts, D.R.; Martin, A.J.; McCarthy, D.P.; Terry, R.L.; Hunter, Z.N.; Yap, W.T.; Getts, M.T.; Pleiss, M.; Luo, X.; King, N.J.C.; et al. Microparticles bearing encephalitogenic peptides induce T-cell tolerance and ameliorate experimental autoimmune encephalomyelitis. *Nat. Biotechnol.* **2012**, *30*, 1217–1224. [CrossRef]
5. Bueyuektimkin, B.; Wang, Q.; Kiptoo, P.; Stewart, J.M.; Berkland, C.; Siahaan, T.J. Vaccine-like Controlled-Release Delivery of an Immunomodulating Peptide to Treat Experimental Autoimmune Encephalomyelitis. *Mol. Pharm.* **2012**, *9*, 979–985. [CrossRef] [PubMed]
6. Hunter, Z.; McCarthy, D.P.; Yap, W.T.; Harp, C.T.; Getts, D.R.; Shea, L.D.; Miller, S.D. A Biodegradable Nanoparticle Platform for the Induction of Antigen-Specific Immune Tolerance for Treatment of Autoimmune Disease. *ACS Nano* **2014**, *8*, 2148–2160. [CrossRef]
7. Bryant, J.; Hlavaty, K.A.; Zhang, X.; Yap, W.-T.; Zhang, L.; Shea, L.D.; Luo, X. Nanoparticle delivery of donor antigens for transplant tolerance in allogeneic islet transplantation. *Biomaterials* **2014**, *35*, 8887–8894. [CrossRef]
8. Yeste, A.; Nadeau, M.; Burns, E.J.; Weiner, H.L.; Quintana, F.J. Nanoparticle-mediated codelivery of myelin antigen and a tolerogenic small molecule suppresses experimental autoimmune encephalomyelitis. *Proc. Natl. Acad. Sci. USA* **2012**, *109*, 11270–11275. [CrossRef]
9. Cappellano, G.; Woldetsadik, A.D.; Orilieri, E.; Shivakumar, Y.; Rizzi, M.; Carniato, F.; Gigliotti, C.L.; Boggio, E.; Clemente, N.; Comi, C.; et al. Subcutaneous inverse vaccination with PLGA particles loaded with a MOG peptide and IL-10 decreases the severity of experimental autoimmune encephalomyelitis. *Vaccine* **2014**, *32*, 5681–5689. [CrossRef]
10. Wiedersberg, S.; Guy, R.H. Transdermal drug delivery: 30+ years of war and still fighting! *J. Control. Release* **2014**, *190*, 150–156. [CrossRef]
11. Prausnitz, M.R. Engineering Microneedle Patches for Vaccination and Drug Delivery to Skin. *Ann. Rev. Chem. Biomol. Eng.* **2017**, *8*, 177–200. [CrossRef] [PubMed]
12. Prausnitz, M.R. Microneedles for transdermal drug delivery. *Adv. Drug Deliv. Rev.* **2004**, *56*, 581–587. [CrossRef] [PubMed]
13. Prausnitz, M.R.; Mikszta, J.A.; Cormier, M.; Andrianov, A.K. Microneedle based vaccines. *Curr. Top. Microbiol. Immunol.* **2009**, *333*, 369–393. [PubMed]
14. Koutsonanos, D.G.; Vassilieva, E.V.; Stavropoulou, A.; Zarnitsyn, V.G.; Esser, E.S.; Taherbhai, M.T.; Prausnitz, M.R.; Compans, R.W.; Skountzou, I. Delivery of subunit influenza vaccine to skin with microneedles improves immunogenicity and long-lived protection. *Sci. Rep.* **2012**, *2*, 357. [CrossRef]
15. Shklovskaya, E.; O'Sullivan, B.J.; Lai Guan, N.; Roediger, B.; Thomas, R.; Weninger, W.; de St. Groth, B.F. Langerhans cells are precommitted to immune tolerance induction. *Proc. Natl. Acad. Sci. USA* **2011**, *108*, 18049–18054. [CrossRef]
16. Pires, L.R.; Amado, I.R.; Gaspar, J. Dissolving microneedles for the delivery of peptides–Towards tolerance-inducing vaccines. *Int. J. Pharm.* **2020**, *586*, 119590. [CrossRef]
17. Ramalheiro, A.; Paris, J.L.; Silva, B.F.B.; Pires, L.R. Rapidly dissolving microneedles for the delivery of cubosome-like liquid crystalline nanoparticles with sustained release of rapamycin. *Int. J. Pharm.* **2020**, *591*, 119942. [CrossRef]
18. Castro, P.M.; Baptista, P.; Madureira, A.R.; Sarmento, B.; Pintado, M.E. Combination of PLGA nanoparticles with mucoadhesive guar-gum films for buccal delivery of antihypertensive peptide. *Int. J. Pharm.* **2018**, *547*, 593–601. [CrossRef]
19. Danhier, F.; Ansorena, E.; Silva, J.M.; Coco, R.; Le Breton, A.; Préat, V. PLGA-based nanoparticles: An overview of biomedical applications. *J. Control. Release* **2012**, *161*, 505–522. [CrossRef]
20. Allahyari, M.; Mohit, E. Peptide/protein vaccine delivery system based on PLGA particles. *Hum. Vaccines Immunother.* **2016**, *12*, 806–828. [CrossRef]

21. Araújo, F.; Shrestha, N.; Shahbazi, M.-A.; Fonte, P.; Mäkilä, E.M.; Salonen, J.J.; Hirvonen, J.T.; Granja, P.L.; Santos, H.A.; Sarmento, B. The impact of nanoparticles on the mucosal translocation and transport of GLP-1 across the intestinal epithelium. *Biomaterials* **2014**, *35*, 9199–9207. [CrossRef]
22. de Vale Morais, A.R.; Alencar, É.D.N.; Xavier Júnior, F.H.; de Oliveira, C.M.; Marcelino, H.R.; Barratt, G.; Fessi, H.; do Egito, E.S.T.; Elaissari, A. Freeze-drying of emulsified systems: A review. *Int. J. Pharm.* **2016**, *503*, 102–114. [CrossRef]
23. Holzer, M.; Vogel, V.; Mäntele, W.; Schwartz, D.; Haase, W.; Langer, K. Physico-chemical characterization of PLGA nanoparticles after freeze-drying and storage. *Eur. J. Pharm. Biopharm.* **2009**, *72*, 428–437. [CrossRef]
24. Franzé, S.; Selmin, F.; Samaritani, E.; Minghetti, P.; Cilurzo, F. Lyophilization of Liposomal Formulations: Still Necessary, Still Challenging. *Pharmaceutics* **2018**, *10*, 139. [CrossRef]
25. Fonte, P.; Soares, S.; Costa, A.; Andrade, J.C.; Seabra, V.; Reis, S.; Sarmento, B. Effect of cryoprotectants on the porosity and stability of insulin-loaded PLGA nanoparticles after freeze-drying. *Biomatter* **2012**, *2*, 329–339. [CrossRef]
26. Starciuc, T.; Malfait, B.; Danede, F.; Paccou, L.; Guinet, Y.; Correia, N.T.; Hedoux, A. Trehalose or Sucrose: Which of the Two Should be Used for Stabilizing Proteins in the Solid State? A Dilemma Investigated by In Situ Micro-Raman and Dielectric Relaxation Spectroscopies During and After Freeze-Drying. *J. Pharm. Sci.* **2020**, *109*, 496–504. [CrossRef]
27. Kim, J.; Cote, L.J.; Kim, F.; Huang, J. Visualizing Graphene Based Sheets by Fluorescence Quenching Microscopy. *J. Am. Chem. Soc.* **2010**, *132*, 260–267. [CrossRef]
28. Doiron, A.L.; Homan, K.A.; Emelianov, S.; Brannon-Peppas, L. Poly(lactic-*co*-glycolic) acid as a carrier for imaging contrast agents. *Pharm. Res.* **2009**, *26*, 674–682. [CrossRef]
29. Abulateefeh, S.R.; Spain, S.G.; Thurecht, K.J.; Aylott, J.W.; Chan, W.C.; Garnett, M.C.; Alexander, C. Enhanced uptake of nanoparticle drug carriers via a thermoresponsive shell enhances cytotoxicity in a cancer cell line. *Biomater. Sci.* **2013**, *1*, 434–442. [CrossRef]
30. Saito, E.; Kuo, R.; Kramer, K.R.; Gohel, N.; Giles, D.A.; Moore, B.B.; Miller, S.D.; Shea, L.D. Design of biodegradable nanoparticles to modulate phenotypes of antigen-presenting cells for antigen-specific treatment of autoimmune disease. *Biomaterials* **2019**, *222*, 119432. [CrossRef]
31. Chereddy, K.K.; Her, C.H.; Comune, M.; Moia, C.; Lopes, A.; Porporato, P.E.; Vanacker, J.; Lam, M.C.; Steinstraesser, L.; Sonveaux, P.; et al. PLGA nanoparticles loaded with host defense peptide LL37 promote wound healing. *J. Control. Release* **2014**, *194*, 138–147. [CrossRef]
32. Kim, H.-G.; Gater, D.L.; Kim, Y.-C. Development of transdermal vitamin D3 (VD3) delivery system using combinations of PLGA nanoparticles and microneedles. *Drug Deliv. Trans. Res.* **2018**, *8*, 281–290. [CrossRef]
33. DeMuth, P.C.; Su, X.; Samuel, R.E.; Hammond, P.T.; Irvine, D.J. Nano-layered microneedles for transcutaneous delivery of polymer nanoparticles and plasmid DNA. *Adv. Mater.* **2010**, *22*, 4851–4856. [CrossRef]
34. Yang, H.W.; Ye, L.; Guo, X.D.; Yang, C.; Compans, R.W.; Prausnitz, M.R. Ebola Vaccination Using a DNA Vaccine Coated on PLGA-PLL/γPGA Nanoparticles Administered Using a Microneedle Patch. *Adv. Healthc. Mater.* **2017**, *6*, 1600750. [CrossRef]

Publisher's Note: MDPI stays neutral with regard to jurisdictional claims in published maps and institutional affiliations.

© 2020 by the authors. Licensee MDPI, Basel, Switzerland. This article is an open access article distributed under the terms and conditions of the Creative Commons Attribution (CC BY) license (http://creativecommons.org/licenses/by/4.0/).

Communication

An Investigation of the Influence of PEG 400 and PEG-6-Caprylic/Capric Glycerides on Dermal Delivery of Niacinamide

Yanling Zhang [1,*], Majella E. Lane [1] and David J. Moore [2]

1. Department of Pharmaceutics, UCL School of Pharmacy, 29-39 Brunswick Square, London WC1N 1AX, UK; m.lane@ucl.ac.uk
2. Tioga Research, Inc., Edinburgh EH31 2BA, UK; dmoore@tiogaresearch.com
* Correspondence: yanling.zhang.15@ucl.ac.uk

Received: 21 November 2020; Accepted: 2 December 2020; Published: 4 December 2020

Abstract: Polyethylene glycols (PEGs) and PEG derivatives are used in a range of cosmetic and pharmaceutical products. However, few studies have investigated the influence of PEGs and their related derivatives on skin permeation, especially when combined with other solvents. Previously, we reported niacinamide (NIA) skin permeation from a range of neat solvents including propylene glycol (PG), Transcutol® P (TC), dimethyl isosorbide (DMI), PEG 400 and PEG 600. In the present work, binary and ternary systems composed of PEGs or PEG derivatives combined with other solvents were investigated for skin delivery of NIA. In vitro finite dose studies were conducted (5 µL/cm^2) in porcine skin over 24 h. Higher skin permeation of NIA was observed for all vehicles compared to PEG 400. However, overall permeation for the binary and ternary systems was comparatively low compared with results for PG, TC and DMI. Interestingly, values for percentage skin retention of NIA for PEG 400:DMI and PEG 400:TC were significantly higher than values for DMI, TC and PG ($p < 0.05$). The findings suggest that PEG 400 may be a useful component of formulations for the delivery of actives to the skin rather than through the skin. Future studies will expand the range of vehicles investigated and also look at skin absorption and residence time of PEG 400 compared to other solvents.

Keywords: niacinamide; polyethene glycol (PEG) 400; solvent; dermal delivery; finite dose; porcine skin

---

1. Introduction

The skin is the largest organ of the human body and serves as a unique interface between humans and the environment. This membrane is a formidable barrier, preventing the egress of water, the ingress of toxins and offers protection against ultraviolet (UV) radiation [1,2]. However, skin also serves as a route for the administration of therapeutic molecules for both local and systemic effects [3]. The primary challenge in the skin penetration process is the passage of molecules through the outermost layer, the stratum corneum (SC). The SC consists of eight to sixteen layers of [4] keratinized corneocytes embedded in a lipid domain. To improve the efficacy of topical and transdermal delivery, various formulation components have been investigated for their potential to facilitate permeation of actives through the SC.

Polyethylene glycols (PEGs) are synthetic polymers of condensed ethylene oxide (EO) and water [5]. These polymers and their derivatives have numerous applications in the food industry and the pharmaceutical and biomedical fields [6,7]. With reference to skin formulations, PEGs and their derivatives are primarily used as solvents, surfactants or stabilizers. Surprisingly, few publications have examined the mechanistic effects of PEGs on the permeation of actives. Sarpotdar and colleagues investigated the effect of PEG 400/water mixtures on the penetration of oxaprozin and guanabenz in cadaver skin using in vitro diffusion cell studies [8]. The flux values of both drugs were found to drop linearly as the concentration of PEG 400 increased in the formulations. Suwanpidokkul et al. compared the penetration of zidovudine in porcine skin from PEG 400/water vehicles with ethanol/water, isopropyl alcohol/water and ethanol/isopropyl myristate (IPM) vehicles. Flux values for the PEG 400/water vehicle were comparable to those for ethanol/water and isopropanol/water, namely 15.52 ± 0.60, 9.96 ± 0.63 and 10.35 ± 0.67 $\mu g/cm^2/h$, respectively. However, both of these studies used infinite dose conditions; therefore, it is difficult to extrapolate the findings for typical "in use" conditions for topical preparations. Aside from the very limited evaluation of PEGs on the skin delivery of actives, comparative studies have not been reported for PEG derivatives, either as neat solvents or as components of simple binary or ternary systems.

Niacinamide (NIA) is the water-soluble form of vitamin B3 and the amide form of nicotinic acid [9]. In 1976, Comaish et al. [10] first reported NIA as an efficient treatment for pellagra, a vitamin deficiency disease with dermatitis symptoms. Since then, topical application of NIA has been shown to improve various skin conditions in humans. The beneficial effects include skin barrier enhancement [9], reduction of hyperpigmentation [11], anti-inflammation [12] and prevention of UV-induced immunosuppression [13]. Favorable compatibility with the active is a crucial criterion when selecting formulation components for dermal delivery of NIA. All solvents investigated in this study have good safety profiles and widespread applications in pharmaceutical and consumer products. Propylene glycol (PG) is the most commonly used glycol in topical and transdermal formulations [2]. Data from both clinical and nonclinical studies support the use of PG as a nontoxic and well-tolerated material [14]. Transcutol® P (TC) and dimethyl isosorbide (DMI) are commonly used vehicles in skin preparations with GRAS status [15,16]. PEG 400 is an FDA approved polymer for use in drug delivery systems because of its safety and tolerance when administered to the body by different routes [17]. PEG-6-CCG was selected as it is currently the most widely used PEG alkyl glyceride derivate in personal care products [18]. According to the 2014 FDA Voluntary Cosmetic Registration Program (VCRP) data, PEG-6-CCG was used in 548 formulations, and the CIR Expert Panel confirmed the GRAS status of the compound [18]. Previously, we reported the dermal delivery of NIA using porcine skin in vitro, from a range of simple solvents for both infinite and finite dose conditions [19]. The vehicles examined included propylene glycol (PG), Transcutol® P (TC), dimethyl isosorbide (DMI), t-butyl alcohol (T-BA) and PEGs 400 and 600. For finite dose studies, the percentage permeation of NIA was ranked as follows: T-BA> DMI > TC > PG; no NIA permeation was observed for PEGs 400 or 600. Corresponding mass balance studies confirmed that NIA was largely deposited on the skin surface with the PEG vehicles. We hypothesized that by combining PEG 400 with neat solvents effective in promoting NIA skin permeation, NIA skin penetration might be enhanced when compared with PEG 400. Thus, the aims of the present study were to (i) design binary and ternary systems of PEG 400 and/or the PEG derivative PEG-6-caprylic/capric glycerides (PEG-6-CCG) and (ii) investigate any synergistic effects of these systems on the permeation of NIA in porcine skin, compared with the neat solvents studied previously.

2. Materials and Methods

2.1. Materials

NIA, PG, PEG 400, high-performance liquid chromatography (HPLC) grade water and methanol were purchased from Sigma-Aldrich, Dorset, UK. TC was a gift from Gattefossé, St. Priest, France, and Croda Ltd., Goole, UK, supplied the DMI. PEG-6-caprylic/capric glycerides (PEG-6-CCG) was a gift from Avon, Suffern, NY, USA. Phosphate buffered saline (PBS) tablets (pH 7.3 ± 0.2 at 25 °C) were purchased from Oxoid, Cheshire, UK.

2.2. Determination of Solubility, Solubility Parameters and HPLC Analytical Method

The method for determining solubility of NIA in PG, TC, DMI, PEG 400, PEG 600 at 32 ± 1 °C has been reported previously [19]. The solubility of NIA in PEG-6-CCG under the same conditions was evaluated by adding an excess amount of NIA to a known volume of PEG-6-CCG. The mixture was continuously stirred with a Teflon coated magnetic bar for 48 h at 32 ± 1 °C. After 48 h, samples were centrifuged (13,200 rpm) for 15 min, and the supernatant was collected and diluted with methanol–water (50:50). All solubility values were reported as % w/v.

The van Krevelen Hoftyzer solubility parameters (δ) for NIA and solvent systems were calculated using Molecular Modeling Pro® (ChemSW Inc., Fairfield, CA, USA). For the binary and ternary systems, as shown in Equations (1) and (2), the solubility parameters (δBinary/Ternary) were calculated according to the respective mole fractions (Φ_1, Φ_2 ...) for the solvents and their respective solubility parameters (δ_1, δ_2 ...) [20].

$$\delta_{Binary} = \frac{\delta_1 \Phi_1 + \delta_2 \Phi_2}{\Phi_1 + \Phi_2} \quad (1)$$

$$\delta_{Ternary} = \frac{\delta_1 \Phi_1 + \delta_2 \Phi_2 + \delta_3 \Phi_3}{\Phi_1 + \Phi_2 + \Phi_3} \quad (2)$$

Samples of NIA were analyzed using HPLC [19]. A Kinetex 5 μm Phenyl-Hexyl 250 × 4.6 mm reverse phase column (Phenomenex, Macclesfield, UK) packed with a SecurityGuard™ cartridge (Phenomenex, UK) was used to achieve separation. The column temperature was set at 30 °C. The mobile phase was water–methanol (80:20) with a flow rate of 1 mL/min. The injection volume was set to 10 μL, and the UV detection wavelength was 263 nm. The HPLC method was validated previously according to ICH guidelines (2005) [21], including accuracy, precision, detection limit, quantitation limit, linearity and robustness.

2.3. Dynamic Vapor Sorption (DVS) Studies

To study the evaporation or hydration of NIA solutions, a Q5000 SA sorption analyzer (TA Instruments, New Castle, DE, USA) was used to examine the mass variation over 24 h (1440 min) following the procedure reported by Iliopoulos et al. [22]. Nitrogen was used as the carrier gas (200 mL/min). To monitor mass differences, 15 mm quartz glass pans connected to a microbalance accurate to 0.00001 mg were employed. Then, 5 μL solutions of 5% NIA (w/v) were added to the pans, and data were recorded using a Universal Analysis 2000 (TA Instrument, New Castle, DE, USA). Temperature and relative humidity (RH) were maintained over the course of the studies at 32 ± 1 °C and 50 ± 2% RH, respectively.

2.4. Permeation Studies and Mass Balance Studies

Binary and ternary solvent systems were prepared for the in vitro permeation studies, namely PEG 400:TC (50:50), PEG 400:PG (50:50), PEG 400:DMI (50:50), PEG 400:PEG-6-CCG (50:50) and PEG 400:PG:DMI (50:25:25). The concentration of NIA typically used in personal care products ranges from 2% to 5% [23]. To be consistent with previous studies, a concentration of 5% (w/v) NIA was therefore selected in the present work.

In vitro permeation studies were performed using Franz diffusion cells [24,25]. Full thickness porcine ear skin has been proposed as an appropriate surrogate for human skin [26]. The in vitro skin permeation procedure was designed based on current Organisation for Economic Co-operation and Development (OECD) guidelines [27]. Porcine tissue was obtained from a local abattoir and prepared on the same day of collection. The skin was surgically separated from the cartilage following a standardized procedure [28]. After preparation, the porcine skin was mounted on aluminum foil and stored at −20 °C until required. Skin integrity was confirmed by assessing electrical resistance before experiments [18,29]. A 2.5 mL degassed freshly prepared PBS solution (pH 7.30 ± 0.20) was used as the receptor medium. To mimic clinically relevant conditions, 5 μL of the NIA solution was applied over a diffusion area of ~1 cm² after the skin temperature had equilibrated to 32 ± 1 °C. The effective diffusion area was measured accurately using a Vernier Caliper (Fisher Scientist, Loughborough, UK). At each sampling point, 200 µL of the receptor medium was withdrawn and refilled with an equal volume of fresh PBS solution for up to 24 h. At the end of the permeation studies, after removing the receptor medium, mass balance studies were performed to determine the total distribution of NIA following a validated procedure [24]. The details of the mass balance method are provided in the Supplementary Materials section of this paper.

2.5. Statistical Analysis

All data are presented as the mean ± standard deviation (SD). The statistical analysis was performed using SPSS® Statistics version 24 (IBM, Feltham, UK). The normality of data was examined using the Shapiro–Wilk test, and the homogeneity of variance was examined using Levene's test. An independent t-test or one-way ANOVA with a post hoc Tukey test was performed for data that met the assumption of normality and homogeneity of variance. The Kruskal–Wallis H test was used for nonparametric data. A p-value lower than 0.05 ($p < 0.05$) was considered a significant difference.

3. Results and Discussion

3.1. Solubility Studies

The solubility parameters of the various vehicles investigated, and corresponding NIA solubility values, are shown in Table 1. The solubility parameters for PEG-6-CCG and the binary systems composed of PEG-6-CCG were not calculated, as this solvent is a mixture of polyethylene glycol derivatives of a range of caprylic/capric glyceride acids [30]. The solubility parameter of NIA was reported as 13.9 $(cal/cm^3)^{1/2}$, and the solubility values for NIA in DMI, TC, PEG 400 and PG were previously reported [19]. The NIA solubility value for PEG 400:PG (50:50) was 25.6 ± 1.1% (w/v). This value was significantly higher (Table 1) than the results obtained for PEG 400:PEG-6-CCG (50:50), PEG 400:DMI (50:50), PEG 400:TC (50:50) and PEG 400:PG:DMI (50:25:25), ($p < 0.05$).

Table 1. Solubility of niacinamide (NIA) in tested single, binary and ternary solvent systems at 32 ± 1 °C (mean ± SD, n = 3). The solubility parameter of NIA was reported as 13.9 $(cal/cm^3)^{1/2}$. Statistical analysis is highlighted in the table (*, $p < 0.05$).

Solvent Systems (v/v)	Solubility Parameter $(cal/cm^3)^{1/2}$	Solubility of NIA ((%, g/100 mL)
PEG-6-CCG	-	5.6 ± 0.3 *
DMI	10.0	8.0 ± 0.6
TC	10.6	13.7 ± 0.4
PEG 400	11.7	18.7 ± 0.9
PG	14.1	28.4 ± 0.9 *
PEG 400:PEG-6-CCG (50:50)	-	15.0 ± 1.0
PEG 400:DMI (50:50)	10.9	17.6 ± 1.2
PEG 400:TC (50:50)	11.2	23.5 ± 0.1
PG:PEG-6-CCG (50:50)	-	25.4 ± 0.1
PEG 400:PG (50:50)	12.8	25.6 ± 1.1
PEG400:PG:DMI (50:25:25)	11.9	21.9 ± 0.8

The NIA solubility values in neat PEG-6-CCG and the binary and ternary systems evaluated here have not previously been reported. The solubility of NIA in PEG-6-CCG was found to be 5.6 ± 0.3%, which was lower than the solubility in neat DMI, PEG 400, TC and PG ($p < 0.05$). For the various binary solvent systems, the NIA solubility ranged from 15.0 to 25.6%. Compared to neat DMI, NIA had a higher solubility in the binary PEG 400:DMI system, determined as 17.6 ± 1.2% ($p < 0.05$). Furthermore, the solubility of NIA in PEG 400:TC (50:50), 23.5 ± 0.1%, was significantly greater compared to the solubility in neat PEG 400 and TC ($p < 0.05$). NIA solubility values in PEG 400:PEG-6-CCG (50:50) and PG:PEG-6-CCG (50:50) were determined as 15.0 ± 1.0 and 25.4 ± 0.1%, respectively. Both values were significantly higher than the result obtained for NIA solubility in neat PEG-6-CCG ($p < 0.05$).

3.2. DVS Studies

Figure 1 shows the DVS results for the various NIA systems over a 24 h period at 32 ± 1 °C and 50 ± 2% RH. Similar trends were observed for NIA in PG, TC and DMI. The initial increase in mass for NIA in PG, TC and DMI reflects the hygroscopic nature of these solvents. After 1 h, the mass of the PG, TC and DMI solutions started to decrease as a result of solvent evaporation. At 24 h, 44.6 ± 0.02, 17.6 ± 0.5 and 76.3 ± 0.1% of the applied weights of the PG, TC and DMI solutions were recovered, respectively. These results are similar to the findings for the evaporation of neat PG and TC reported by Haque et al. [31], and the mass loss for the DMI solution is also consistent with the findings of Iliopoulos et al. [22]. For PEG 400 and PEG-6-CCG, the mass increase likely reflects water retention by these solvents with 110.7 ± 1.2 and 103.3 ± 0.4% recovered at 24 h, respectively. Statistical differences were evident between the percentage recovery values determined at 24 h for the PEG 400 and the PEG-6-CCG formulations ($p < 0.05$). To our knowledge, DVS has not been previously used to characterize the behavior of PEG-6-CCG under controlled conditions of temperature and humidity.

Figure 1B shows the DVS results for the binary and ternary systems of NIA over 24 h. As for the results observed for neat PEG 400 and PEG-6-CCG, an initial increase in weight was evident for all binary and ternary systems, again reflecting the hygroscopic nature of PEG 400 and PEG-6-CCG. For the ternary PEG 400:PG:DMI system, 76.5 ± 1.0% of the initial applied mass was recovered at 24 h. This value was significantly higher than the results for neat PG ($p < 0.05$). As expected, after the initial increase in weight at the start of the experiment, no further mass changes were observed for the PEG 400:PEG-6-CCG NIA system. For the remaining binary systems composed of PG, TC or DMI, the mass decreased after 1 h, resulting from evaporation of these solvents. At 24 h, 58.1 ± 0.8 and 71.1 ± 1.5% of the applied amounts of PG:PEG-6-CCG (50:50) and PEG 400:PG (50:50) NIA solutions were recovered, respectively. These values were higher than the corresponding results for neat PG ($p < 0.05$). At 24 h, 69.1 ± 1.0 and 91.4 ± 0.8% of the applied PEG 400:TC (50:50) and PEG 400:DMI (50:50) mass were recovered, respectively, and these values were also significantly higher than the values for neat TC or DMI ($p < 0.05$).

3.3. In Vitro Permeation Studies

The permeation profiles for all vehicles are shown in Figure 2. All experiments were performed using Franz diffusion cells and porcine skin under finite dose conditions. For the binary PG:PEG-6-CCG (50:50) system, permeation of NIA was not evident until the 8 h sampling point; for all other systems, NIA was only detected at the end of the permeation study. At 24 h, NIA permeation from PG:PEG-6-CCG (50:50) and PEG 400:TC (50:50) was determined as 8.0 ± 3.7 and 5.8 ± 3.4 µg/cm^2. The corresponding values for PEG 400:PEG-6-CCG (50:50), PEG 400:DMI (50:50), PEG 400:PG (50:50) and PEG 400:PG:DMI (50:25:25) were 3.0 ± 0.3, 2.3 ± 0.7, 1.9 ± 0.5 and 1.3 ± 0.4 µg/cm^2, respectively. Statistical analysis confirmed a significant difference between NIA permeation for PG:PEG-6-CCG and PEG 400:PG:DMI ($p < 0.05$). As reported previously, the corresponding cumulative permeation values of NIA for neat PG, DMI and TC solutions were 46.0, 103.6 and 95.1 µg/cm^2 at 24 h, respectively; no permeation of NIA was observed for neat PEG 400 [19].

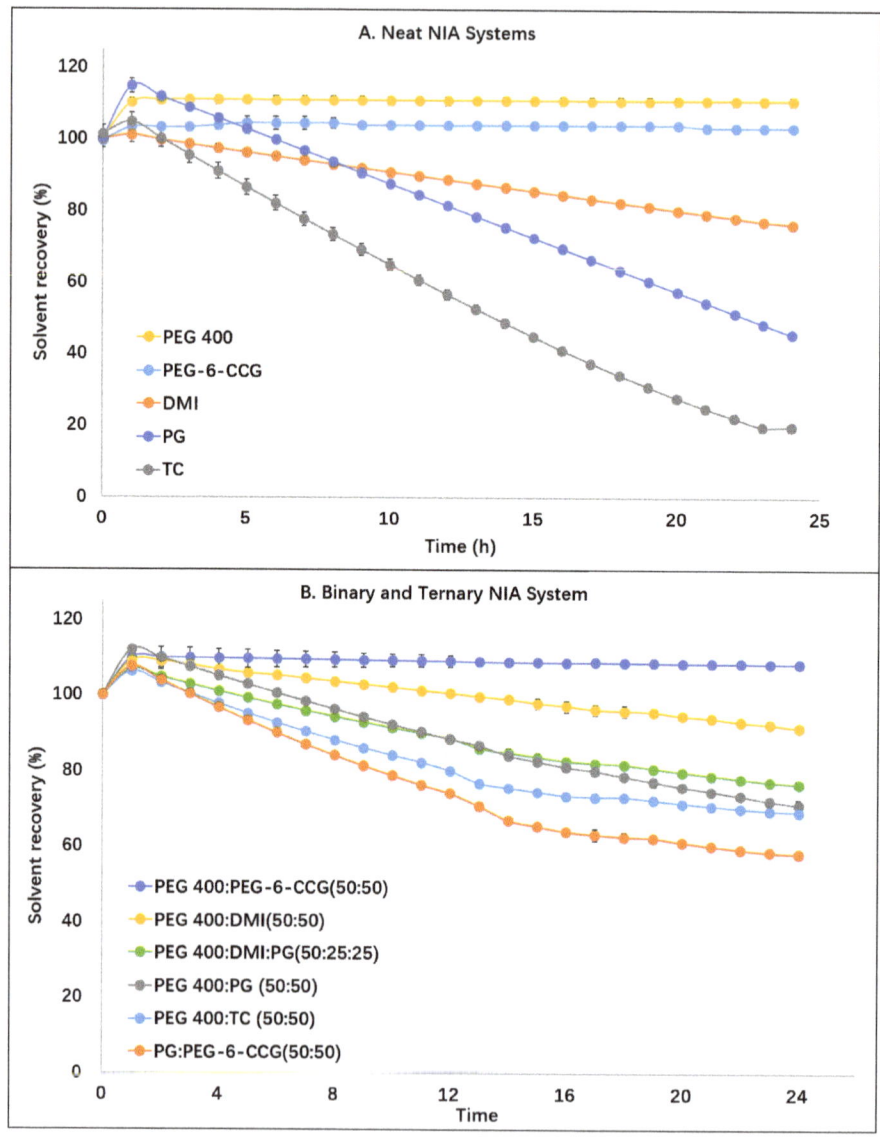

Figure 1. Dynamic Vapor Sorption (DVS) studies of NIA (5%) in propylene glycol (PG), Transcutol®P (TC), dimethyl isosorbide (DMI) and PEG400 and PEG-6-CCG solutions (**A**) and PEG:PEG-6-CCG (50:50), PEG400:DMI (50:50), PEG400:PG (50:50), PEG400:TC (50:50), PG:PEG-6-CCG and PEG400:PG:DMI (50:25:25) (**B**) over 24 h at 32 ± 1 °C and 50 ± 2% RH (n = 3, mean ± SD).

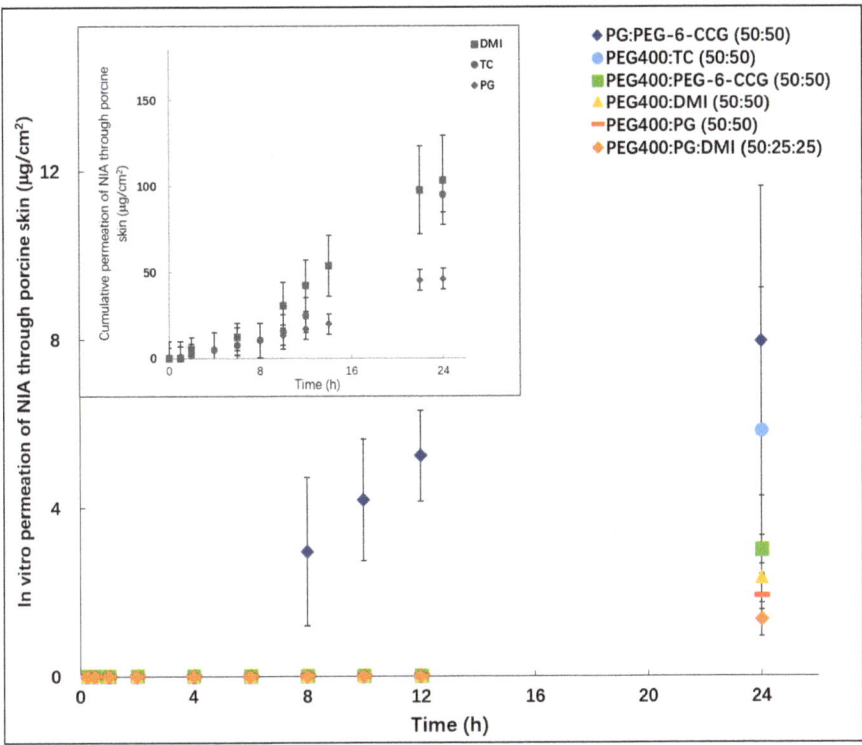

Figure 2. In vitro permeation of NIA from binary and ternary solvent systems under finite dose conditions (5 µL/cm^2) (n = 4, mean ± SD). The permeation profiles of NIA from single solvents (DMI, TC and PG) were adopted from Zhang et al. [19].

The results of the mass balance studies are summarized in Figure 3A–C. The corresponding percentage values determined for neat PEG 400, PG, TC and DMI, reported previously [19], are also included in the figures for comparison. For the binary and ternary vehicles, the values for total recovery of NIA were within the recommended ranges as published in the Scientific Committee on Consumer Safety (SCCS) guidelines for dermal absorption studies (85–115%) [32]. For the binary and ternary systems, 0.6 to 2.7% of the applied NIA penetrated through the skin membrane after 24 h permeation. Lower NIA permeation percentage values were still evident for the binary/ternary systems compared to neat PG, TC and DMI ($p < 0.05$).

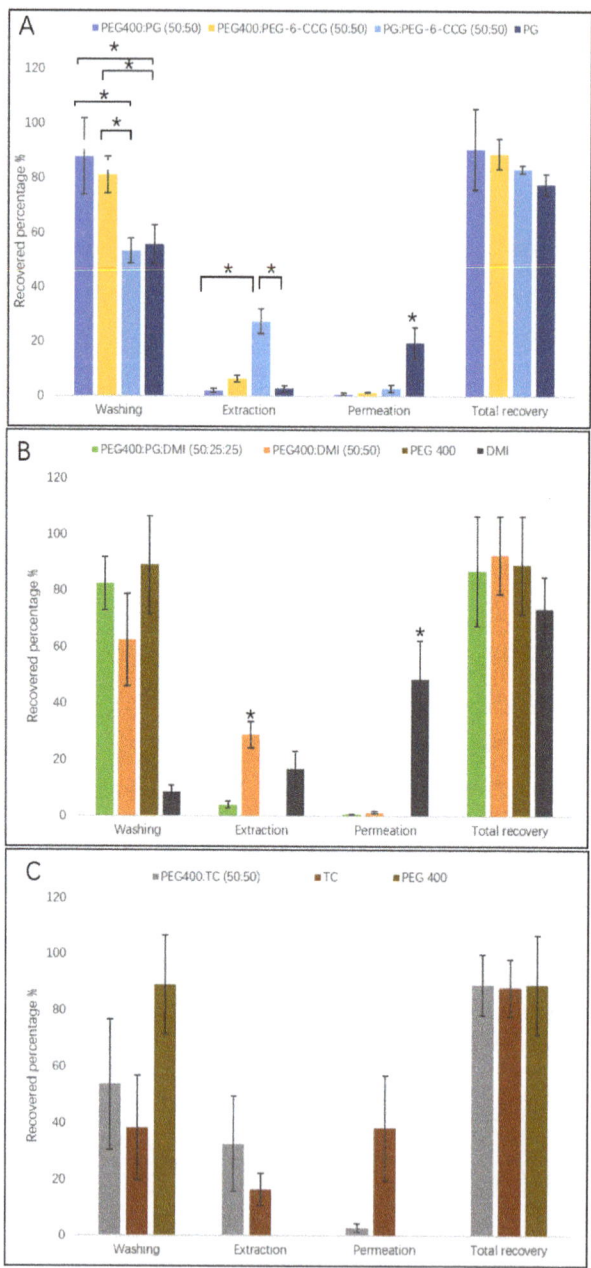

Figure 3. Mass balance results of the applied NIA from binary and ternary systems after 24 h permeation studies (n = 4, mean ± SD). (**A**) summarizes the mass balance results from PEG400:PG (50:50), PEG400:PEG-6-CCG (50:50), PG:PEG-6-CCG (50:50) and neat PG; (**B**) summarizes the results from PEG400:PG:DMI (50:25:25), PEG400:DMI (50:50), PEG400 and DMI; (**C**) summarizes the results from PEG400:TC (50::50), TC and PEG400. The mass balance results of NIA from neat TC, PG, PEG 400 and DMI were adopted from Zhang et al. [19]. Statistical analysis was highlighted in the figure (*, $p < 0.05$).

The percentage of the applied dose extracted from the skin membrane ranged from 2 to 33% for all assessed binary vehicles. The percentage retention of NIA was 32.6 ± 16.9% for PEG 400:TC (50:50), followed by PEG 400:DMI (50:50) and PG:PEG-6-CCG (50:50), with corresponding values of 28.9 ± 4.7 and 27.4 ± 4.6%, respectively. In comparison to neat PG, a higher NIA skin retention was evident for PG:PEG-6-CCG (50:50) ($p < 0.05$) (Figure 3A). Previously, we reported skin retention of NIA as 16.7 ± 6.3% for neat DMI [19]. Hence, the binary PEG 400-DMI system also significantly improved NIA skin retention compared with neat DMI ($p < 0.05$) (Figure 3B). The percentage of NIA extracted from skin for PEG 400:TC was 32.6 ± 16.9%, but no significant difference was evident in comparison to neat TC (16.4 ± 5.8%, $p > 0.05$) (Figure 3C).

PG has been used in a range of topical and transdermal formulations [2]. Hoelgaard et al. [33] reported a "carrier-solvent" effect for PG enhancement of metronidazole permeation. Haque et al. [31] investigated the skin delivery of anthramycin from a PG solution and noted that anthramycin appeared to "track" the permeation of PG. The authors suggested that PG might enhance skin permeability by increasing the partition or solubility of the drug in skin. Recently, Kung et al. [34] investigated the skin penetration of methadone from binary systems composed of PG and other solvents. Increased permeation of methadone corresponded to high skin uptake of PG. These works suggest that PG may interact with skin lipids and influence the barrier function of the SC. The results reported here indicate that combining PG with PEG 400 does not promote enhanced permeation or skin retention compared with PG. However, the combination of PG with PEG-6-CCG improved the percent skin retention of NIA compared to PG alone ($p < 0.05$). NIA skin retention was comparable for PEG 400:PG ($p > 0.05$) and neat PG ($p < 0.05$). As the solubility values for NIA in PG, PEG 400:PG and PG:PEG-6-CCG are comparable (Table 1), other properties of the solvents appear to influence the skin permeation of NIA.

Reports in the literature suggest that TC might increase percutaneous absorption by changing the solubility of drugs in the skin rather than disrupting skin lipids [35]. The comparatively rapid skin permeation of TC compared with PG, 1,3-butanediol and dipropylene glycol was reported by Haque et al. [31]. The solubility of NIA in TC was previously reported as 8.0 ± 0.6% [19]. In the present work of combining PEG 400 with TC, significantly higher solubility values for NIA were evident compared with those for neat TC or PEG 400 ($p < 0.05$). Combining DMI with PEG 400 resulted in a two-fold increase in NIA solubility. As for the PG results, high solubility in the binary DMI and TC vehicles did not promote skin permeation of NIA compared to TC or DMI alone.

4. Conclusions

PEGs and their derivatives are commonly used in topical and transdermal formulations. Previously, we reported no permeation or skin retention of NIA from a neat PEG 400 solution in porcine skin. The present work examined NIA skin delivery from vehicles composed of PEG 400 and the PEG derivative PEG-6-CCG. Permeation of NIA was increased for the PEG 400 binary systems compared to PEG 400 alone, but overall, very low permeation of NIA was observed. On the other hand, high skin retention was observed for these vehicles when compared with the neat solvents investigated. Depending on the active of interest, skin retention rather than permeation may be more desirable. An assessment of the residence time of PEGs on and in the skin should also allow for a better understanding of how PEGs can be utilized when targeting actives to the skin. This will be the focus of future work, as well as an investigation of the influence of other PEGs and their derivatives on skin delivery of actives.

Supplementary Materials: The following are available online at http://www.mdpi.com/2073-4360/12/12/2907/s1, File S1: The Method of Mass Balance Study.

Author Contributions: Conceptualization, M.E.L.; methodology, M.E.L., Y.Z.; software, M.E.L. and Y.Z.; validation, M.E.L. and Y.Z.; formal analysis, M.E.L. and Y.Z.; investigation, Y.Z.; resources, M.E.L. and Y.Z.; data curation, Y.Z.; writing—original draft preparation, Y.Z.; writing—review and editing, M.E.L., Y.Z. and D.J.M.; visualization, M.E.L., Y.Z. and D.J.M.; supervision, M.E.L.; project administration, M.E.L.; funding acquisition, M.E.L. and Y.Z. All authors have read and agreed to the published version of the manuscript.

Funding: This research received no external funding.

Conflicts of Interest: The authors declare no conflict of interest.

References

1. Hadgraft, J. Skin, the final frontier. *Int. J. Pharm.* **2001**, *224*, 1–18. [CrossRef]
2. Lane, M.E. Skin penetration enhancers. *Int. J. Pharm.* **2013**, *447*, 12–21. [CrossRef]
3. Benson, H.A.E.; Grice, J.E.; Mohammed, Y.; Namjoshi, S.; Roberts, M.S. Topical and transdermal drug delivery: From simple potions to smart technologies. *Curr. Drug Deliv.* **2019**, *16*, 444–460. [CrossRef]
4. Menon, G.K.; Cleary, G.W.; Lane, M.E. The structure and function of the stratum corneum. *Int. J. Pharm.* **2012**, *435*, 3–9. [CrossRef]
5. Jang, H.-J.; Shin, C.Y.; Kim, K.-B. Safety evaluation of polyethylene glycol (PEG) compounds for cosmetic use. *Toxicol. Res.* **2015**, *31*, 105–136. [CrossRef]
6. Casiraghi, A.; Selmin, F.; Minghetti, P.; Cilurzo, F.; Montanari, L. Nonionic surfactants: Polyethylene glycol (peg) ethers and fatty acid esters as penetration enhancers. *Percutaneous Penetration Enhanc. Chem. Methods Penetration Enhanc.* **2015**, *10*, 251–271.
7. Fruijtier-Pölloth, C. Safety assessment on polyethylene glycols (PEGs) and their derivatives as used in cosmetic products. *Toxicology* **2005**, *214*, 1–38. [CrossRef]
8. Sarpotdar, P.P.; Gaskill, J.L.; Giannini, R.P. Effect of polyethylene glycol 400 on the penetration of drugs through human cadaver skin in vitro. *J. Pharm. Sci.* **1986**, *75*, 26–28. [CrossRef] [PubMed]
9. Mohammed, D.; Crowther, J.M.; Matts, P.J.; Hadgraft, J.; Lane, M.E. Influence of niacinamide containing formulations on the molecular and biophysical properties of the stratum corneum. *Int. J. Pharm.* **2013**, *441*, 192–201. [CrossRef] [PubMed]
10. Comaish, J.S.; Felix, R.H.; McGrath, H. Topically applied niacinamide in isoniazid-induced pellagra. *Arch. Dermatol.* **1976**, *112*, 70–72. [CrossRef] [PubMed]
11. Hakozaki, T.; Minwalla, L.; Zhuang, J.; Chhoa, M.; Matsubara, A.; Miyamoto, K.; Greatens, A.; Hillebrand, G.; Bissett, D.; Boissy, R. The effect of niacinamide on reducing cutaneous pigmentation and suppression of melanosome transfer. *Br. J. Dermatol.* **2002**, *147*, 20–31. [CrossRef] [PubMed]
12. Shalita, A.R.; Smith, J.G.; Parish, L.C.; Sofman, M.S.; Chalker, D.K. Topical nicotinamide compared with clindamycin gel in the treatment of inflammatory acne vulgaris. *Int. J. Dermatol.* **1995**, *34*, 434–437. [CrossRef] [PubMed]
13. Damian, D.L.; Halliday, G.M.; Taylor, C.A.; Barnetson, R.S. Ultraviolet radiation induced suppression of Mantoux reactions in humans. *J. Investig. Dermatol.* **1998**, *110*, 824–827. [CrossRef] [PubMed]
14. Fiume, M.M.; Bergfeld, W.F.; Belsito, D.V.; Hill, R.A.; Klaassen, C.D.; Liebler, D.; Marks, J.G.; Shank, R.C.; Slaga, T.J.; Snyder, P.W.; et al. Safety assessment of propylene glycol, tripropylene glycol, and PPGs as used in cosmetics. *Int. J. Toxicol.* **2012**, *31*, 245S–260S. [CrossRef] [PubMed]
15. Sullivan, D.W.; Gad, S.C.; Julien, M. A review of the nonclinical safety of Transcutol®, a highly purified form of diethylene glycol monoethyl ether (DEGEE) used as a pharmaceutical excipient. *Food Chem. Toxicol.* **2014**, *72*, 40–50. [CrossRef]
16. Osborne, D.W.; Musakhanian, J. Skin penetration and permeation properties of Transcutol®—Neat or diluted mixtures. *AAPS PharmSciTech* **2018**, *19*, 3512–3533. [CrossRef]
17. Hoang Thi, T.T.; Pilkington, E.H.; Nguyen, D.H.; Lee, J.S.; Park, K.D.; Truong, N.P. The importance of poly (ethylene glycol) alternatives for overcoming peg immunogenicity in drug delivery and bioconjugation. *Polymers* **2020**, *12*, 298. [CrossRef]

18. Fiume, M.M.B.; Bergfeld, W.F.; Belsito, D.V.; Hill, R.A.; Klaassen, C.D.; Liebler, D.C.; Marks, J.G.; Shank, R.C.; Slaga, T.J.; Snyder, P.W. Safety assessment of pegylated alkyl glycerides as used in cosmetics. *Int. J. Toxicol.* **2020**, *39*, 26S–58S. [CrossRef]
19. Zhang, Y.; Lane, M.E.; Hadgraft, J.; Heinrich, M.; Chen, T.; Lian, G.; Sinko, B. A comparison of the in vitro permeation of niacinamide in mammalian skin and in the parallel artificial membrane permeation assay (PAMPA) model. *Int. J. Pharm.* **2019**, *556*, 142–149. [CrossRef]
20. Braon, T.L. Determination of solubility parmeter values for pure solvents and binary mixtures. *Drug Dev. Ind. Pharm.* **1980**, *6*, 87–98. [CrossRef]
21. ICH. *Validation of Analytical Procedures: Text and Methodology*; ICH Harmonised Tripartite Guideline: Geneva, Switzerland, 2005; pp. 1–13.
22. Iliopoulos, F.; Sil, B.C.; Monjur Al Hossain, A.S.M.; Moore, D.J.; Lucas, R.A.; Lane, M.E. Topical delivery of niacinamide: Influence of neat solvents. *Int. J. Pharm.* **2020**, *579*, 119137. [CrossRef] [PubMed]
23. Davis, E.C.; Callender, V.D. Postinflammatory hyperpigmentation: A review of the epidemiology, clinical features, and treatment options in skin of color. *J. Clin. Aesthet. Dermatol.* **2010**, *3*, 20–31. [PubMed]
24. Haque, T.; ME, L.; BS, S.; Crowther, J.M.; Moore, D.J. In vitro permeation and disposition of niacinamide in silicone and porcine skin of skin barrier-mimetic formulations. *Int. J. Pharm.* **2017**, *520*, 158–162. [CrossRef] [PubMed]
25. Zhang, Y.; Kung, C.P.; Sil, B.C. Topical delivery of niacinamide: Influence of binary and ternary solvent systems. *Pharmaceutics* **2019**, *11*, 668. [CrossRef] [PubMed]
26. Flaten, G.E.; Palac, Z.; Engesland, A.; Filipović-Grčić, J.; Vanić, Ž.; Škalko-Basnet, N. In vitro skin models as a tool in optimization of drug formulation. *Eur. J. Pharm. Sci.* **2015**, *75*, 10–24. [CrossRef]
27. OECD. *Test No. 428: Skin Absorption: In Vitro Method*; OECD Publishing: Paris, France, 2004.
28. Santos, P.; Watkinson, A.C.; Hadgraft, J.; Lane, M.E. Oxybutynin permeation in skin: The influence of drug and solvent activity. *Int. J. Pharm.* **2010**, *384*, 67–72. [CrossRef]
29. Davies, D.J.; Ward, R.J.; Heylings, J.R. Multi-species assessment of electrical resistance as a skin integrity marker for in vitro percutaneous absorption studies. *Toxicol. Vitr.* **2004**, *18*, 351–358. [CrossRef]
30. Zhang, G.; Bao, C.; Fu, K.; Lin, Y.; Li, T.; Yang, H. Synthesis, characterization, self-assembly, and irritation studies of polyglyceryl-10 caprylates. *Polymers* **2020**, *12*, 294. [CrossRef]
31. Haque, T.; Rahman, K.M.; Thurston, D.E.; Hadgraft, J.; Lane, M.E. Topical delivery of anthramycin I. Influence of neat solvents. *Eur. J. Pharm. Sci.* **2017**, *104*, 188–195. [CrossRef]
32. SCCS (The Scientific Committee on Concumer Safety). *Opinion on Basic Criteria for the In Vitro Assessment of Dermal Absorption of Cosmetic Ingredients*; SCCS: Brussels, Belgium, 2010.
33. Hoelgaard, A.; Møllgaard, B.; Baker, E. Vehicle effect on topical drug delivery. IV. Effect of N-methylpyrrolidone and polar lipids on percutaneous drug transport. *Int. J. Pharm.* **1988**, *43*, 233–240. [CrossRef]
34. Kung, C.-P.; Zhang, Y.; Sil, B.C.; Hadgraft, J.; Lane, M.E.; Patel, B.; McCulloch, R. Investigation of binary and ternary solvent systems for dermal delivery of methadone. *Int. J. Pharm.* **2020**, *586*, 119538. [CrossRef] [PubMed]
35. Harrison, J.E.; Watkinson, A.C.; Green, D.M.; Hadgraft, J.; Brain, K. The relative effect of Azone® and Transcutol® on permeant diffusivity and solubility in human stratum corneum. *Pharm. Res.* **1996**, *13*, 542–546. [CrossRef] [PubMed]

Publisher's Note: MDPI stays neutral with regard to jurisdictional claims in published maps and institutional affiliations.

© 2020 by the authors. Licensee MDPI, Basel, Switzerland. This article is an open access article distributed under the terms and conditions of the Creative Commons Attribution (CC BY) license (http://creativecommons.org/licenses/by/4.0/).

Article

Development and Evaluation of Polymeric Nanosponge Hydrogel for Terbinafine Hydrochloride: Statistical Optimization, In Vitro and In Vivo Studies

Aditee Ghose [1,†], Bushra Nabi [1,†], Saleha Rehman [1], Shadab Md [2,3], Nabil A. Alhakamy [2,3], Osama A. A. Ahmad [2,3], Sanjula Baboota [1] and Javed Ali [1,*]

1. Department of Pharmaceutics, School of Pharmaceutical Education and Research, Jamia Hamdard, New Delhi 110062, India; aditeeghose@gmail.com (A.G.); nabibushra79@gmail.com (B.N.); saleharehman90@gmail.com (S.R.); sbaboota@jamiahamdard.ac.in (S.B.)
2. Department of Pharmaceutics, Faculty of Pharmacy, King Abdulaziz University, Jeddah 21589, Saudi Arabia; shaque@kau.edu.sa (S.M.); nalhakamy@kau.edu.sa (N.A.A.); oaahmed@kau.edu.sa (O.A.A.A.)
3. Center of Excellence for Drug Research & Pharmaceutical Industries, King Abdulaziz University, Jeddah 21589, Saudi Arabia
* Correspondence: jali@jamiahamdard.ac.in or javedaali@yahoo.com; Tel.: +91-9811312247; Fax: +91-11-2605-9663
† The authors have contributed equally.

Received: 14 October 2020; Accepted: 1 December 2020; Published: 3 December 2020

Abstract: Terbinafine hydrochloride, although one of the prominent antifungal agents, suffers from low drug permeation owing to its hydrophobic nature. The approach of nanosponge formulation may thus help to resolve this concern. Thus, the present research was envisioned to fabricate the nanosponge hydrogel of terbinafine hydrochloride for topical delivery since nanosponge augments the skin retentivity of the drug. The optimized formulation was obtained using Box Behnken Design. The dependent and independent process parameters were also determined wherein polyvinyl alcohol (%), ethylcellulose (%), and tween 80 (%) were taken as independent process parameters and particle size, polydispersity index (PDI), and entrapment efficiency (EE) were the dependent parameters. The nanosponge was then incorporated into the hydrogel and characterized. In-vitro drug release from the hydrogel was 90.20 ± 0.1% which was higher than the drug suspension and marketed formulation. In vitro permeation potential of the developed formulation through rat skin showed a flux of 0.594 ± 0.22 µg/cm^2/h while the permeability coefficient was 0.059 ± 0.022 cm/s. Nanosponge hydrogel was evaluated for non-irritancy and antifungal activity against *C. albicans* and *T. rubrum* confirming the substantial outcome. Tape stripping studies exhibited ten times stripping off the skin quantified 85.6 ± 0.21 µg/cm^2. The confocal analysis justified the permeation potential of the prepared hydrogel. The mean erythemal score was 0.0, confirming that the prepared hydrogel did not cause erythema or oedema. Therefore, based on results obtained, nanosponge hydrogel formulation is a potential carrier for efficient topical delivery of terbinafine hydrochloride.

Keywords: nanosponge; hydrogel; Box–Behnken design; pharmacokinetic; terbinafine hydrogel

1. Introduction

Fungal infections currently account for about the fourth common disease in the world that affects millions of people every year, about 25% of the world's population, mostly adults. The enhanced manifestation of the infections has led to increased cases of immunocompromised patients suffering from malignancies, HIV, and diseases of the similar sought [1–3].

The skin serves as a defence system for the human body when exposed to ultraviolet radiation that induces oxidative stress. Oxidative stress may lead to lipid peroxidation and DNA breakage on some occasion. However, there are certain self-defence mechanisms, such as superoxide dismutase, available for protection [4,5]. Furthermore, in the current scenario, skin and specifically stratum corneum are commonly used for drug delivery. The stratum corneum is considered as the target organ for the delivery of antimycotic drugs because the desirous concentration of drugs to render the therapeutic effect could be easily achieved post topical delivery. However, to achieve the same peroral effect, a higher dose is needed, resulting in a higher incidence of adverse effects [6,7].

Topical drug delivery systems have established a reputation for themselves and have the potential for efficient drug delivery due to the vehicles used in their preparation that ultimately affect the rate of drug permeation across the skin [6]. Topical delivery provides great advantages such as patient compliance due to ease of applicability and non-invasive design, on-site delivery minimizing systemic side effects and effective targeting ability. Certain drawbacks are associated with the conventional formulations. Therefore, to overcome the lacunas, hydrogels are tailored and widely used for topical application. The fact that they are widely gaining impetus is attributable to their swelling property in association with their adhesiveness and potential to modulate the drug release [8].

Nanosponges (NS) are a novel formulation, a sponge-like structure used to encapsulate nanoparticles with a non-collapsible and porous structure. It is primarily used for pharmaceutical and cosmeceutical approaches, as it blends the advantages of microsponges and nanosized vesicular structure. The porous structure not only enables us to entrap a wide range of active ingredients but also modulates the release pattern. NS, if incorporated in hydrogel offers remarkable perks, the most important being improved skin retention [9,10]. NS offers remarkable advantages including higher entrapment efficiency, improving the drug profile, economical method of preparation, and ease of drug release owing to three-dimensional porous structures. Different preparation methods are used: solvent method, ultra-assisted synthesis, emulsion solvent diffusion method, and melting method to formulate stable NS in different categories. Software-based optimization techniques are employed to derive optimized product of superior attributes and quality. Moreover, 3D printing techniques are now being considered to ease the production of NS. Different routes and modes of drug administration e.g., aerosols, capsules, parenteral, tablets, topicals are now being exhausted for NS delivery [11].

Terbinafine hydrochloride (TH) is the fungicidal allylamine drug that kills the fungal organisms. It acts as a non-competitive inhibitor of squalene epoxidase lodging within the fungi cell membrane. Furthermore, it plays an active role in diminishing the level of ergosterol along with the lowering of the intracellular buildup of squalene which leads to fungal cell death. Fungi often infect the skin surface and eventually invade the stratum corneum, and desquamation from the skin surface is not possible [12]. The drug is hydrophobic thus exhibiting poor drug permeation. Hence to overcome the drawbacks of the conventional topical formulations, a novel approach has been envisaged to demonstrated promising results. Cerebi et al. and Vagashiya et al. developed TH microemulsion-based hydrogel and solid lipid nanoparticles, which were instrumental in reducing the fungal load [8,13]. Barot and colleagues developed the microemulsion of TH. The cumulative amount of drug permeation was enhanced 3-folds when compared with conventionally available cream [14]. Mahaparale and colleagues fabricated TH polymeric microsponge using a quasi emulsion solvent diffusion method. The studies established that the sustained release mechanism was followed by the developed formulation. Satisfactory drug deposition was also deduced [15]. Amer et al. showed the importance of developing TH nanosponge that exhibited 90% of drug release within 8 h. Furthermore, the highest in vivo skin deposition and antifungal activity were also demonstrated in the developed formulation [16]. However, the study conducted on NS of TH was not exhaustive enough and lacked an in-depth analysis and discussion section. Thus, the authors intended to conduct extensive analysis, thereby, providing a novelty quotient to the research envisaged.

The purpose of the research undertaken is to formulate and optimize NS of TH and further incorporate it into the hydrogel. This step would be superseded by the in vitro release and permeation studies followed by an antifungal activity. This aids in determining the in vivo prospect of TH.

2. Experimental Methodologies

2.1. Materials and Animals Used

2.1.1. Materials

Terbinafine Hydrochloride (TH) was received from A.S Lifesciences, (Haryana, India). Other ingredients used were Polyvinyl Alcohol (PVA) (GS Chemical Testing & Chemical Industries, New Delhi, India); Ethyl Cellulose (EC) (Titan biotech Ltd., Rajasthan, India); Dichloromethane, Tween 80, Carbopol 940, and Ethanol (S.D. Fine Chemicals Ltd., Mumbai, India); Triethanolamine (Loba Chemie, Mumbai; India). Analytical grade reagents were used for the research.

2.1.2. Animals

Albino Wistar rats (weighing 200–250 g) were chosen as the animal model which was obtained by Central Animal House, Jamia Hamdard, New Delhi upon approval from the IAEC, Jamia Hamdard (173/GO/Re/S/2000/CPCSEA)(Protocol Number: 1404) and CPCSEA (Registration number 173/CPCSEA), Government of India.

2.2. Development and Characterization of TH-NS

2.2.1. Optimization of TH-NS

Significant endeavours were made for the fabrication of the optimal NS for which the quality target product profile (QTPP) was determined. This step was instrumental in establishing the plausible attributes of the end product. Thus, it is a vital stage that needs to be emphasized to deduce the finished product of high standards concerning quality, safety, and efficacy. Nevertheless, this could be accomplished only by focusing on critical process parameters (CPP) and critical material attributes (CMA) that might, in turn, affect the CQA (critical quality attributes) [17]. The different essentials of QTPP with their validation for the fabrication of TH-NS are mentioned in Table 1. These QTPP have specific targets that need to be addressed for the research. For instance, the product developed should have optimal features to allow the route of administration to be topical. The Ishikawa plot as represented in Figure 1 elaborates on the various risk factors, conditions required, along with CQA in the development process.

Table 1. QTPP with its target and justification for the fabrication of TH-NS.

QTPP	Target	Justification
Drug delivery system	NS	The system offers augmented skin retentivity in comparison to other nanosystems
Dosage type	Controlled release	This will enable amplified drug absorption profile
Route of administration	Topical	They offer ease of applicability, non-invasive nature, and on-site delivery
Drug release (%)	More than 80%	It is a pre-requisite for optimal therapeutic and pharmacological activity

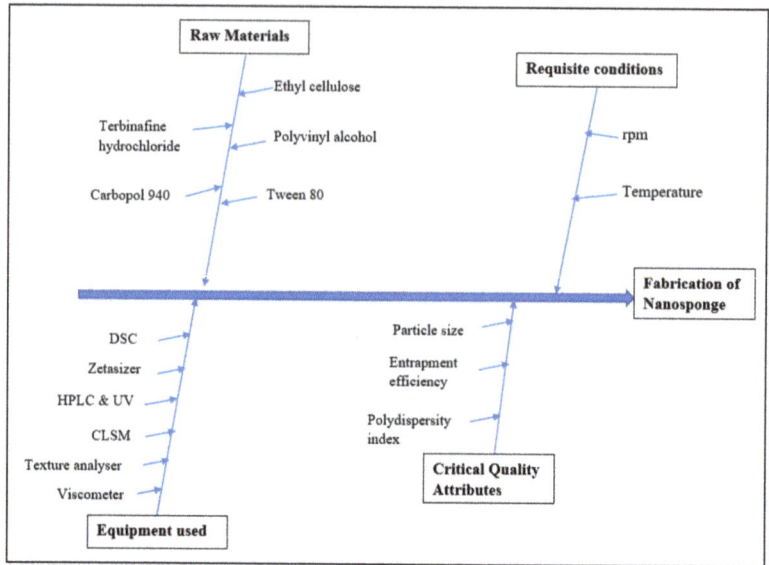

Figure 1. Ishikawa diagram used as a visualization tool for the fabrication of TH-NS.

2.2.2. Fabrication of TH-NS

The NS was developed by the emulsion diffusion method using different proportions of excipients i.e., EC, PVA, and Tween-80. The continuous phase comprised of 17.5 mL aqueous solution of PVA in different ratios, with varying amounts of surfactant, and the dispersed phase consisting of TH (100 mg) and in 2 mL of dichloromethane optimum quantity of EC is dissolved. The organic phase was incorporated in the aqueous phase maintaining the requisite conditions (temperature: 35 °C, 1000 rpm for 2 h). The resultant NS was collected and dried at 40 °C for 24 h [18].

The optimization process was carried out using a Box–Behnken design wherein PVA (%), EC (%), and Tween 80 (%) were taken as independent process parameters. The dependent parameters selected for the optimization process were particle size, polydispersity index (PDI), and entrapment efficiency (EE) [19]. The relation between variables and expected goals is depicted in Table 2. The various independent and dependent variables are demonstrated in Table 2. The levels of independent variables are also mentioned. The three-factor 3-level design employed led to 17 experimental runs, yielding a quadratic equation defining the model, as shown in the equation below.

$$Y = b_0 + b_1x_1 + b_2x_2 + b_3x_3 + b_4x_4 + b_{12}x_1x_2 + b_{13}x_1x_3 + b_{23}x_2x_3 + b_{11}x_1^2 + b_{22}x_2^2 + b_{33}x_3^2 \quad (1)$$

Table 2. Dependent and independent variables of Box-Behnken Design with their respective levels and goals.

Variables	Levels of Variables	
Independent variables	Low	High
PVA (%)	1	3
EC (%)	1	3
T80 (%)	1	5
Dependent variables	Goals	
Particle mean size (nm)	Minimize	
PDI	Minimize	
EE (%)	Maximize	

Y: effect observed from the combination of different levels of process parameters; b_0: intercept, b_1–b_3: regression coefficients; x_1–x_3: independent variables taken for study; b_{12}, b_{13}, b_{23}: interaction coefficients; b_{11}, b_{22}, b_{33}: quadratic coefficients of the experiment conducted.

2.2.3. Characterization of Prepared TH-NS

Particle size, polydispersity index (PDI), and entrapment efficiency (EE).

The particle size and PDI were measured using Malvern zeta sizer (Malvern Instruments Ltd., Worcestershire, UK). Sample dilution was performed using distilled water at 25 °C to yield dispersion of sufficient scattering intensity, and analysis was performed at 372.0 (kcps) for 20 s.

TH-NS was weighted and centrifuged at 10,000 rpm for 30 min. It was diluted in 10 mL methanol. This step is envisaged to break the aggregates if formed. The mixture was sonicated for 5 min. It was filtered and analyzed at 223 nm by UV spectrophotometric technique [1].

$$\% \text{ Drug entrapment efficiency} = \frac{W(\text{encapsulated NS}) \times 100}{W(\text{total drug})}$$

Again, the particle size determination was further validated using transmission electron microscopy (TEM). The sample for analysis was diluted 10 folds. It was then placed on the copper grid immersed within a copper film. This step was then followed by staining with phosphotungstic acid (1%) to provide a negative charge. Finally, the preparation was air-dried before viewing [20].

The pure drug and TH-NS (5 mg each) were utilized for carrying out DSC analysis wherein Perkin Elmer differential scanning calorimeter was used. The requisite conditions were maintained (heating rate: 10 °C/min, temperature: 30–400 °C, and nitrogen flow: 60 mL/min) [20].

2.3. Fabrication of Topical Hydrogel Integrating TH-NS

The TH-NS was dispersed in 1% w/w Carbopol 940 to fabricate hydrogel (HG). Various TH concentrations were measured at 0.2, 0.5, 0.8, 1 and 1.5% w/w. The 2-h polymer dispersion in water allowed it to act as a gel-forming agent. The external force was employed by agitating the established system at 600 rpm, followed by a 15-min stagnation process. This move is critical in removing trapped air. Additionally, 2% v/v of aqueous triethanolamine was added to the prepared formulation by holding it at 100 rpm.

2.4. Characterisation of TH-NS HG

2.4.1. Visual Observation

The prepared HG was inspected visually for its colour, homogeneity, grittiness, and syneresis [21].

2.4.2. Viscosity Determination

The prepared gel viscosity was determined using a programmable Viscometer. The HG prepared was taken in a beaker wherein the T-bar spindle (spindle-C, S-96) was immersed at 90° such that the spindle and the base of the beaker are not in contact. The speed of rotation of the spindle was maintained at 50 rpm. The evaluation is made after a duration of 30 s post which the HG system prepared is stabilized [12,22].

2.4.3. pH Determination

The pH meter was used to determine the formulation pH. In a known quantity of distilled water (100 mL), the prepared HG (1 g) was dissolved and kept for 2 h. The electrode was then immersed in the mixture produced and observed at room temperature in triplicate [21].

2.4.4. Spreadability

Spreadability (g·cm/s) is referred to the time in sec required by two slides to slip over the HG placed between them upon the influence of external stimuli. The glass slides of 7.5 cm in length were employed along with the load on the upper plate being 20 g [23]. Spreadability was evaluated using the formula below mentioned:

$$\text{Spreadability} = \frac{\text{Weight (g)} \times \text{Lenght (cm)}}{\text{Time (s)}}$$

2.4.5. Texture Analysis

The mechanical property of HG was evaluated with the help of software-texture analyzer TA (XT *Plus* Stable Micro System, UK equipped with 5 kg load cell). The sample weighing 100 mg was placed in the beaker cautiously to avoid air bubbles in it. The speed of the probe for analysis was fixed at 1.0 mm/s for pre-test, 2.0 mm/s for the test, and 10.0 mm/s for post-test analysis.

The probe was immersed with a load cell capacity of 5.0 g and a distance of 10.0 mm. Data interpretation was undertaken using the Texture Exponent software installed within the equipment. The resultant force-time plot gave the values of different mechanical parameters [24,25].

2.4.6. Determination of Drug Content

The prepared HG (50 mg) was dissolved in 100 mL phosphate buffer pH 5.5 and shaken for 2 h. This move is intended to ensure optimum drug solubility during mechanical shaking. The solution was purified and spectrophotometrically quantified at 223 nm [6,21].

2.5. In Vitro Release Study

The in vitro drug release study of TH from the drug suspension (DS), TH-NS, HG, and the marketed formulation was determined using an activated dialysis bag in a dissolution chamber where the system was maintained at 100 rpm and 37 ± 0.5 °C. The release profile was determined in phosphate buffer pH 5.5 (pH of normal skin) and phosphate buffer pH 7.4 (physiological pH) respectively. Sample equivalent to 3 mL was removed at specified time intervals and an equal quantity of fresh dissolution media was replenished in the assembly to maintain sink condition. Filtered and quantitatively determined at 223 nm. The data obtained were fitted to different kinetic models to interpret drug release mechanisms and kinetics [12,26].

2.6. In Vitro Permeation Studies

The rat's skin was excised, washed using distilled water before mounting on the Franz diffusion cell. The skin was so positioned that the donor compartment faced the stratum corneum (SC) and the dermis side faced the receptor. The HG was taken as necessary and put on the skin surface. The receptor compartment consisted of phosphate buffer pH 5.5, taken as the diffusion media. The appropriate condition was maintained at 37 ± 1 °C along with constant aeration. Sample (1 mL) was removed at a predetermined time and reloaded with fresh media. The cumulative amount of drug permeation as a function of time along with flux (J) was statistically determined using HPLC analysis [1,6,27].

2.7. Confocal Laser Microscopy

Confocal microscopic examination is essential to the drug's in vivo prospects. The HG prepared by labelling it with Rhodamine 123 dye, applied to Albino Wistar rats' dorsal skin and allowed for 8 h. Then the animal was slaughtered, skin excised and washed with phosphate buffer pH 5.5. Prepared skin section was placed on a slide and analyzed microscopically using CLSM (Leica microsystems). For rhodamine 123 operation, optical excitation was performed with a laser beam of 488 nm and fluorescence emission was detected above 560 nm [22,28].

2.8. Tape Stripping Technique

Post conducting the in vitro permeation investigation, the surplus quantity of the developed formulation was scrapped off from the skin surface. It was then washed thrice with the phosphate buffer (pH 5.5) and dried gently using a cotton swab. The final process envisaged the use of a serial tape-stripping method whereby, the removal of 15 strips using adhesive tape was undertaken. Each tape was carefully weighed before and after the procedure. The adhesive tapes were collected using a methanol-water mixture (95:5 *v/v*) and further analyzed using the HPLC technique [28,29].

2.9. Skin Irritation Study

The test of skin irritation potential was carried out with the HG prepared in comparison to placebo HG and performed by the Draize patch test. The animals were divided into the following groups ($n = 3$)

- Group I: Control (no drug treatment);
- Group II: DS;
- Group III: Placebo HG;
- Group IV: TH-NS HG.

The animal's dorsal side was shaved 24 h before drug application. HG (0.5 g) was applied evenly and homogenously to hair-free rat skin covering 4 cm^2. Any visual improvement was monitored on rat skin after 24, 48, and 72 h post HG applicability. The mean erythema scores of the different formulations were graded from 0 to 4 [10,30].

2.10. Antifungal Activity Study (Cylinder Plate Method)

Trichophyton rubrum and *Candida albicans* species were used to determine HG antifungal inhibitory activity. For the analysis of antifungal potential, the cylinder plate method was selected using sabouraud dextrose media (Himedia, India). The requisite conditions are 25 °C for 7 days and 30 °C for 2 days for *Trichophyton rubrum* and *Candida albicans* respectively. The spores were then harvested and suspended into the media of amount 20 mL. The inoculated media (1 mL) was incorporated into 100 mL of sabouraud dextrose agar at 37 ± 1 °C. From the mixture thus obtained, 10 mL was inserted in the petri dish in solidified agar media. Four wells were bored in a media-containing petri dish and filled with 0.1 g HG, placebo HG, advertised formulation and control (distilled water). Different Petri dishes contained different microbial strains from which zone of inhibition was determined using a hemocytometer [6,14].

2.11. Statistical Analysis

The results obtained in different studies were statistically analyzed. The experiments were conducted in triplicate and expressed in the form of mean ± SD. One-way analysis of variance (ANOVA) (GraphPad Software Inc., San Diego, CA, USA) was employed for the determination.

3. Results and Discussion

3.1. Optimization of TH-NS

The primary target was to tailor TH-NS with enhanced skin retentivity. The prime concept behind predefining the QTPP was to get a patient-centric formulation of the utmost quality that would offer maximum therapeutic outcome. Therefore, the attributes, i.e., particle size, EE, and PDI, were selected as CQA which offers its respective advantages towards the developed systems.

The determination of process parameters in the optimization process is of prime concern to yield effective and economical outcome. The conventional "one-factor-at-a-time approach" is time-consuming along with avoidance of interactions between independent variables, thereby paving way for novel optimization techniques. Response surface methodology (RSM) is a potential optimization tool wherein, multiple factors and their interactions could be ascertained with a minimal number of experimental runs. In this method, the response of interactions of statistically designed combinations coupled with the determination of the coefficients of the best fit model has endeavoured to predict the adequacy of the model [31]. The outcome of the trails was accessed using Design-Expert software which validated the substantial use of statistical design methodology. The independent process attributes chosen were PVA, EC concentration, and Tween80 concentration for which the dependent parameters were particle size (nm), PDI, and EE (%). Polynomial quadratic equations along with respective correlations were derived based on the interactions of the dependent and independent variables. The contour plots thereby obtained from the analysis demonstrated the qualitative effect of each variable on each response.

On applying the design, 17 runs were obtained. The result obtained is shown in Table 3 and the contour plots are shown in Figure 2. The result obtained from the experimental runs as evident from Table 3 reveals that the particle size of the different formulations was in the range of 425.7–571.4 nm. The PDI was in the range of 0.30–0.564 while EE lies between 50.5% and 85.45%.

Table 3. Experimental runs of BBD with obtained results.

Formulation Code	PVA Conc (X1)	EC Conc (X2)	T-80 Conc (X3)	Particle Size (nm) (Y1)	PDI (Y2)	EE (Y3)
N1	3	2	1	448.4	0.3	85.45
N2	1	3	3	440.7	0.425	50.5
N3	2	1	5	520.4	0.47	73.6
N4	2	1	1	560.8	0.509	65.9
N5	1	1	3	425.7	0.42	54.72
N6	1	2	1	476.89	0.411	58.2
N7	1	2	5	490.1	0.48	59.6
N8	2	3	5	530.25	0.48	81.4
N9	2	2	3	571.4	0.482	77.2
N10	2	2	3	573.7	0.501	77.9
N11	3	3	3	360.9	0.41	80.1
N12	2	3	1	452.62	0.32	83.77
N13	3	2	5	555.2	0.564	76
N14	2	2	3	571.4	0.501	77.2
N15	3	1	3	509.12	0.51	78.61
N16	1	2	3	500.2	0.43	72.21
N17	3	1	1	489.12	0.5	75.67

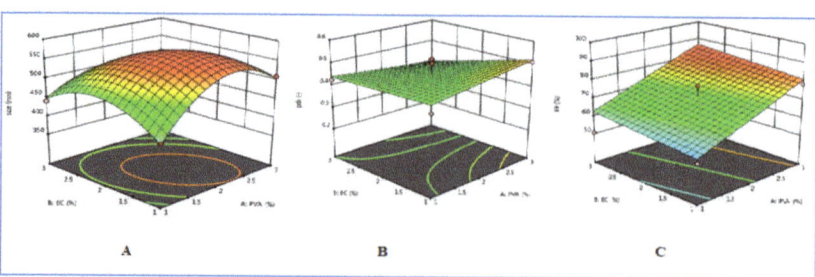

Figure 2. (**A**) Contour plot depicting size based on the concentration of PVA and EC% (**B**) Contour plot depicting PDI based on the concentration of PVA and EC% (**C**) Contour plot depicting entrapment efficiency based on the concentration of PVA and EC%.

The second-order polynomial equation relating the response of particle size (Y1), PDI (Y2), and EE% (Y3) respectively are given below:

Y1 = 28.94B + 19.66C − 40.80AB + 29.51BC

Y2 = 0.0060A − 0.0340B + 0.0570C + 0.0262AB + 0.0502AC + 0.0487BC

Y3 = 71.96A + 12.14B + 2.87C

The positive sign in the quadratic relation depicts a synergism while on the contrary, the negative sign shows an antagonistic relation. A backward elimination process was adopted to allow the data to fit within the quadratic model obtained. It could be observed that PVA concentration had a major effect on particle size followed by PDI while the minimum effect on EE [10].

3.2. Characterisation of Optimised TH-NS

3.2.1. Particle Size Determination, Polydispersity Index, Entrapment Efficiency

The particle size of prepared TH-NS was 448.4 ± 12.6 nm, while PDI was 0.3 ± 0.04. The findings observed were based on earlier literature [12]. The entrapment efficiency of the TH-NS was found to be 85.45 ± 3.7% which was further corroborated from the given literature of loaded NS [10,18]. TEM image showed that the prepared TH-NS morphology was spherical with a size of 440 nm as shown in Figure 3. The result obtained was consistent with the zeta sizer analysis [10,12].

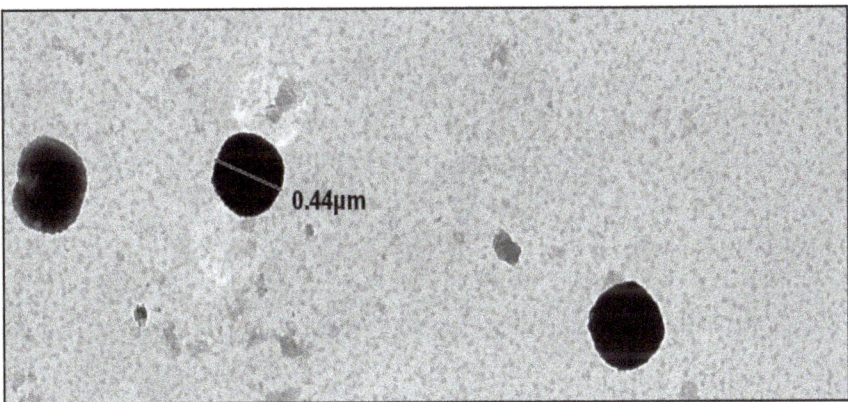

Figure 3. TEM image of TH-NS.

3.2.2. Thermoanalytical Technique (DSC)

Pure TH gives a sharp endothermic peak at 214 °C evaluated by DSC spectra as depicted in Figure 4. However, when the DSC spectra of TH loaded NS formulated was compared with spectra of pure drug, it showed the difference in absence of peak formations at the temperature as depicted in Figure 4. The absence of peak demonstrated that the TH loaded NS has been developed which leads to the reduction in crystallinity of the drug. The change in the structure from crystalline to amorphous established the fact that NS has been developed [1]. The molecularly dispersed phase of TH within the NS structure leads to the broadening of the peak. The process seems to be following the Gibbs-Thompson equation [32]. The amorphous structure of the TH-NS established by the result obtained is desirable for enhanced drug entrapment within the NS structure [33].

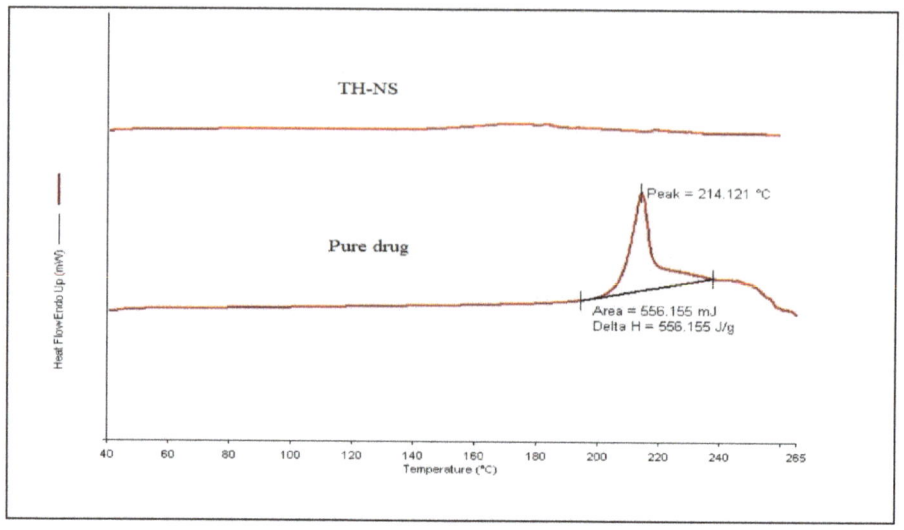

Figure 4. DSC thermogram of pure drug and TH-NS.

3.3. Preparation and Evaluation of TH-NS HG

Preliminary tests were conducted to determine the best formulation to prepare the optimum gel. Table 4 indicates a different HG composition. The Carbopol 90 composition ranged from 0.2% to 1.5% w/w. However, the compositions of other constituents remained constant at a fixed concentration. The effects of the compositions on the following parameters were tested. These properties are instrumental in establishing the physical stability of the prepared hydrogel [8]. The purpose of following such objectives was to optimize the formulation on the basis of experimental demonstrations. Table 5 demonstrates the results obtained for different HG preparations.

Table 4. Composition amount of HG preparation.

Compositions (% w/w)	Formulation Code					
	G1	G2	G3	G4	G5	G6
Carbopol 940	0.2	0.5	0.8	1	1.2	1.5
TH-loaded NS	2	2	2	2	2	2
Ethanol (95% v/v)	10	10	10	10	10	10
TEA (2%)	2	2	2	2	2	2
Distilled water	q.s	q.s	q.s	q.s	q.s	q.s

Table 5. Physicochemical characteristics for evaluation of HG.

Formulation Code	Visual Characteristics			Viscocity (Centipoises) ($n = 20$) ± SD	pH ($n = 3$) ± SD	Spreadability (g·cm/s) ($n = 3$) ± SD	Texture Analysis				Drug Content (%) ± SD
	Colour	Homogenity	Grittiness				Hardness	Adhesiveness	Elasticity	Cohesiveness	
G1	Whitish	-	-	0.384 ± 0.0004	5.72 ± 0.196	2.00 ± 0.04	198.10	−210.233	0.372	0.563	82.6 ± 0.002
G2	Whitish	-	-	0.465 ± 0.0003	5.76 ± 0.05	1.8 ± 0.03	222.283	−228.233	0.793	0.762	83.4 ± 0.002
G3	Whitish	-	-	0.789 ± 0.0002	5.90 ± 0.005	1.8 ± 0.02	240.10	−232.22	0.801	0.77	84.4 ± 0.002
G4	Whitish	-	-	1.177 ± 0.0002	5.91 ± 0.02	1.00 ± 0.02	254.999	−247.275	0.822	0.8	86.5 ± 0.002
G5	Whitish	Slightly clumpy	-	1.878 ± 0.0004	5.95 ± 0.02	0.8 ± 0.05	260.002	−259.087	0.822	0.801	84.1 ± 0.001
G6	Whitish	clumpy	-	2.999 ± 0.0003	5.99 ± 0.05	0.4 ± 0.03	286.666	−272.794	0.824	0.802	84.4 ± 0.002

3.3.1. Visual Examination

The prepared HGs were white, uniform, and clear homogeneity, missing lumps and syneresis, and showed no signs of gritty. The HGs had a glossy appearance except for the G6 formulation showing lumps in it. G5's formulation was also slightly clumpy. Table 5 shows the result.

3.3.2. Viscosity

Viscosity was testified as an important rheological physical parameter for topical formulations, whichever affects the rate of drug release into the skin. The prepared hydrogels had a viscosity of 0.84–2.99 centipoise. The lowest viscosity was observed in a lower polymer containing hydrogel. The shear rate increases with the decrease in viscosity suggesting the shear-thinning pseudoplastic nature of the formulation. The finding was corroborated with previous literature [6].

3.3.3. pH Determination

The pH value of the prepared HG was found to be 5–6, i.e., within the appropriate limits, and does not cause skin irritation upon application. Additionally, the pH values of different formulations did not change significantly in time.

3.3.4. Spreadability

Spreadability is the degree to which a gel spreads upon application. It was founded that therapeutic effectiveness depends largely on any hydrogel's spreading value. The result showed spreadability within 0.4–2.0 g·cm/sec. Good spreadability is a requirement for ideal gel formulation. The result verified that increased polymer concentration reduces HG's spreadability.

3.3.5. Texture Analysis

Texture can be considered as an index for the product's rheological attribute. For topical delivery systems, the key criteria controlling therapeutic outcome are minimum firmness and maximum adhesiveness. Firmness addresses the product removal from the container and ease of applicability while, on the contrary, adhesiveness pertains to bioadhesion. Hardness was found to lie between 198.10–286.66, adhesiveness between 210.233–286.66, elasticity between 0.372–0.824 and cohesiveness between 0.563–0.802. The results obtained validate that the formed HG has the property of adhering firmly but gently to the skin surface because they have low firmness and high adhesivity.

3.3.6. Drug Content

The percentage of drug content for the prepared HGs were ranging from 82% to 86.5%. Thus, indicating the drug was uniformly distributed throughout the gel.

According to previous studies, the hardness and compressibility of prepared gel should be strong should be low. Furthermore, high cohesiveness is a desirous attribute. Therefore, based on the results obtained from the above-performed studies it could be concluded that G4 with 1% Carbopol is the optimized formulation.

3.4. In Vitro Release Study

The cumulative % drug release was found to be 83.92% ± 0.22%, 63.06% ± 0.2%, 90.20% ± 0.1%, and 82.83% ± 0.29% respectively in DS, TH-NS, HG, and marketed formulation (Phyte gel) at pH 5.5 (as shown in Figure 5) The difference between the cumulative % drug release in DS and the marketed formulation was not found to be statistically significant. However, statistical significance ($p < 0.001$) was observed when the result obtained for TH-NS was compared with that of DS, HG, and marketed formulation. It was evident that Korsmeyer Peppas kinetic model was followed with R^2 0.995, 0.998, 0.998, and 0.989 respectively indicating an anomalous diffusion. The result could be due to the matrix's relatively slow diffusion of the trapped drug. It was apparent that HG showed up superior sustained release potential. As hydrophilic polyacrylic acid polymer, i.e., Carbopol 940 has carboxyl functional groups that get ionized after reaction with TEA. This develops a gel-like structure attributable to the electrostatic repulsion among charged polymer chains. This step leads to an increase in the pH of the HG developed rendering it appropriate for skin applicability. There was no time lag observed in the drug release from the gel observed. The result obtained was following the previously established literature [8,26].

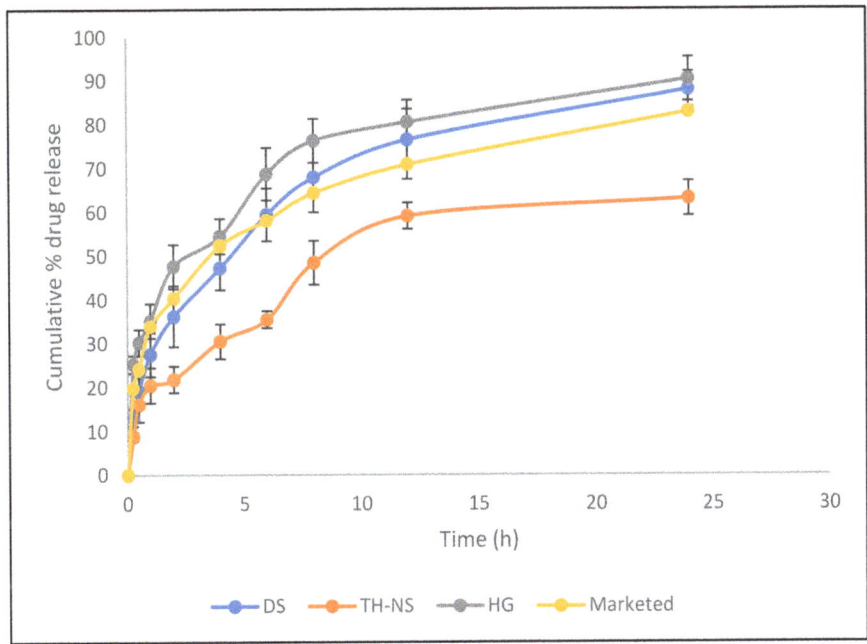

Figure 5. % cumulative in vitro drug release.

3.5. In Vitro Permeation Studies

Permeation study is the passive diffusional study where the drug permeates the entire excised skin (Moreno et al., 2019). The Flux ($\mu g/cm^2/h$) of DS, HG and marketed gel was found to be 0.10 ± 0.12, 0.594 ± 0.22 and 0.334 ± 0.23 while permeability coefficient (cm/s) was 0.01 ± 0.012, 0.059 ± 0.022 and 0.033 ± 0.023 respectively. There was statistical significance ($p < 0.05$) when the flux and permeability coefficient of HG was compared with that of DS and marketed formulation. The permeation curve was non-linear followed by a controlled release profile (Figure 6). However, an initial faster release where approximately 20% of the drug was released from the HG in the first 2 h of the experiment. The result was consistent with Malakar and colleagues' analysis [34]. There was an enhanced drug permeation compared to DS and marketed gel due to TH-NS encapsulation in HG. Another related reason for retaining HG on the skin surface may be TH's gradual release pattern over time. Additionally, due to the loose gel structure, the drug release was increased [6,18,35].

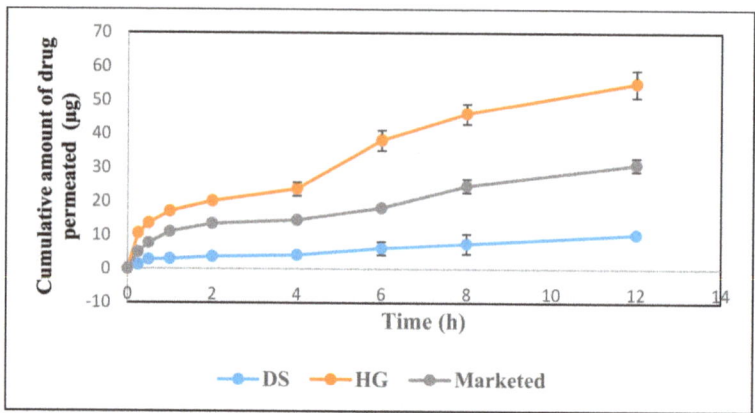

Figure 6. Cumulative amount of drug permeated vs. time profile.

3.6. Confocal Study

CLSM analysis was used to confirm HG's permeation potential and drug deposition on the skin for topical application. The skin was examined microscopically after the in vitro permeation of HG and the control solutions. Figure 7 demonstrates the CLSM photomicrographs where Figure 7A is the photomicrographs of cross-sections of hairless viable rat skin, sectioned 0.5 mm. (as standard for comparison), Figure 7B is the skin treated with HG, and Figure 7C is the skin treated with marketed gel 1%. The skin surface was referred to as the brightest autofluorescence of the imaging plane with characteristic morphology of the SC surface. In the images of the SC surface, the corneocyte groups form distinct "island-like" structures forming dark furrows. Increased fluorescence intensity was exhibited by HG. Furthermore, the fluorescence persisted on the upper layers of the skin while diminished rapidly with the depth. The increased fluorescence on the upper layer of the skin due to the retention of hydrogels. It was established that the dye probe distribution was dramatically increased in the topmost dermal layer from the prepared formulations. This might be due to the matrix-forming potential of NS upon integration in hydrogel and probable fusion with the skin membrane [36].

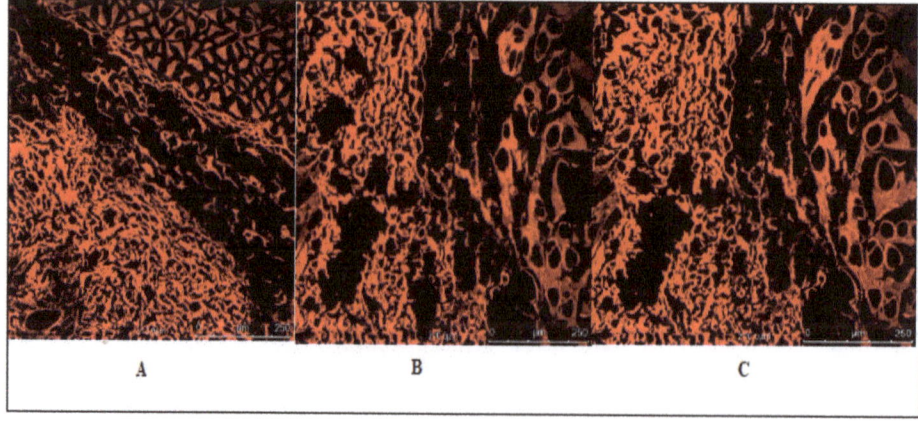

Figure 7. (**A**) CLSM photomicrographs of cross-sections of hairless viable rat skin, sectioned 0.5 mm (as standard for comparison) (**B**) Skin treated with HG (**C**) Skin treated with TH marketed gel 1%.

3.7. Tape Stripping Study

The tape stripping method is a dermatokinetic method to evaluate the distribution of formulation on the skin and in the stratum corneum segregating them into different layers when stripped off. As reported by Morena and colleagues, the concentration of drugs decreases from upper to lower layers. Furthermore, in the tape stripping method, there were chances of the drug being accumulated on the subsequent layers upon an administration which was stripped off leading to a decrease in the drug concentration in the subsequent layers [37,38].

On conducting the method of tape stripping for 10 times after 12 h, it was observed that HG collected in tape stripping was highest 85.6 ± 0.21 µg/cm^2. In the case of TH-NS and marketed gel, it was 6.54 ± 0.42 µg/cm^2 and 24.83 ± 0.1 µg/cm^2 respectively. Further by 15 times stripping observed HG was reduced to amount 20.64 ± 0.32 µg/cm^2 in comparison to 10 times while TH-NS and marketed were 2.75 ± 0.31 µg/cm^2 and 9.89 ± 0.18 µg/cm^2 respectively (Figure 8A). The result obtained for HG in the tape stripping method was statistically significant ($p < 0.001$) when compared with TH-NS and marketed formulation after 10 times stripping. However, after 15 times stripping, the statistical significance ($p < 0.05$) was observed when the HG result was compared to TH-NS and marketed formulation (Figure 8A). The outcome obtained further validated that enhanced drug level could be achieved on the subcutaneous layer with lesser penetration into deeper skin layers from the HG developed, a prerequisite for topical drug delivery system. Here, the subcutaneous layer acts as a reservoir to deliver drug progressively to the viable dermis layer. The slower penetration of the developed formulation will also lead to reduced systemic toxicity [33,35]. The study finding that the drug resides on the top of the dermal layer confirmed an enhanced therapeutic drug profile to treat fungal infection [39,40]. The HG demonstrated better results compared to the marketed formulation in both the studies, i.e., the permeation study and the tape stripping study.

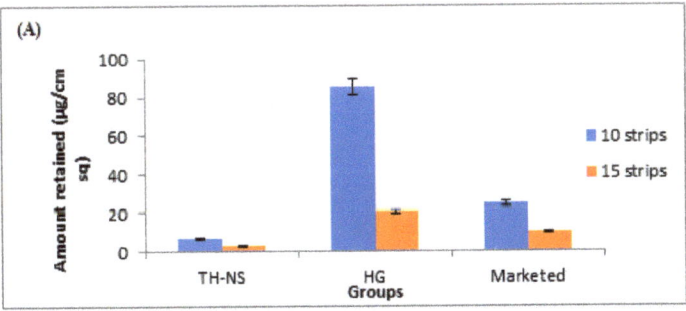

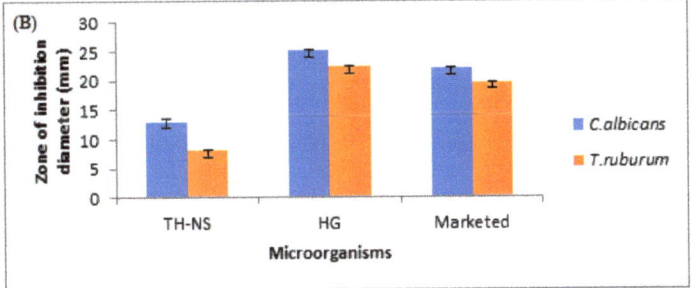

Figure 8. (**A**) Comparative amount retained by tape stripping (10 and 15 strips) and (**B**) Comparison of the zone of inhibition for the microorganisms of three different groups TH solution, TH-NS gel, and marketed gel.

3.8. Skin Irritation Study

A control group to which no drug was administered was used in the experiment. The experiment was performed according to the study reported by Iqbal et al., 2018 [41]. The erythemal scores were recorded for all rats at different time intervals of 24, 48, and 72 h. The TH-NS tailored demonstrated the mean erythemal score of 0.00 confirming no erythema or edema on the shaved rat skin, thereby exhibiting minimized skin irritation. However, the drug suspension had a score of 1 at 24 h and post that. As per the degree of erythema, different grades have been allocated no erythema = 0; slight erythema = 1; moderate = 2; more severe = 3; and most severe = 4. The commercially available convention treatments have a considerable drawback as they cause skin irritation, hence the prepared formulation provides an advantage over the conventionally available preparations. The result agreed with the previous literature [3,27].

3.9. Antifungal Studies

The cylinder plate method was used for the antifungal potential screening wherein the result demonstrated a slight difference in the zone of inhibition between TH-NS, HG, and marketed gel. The diameter of zone of inhibition shown in C. albicans and T. ruburum were 13 ± 0.42 and 8 ± 0.31 mm by TH-NS, 25.01 ± 0.21 and 22.2 ± 0.32 mm by HG while that from the marketed gel was 22 ± 0.1 and 19.5 ± 0.18 mm respectively. The result obtained for the antifungal study is exhibited in Figure 8B. The result obtained for the zone of inhibition for HG and marketed formulation was found to be statistically insignificant. Although, the result for TH-NS was statistically significant ($p < 0.05$) when compared with HG and the marketed formulation. The HG prepared demonstrated proximity towards the killing of fungal pathogens and demonstrates acceptable susceptibility values against C. albicans and T. ruburum. Because of the low viscosity of the prepared hydrogel, its penetration was increased compared to the drug solution and marketed formulation [6,42]. Therefore, the TH-NS formulation showed better fungal burden results.

4. Conclusions

The research undertaken establishes the successful incorporation of TH-NS into hydrogel which was further optimized by the QbD approach (Box–Behnken statistical design). On further in vitro and in vivo study, the prepared formulation provided considerable results, revealing enhanced drug permeability after topical administration. Thus, based on the experimental results obtained, the hydrogel approach is a potential carrier to deliver TH. However, it warrants further elaborate evaluation for commercial applicability.

Author Contributions: Conceptualization, A.G., S.B., and J.A.; Methodology, A.G., B.N., and S.R.; Software and Formal Analysis, S.M., O.A.A.A.; Validation, N.A.A. Investigation, J.A. and S.B.; Resources, J.A. and S.M.; Data Curation, B.N. and S.R.; Writing—Original Draft Preparation, B.N., N.A.A. and O.A.A.A.; Writing—Review and Editing, J.A., S.M. and S.B.; Visualization, J.A. and S.B.; Supervision, J.A. and S.B.; Funding Acquisition, S.M. and N.A.A. All authors have read and agreed to the published version of the manuscript.

Funding: The Deanship of Scientific Research (DSR) at King Abdulaziz University, Jeddah, Saudi Arabia has funded this project, under grant no. (FP-134-42).

Acknowledgments: The authors are thankful to Jamia Hamdard for providing facilities to carry out the research work.

Conflicts of Interest: The authors declare no conflict of interest.

References

1. Abdellatif, M.M.; Khalil, I.A.; Khalil, M.A. Sertaconazole nitrate loaded nanovesicular systems for targeting skin fungal infection: In-vitro, ex-vivo and in-vivo evaluation. *Int. J. Pharm.* **2017**, *527*, 1–11. [CrossRef] [PubMed]

2. Voltan, A.R.; Quindós, G.; Alarcón, K.P.M.; Fusco-Almeida, A.M.; Mendes-Gianinni, M.J.; Chorilli, M. Fungal diseases: Could nanostructured drug delivery systems be a novel paradigm for therapy? *Int. J. Nanomed.* **2016**, *11*, 3715–3730. [CrossRef] [PubMed]
3. Radwan, S.A.A.; ElMeshad, A.N.; Shoukri, R.A. Microemulsion loaded hydrogel as a promising vehicle for dermal delivery of the antifungal sertaconazole: Design, optimization and ex vivo evaluation. *Drug Dev. Ind. Pharm.* **2017**, *43*, 1351–1365. [CrossRef] [PubMed]
4. Lv, X.; Liu, T.; Ma, H.; Tian, Y.; Li, L.; Li, Z.; Gao, M.; Zhang, J.; Tang, Z. Preparation of Essential Oil-Based Microemulsions for Improving the Solubility, pH Stability, Photostability, and Skin Permeation of Quercetin. *AAPS PharmSciTech* **2017**, *18*, 3097–3104. [CrossRef]
5. Fachinetti, N.; Rigon, R.B.; Eloy, J.O.; Sato, M.R.; Dos Santos, K.C.; Chorilli, M. Comparative Study of Glyceryl Behenate or Polyoxyethylene 40 Stearate-Based Lipid Carriers for Trans-Resveratrol Delivery: Development, Characterization and Evaluation of the In Vitro Tyrosinase Inhibition. *AAPS PharmSciTech* **2018**, *19*, 1401–1409. [CrossRef]
6. Özcan, I.; Güneri, T.; Abacı, O.; Özer, O.; Uztan, A.H.; Aksu, B.; Hayal Boyacıoğlu, H. Enhanced Topical Delivery of Terbinafine Hydrochloride with Chitosan Hydrogels. *AAPS PharmSciTech* **2009**, *10*, 1024–1031. [CrossRef]
7. Kahraman, E.; Güngör, S.; Ozsoy, Y. Potential enhancement and targeting strategies of polymeric and lipid-based nanocarriers in dermal drug delivery. *Ther. Deliv.* **2017**, *8*, 967–985. [CrossRef]
8. Celebi, N.; Ermiş, S.; Özkan, S. Development of topical hydrogels of terbinafine hydrochloride and evaluation of their antifungal activity. *Drug Dev. Ind. Pharm.* **2014**, *41*, 631–639. [CrossRef]
9. Anandam, S.; Selvamuthukumar, S. Fabrication of cyclodextrin nanosponges for quercetin delivery: Physicochemical characterization, photostability, and antioxidant effects. *J. Mater. Sci.* **2014**, *49*, 8140–8153. [CrossRef]
10. Badr-Eldin, S.M.; Aldawsari, H.M.; Labib, G.S.; El-Kamel, A.H. Design and formulation of a topical hydrogel integrating lemongrass-loaded nanosponges with an enhanced antifungal effect: In vitro/in vivo evaluation. *Int. J. Nanomed.* **2015**, *10*, 893–902. [CrossRef]
11. Jain, A.; Prajapati, S.K.; Kumari, A.; Mody, N.; Bajpai, M. Engineered nanosponges as versatile biodegradable carriers: An insight. *J. Drug Deliv. Sci. Technol.* **2020**, *57*, 101643. [CrossRef]
12. Akhtar, N.; Pathak, K. Cavamax W7 Composite Ethosomal Gel of Clotrimazole for Improved Topical Delivery: Development and Comparison with Ethosomal Gel. *AAPS PharmSciTech* **2012**, *13*, 344–355. [CrossRef] [PubMed]
13. Vaghasiya, H.; Kumar, A.; Sawant, K. Development of solid lipid nanoparticles based controlled release system for topical delivery of terbinafine hydrochloride. *Eur. J. Pharm. Sci.* **2013**, *49*, 311–322. [CrossRef] [PubMed]
14. Barot, B.S.; Parejiya, P.B.; Patel, H.K.; Mehta, D.M.; Shelat, P.K. Microemulsion-based antifungal gel delivery to nail for the treatment of onychomycosis: Formulation, optimization, and efficacy studies. *AAPS PharmSciTech* **2012**, *13*, 184–192. [CrossRef] [PubMed]
15. Mahaparale, P.R.; Ikam, S.A.N.; Chavan, M.S. Development and Evaluation of Terbinafine Hydrochloride Polymeric Microsponges for Topical Drug Delivery. *Indian J. Pharm. Sci.* **2018**, *80*, 1086–1092. [CrossRef]
16. Amer, R.I.; El-Osaily, G.H.; Gad, S.S. Design and optimization of topical terbinafine hydrochloride nanosponges: Application of full factorial design, in vitro and in vivo evaluation. *J. Adv. Pharm. Technol. Res.* **2020**, *11*, 13–19. [CrossRef]
17. Cunha, S.; Costa, C.P.; Loureiro, J.A.; Alves, J.; Peixoto, A.F.; Forbes, B.; Sousa-Lobo, J.; Silva, A. Double Optimization of Rivastigmine-Loaded Nanostructured Lipid Carriers (NLC) for Nose-to-Brain Delivery Using the Quality by Design (QbD) Approach: Formulation Variables and Instrumental Parameters. *Pharmaceutics* **2020**, *12*, 599. [CrossRef]
18. Sharma, R.; Pathak, K. Polymeric nanosponges as an alternative carrier for improved retention of econazole nitrate onto the skin through topical hydrogel formulation. *Pharm. Dev. Technol.* **2010**, *16*, 367–376. [CrossRef]
19. Bachir, Y.N.; Bachir, R.N.; Hadj-Ziane-Zafour, A. Nanodispersions stabilized by β-cyclodextrin nanosponges: Application for simultaneous enhancement of bioactivity and stability of sage essential oil. *Drug Dev. Ind. Pharm.* **2018**, *45*, 333–347. [CrossRef]
20. Anandam, S.; Selvamuthukumar, S. Optimization of microwave-assisted synthesis of cyclodextrin nanosponges using response surface methodology. *J. Porous Mater.* **2014**, *21*, 1015–1023. [CrossRef]

21. Zakaria, A.S.; Afifi, S.A.; Elkhodairy, K.A. Newly Developed Topical Cefotaxime Sodium Hydrogels: Antibacterial Activity andIn VivoEvaluation. *BioMed Res. Int.* **2016**, *2016*, 6525163. [CrossRef] [PubMed]
22. Hussain, A.; Samad, A.; Ramzan, M.; Ahsan, M.N.; Rehman, Z.U.; Ahmad, F.J. Elastic liposome-based gel for topical delivery of 5-fluorouracil: In vitro and in vivo investigation. *Drug Deliv.* **2016**, *23*, 1115–1129. [CrossRef] [PubMed]
23. Chen, M.X.; Alexander, K.S.; Baki, G. Formulation and Evaluation of Antibacterial Creams and Gels Containing Metal Ions for Topical Application. *J. Pharm.* **2016**, *2016*, 5754349. [CrossRef] [PubMed]
24. Karavana, Y.S.; Güneri, P.; Ertan, G. Benzydamine hydrochloride buccal bioadhesive gels designed for oral ulcers: Preparation, rheological, textural, mucoadhesive and release properties. *Pharm. Dev. Technol.* **2009**, *14*, 623–631. [CrossRef]
25. Pons, M.; Fiszman, S. Instrumental Texture Profile Analysis with Particular Reference to Gelled Systems. *J. Texture Stud.* **1996**, *27*, 597–624. [CrossRef]
26. Chauhan, S.; Gulati, N.; Nagaich, U. Fabrication and evaluation of ultra deformable vesicles for atopic dermatitis as topical delivery. *Int. J. Polym. Mater.* **2018**, *68*, 266–277. [CrossRef]
27. Lee, S.G.; Kang, J.B.; Kim, S.R.; Kim, C.J.; Yeom, D.W.; Yoon, H.Y.; Kwak, S.S.; Choi, Y.W. Enhanced topical delivery of tacrolimus by a carbomer hydrogel formulation with Transcutol P. *Drug Dev. Ind. Pharm.* **2016**, *42*, 1–29. [CrossRef]
28. Alvarez-Román, R.; Naik, A.; Kalia, Y.N.; Guy, R.H.; Fessi, H. Enhancement of Topical Delivery from Biodegradable Nanoparticles. *Pharm. Res.* **2004**, *21*, 1818–1825. [CrossRef]
29. Jacobs, A.G.; Gerber, M.; Malan, M.M.; Preez, L.J.; Lizelle, T. Topical delivery of acyclovir and ketoconazole. *Drug Deliv.* **2016**, *23*, 631–641. [CrossRef]
30. Gupta, M.; Vyas, S.P. Development, characterization and in vivo assessment of effective lipidic nanoparticles for dermal delivery of fluconazole against cutaneous candidiasis. *Chem. Phys. Lipids* **2012**, *165*, 454–461. [CrossRef]
31. Venugopal, V.; Kumar, K.J.; Muralidharan, S.; Parasuraman, S.; Raj, P.V.; Kumar, K.V. Optimization and in-vivo evaluation of isradipine nanoparticles using Box-Behnken design surface response methodology. *OpenNano* **2016**, *1*, 1–15. [CrossRef]
32. Rehman, S.; Nabi, B.; Baboota, S.; Ali, J. Tailoring lipid nanoconstructs for the oral delivery of paliperidone: Formulation, optimization and in vitro evaluation. *Chem. Phys. Lipids* **2020**, 105005. [CrossRef] [PubMed]
33. Fatima, N.; Rehman, S.; Nabi, B.; Baboota, S.; Ali, J. Harnessing nanotechnology for enhanced topical delivery of clindamycin phosphate. *J. Drug Deliv. Sci. Technol.* **2019**, *54*, 101253. [CrossRef]
34. Malakar, J.; Sen, S.O.; Nayak, A.K.; Sen, K.K. Formulation, optimization and evaluation of transferosomal gel for transdermal insulin delivery. *Saudi Pharm. J.* **2012**, *20*, 355–363. [CrossRef]
35. Abdel-Salam, F.S.; Mahmoud, A.A.; Ammar, H.O.; Elkheshen, S.A. Nanostructured lipid carriers as semisolid topical delivery formulations for diflucortolone valerate. *J. Liposome Res.* **2016**, *27*, 41–55. [CrossRef]
36. Verma, A.; Pathak, K. Topical Delivery of Drugs for the Effective Treatment of Fungal Infections of Skin. *Curr. Pharm. Des.* **2015**, *21*, 2892–2913. [CrossRef]
37. Lademann, J.; Jacobi, U.; Surber, C.; Weigmann, H.-J.; Fluhr, J. The tape stripping procedure—Evaluation of some critical parameters. *Eur. J. Pharm. Biopharm.* **2009**, *72*, 317–323. [CrossRef]
38. Moreno, E.; Calvo, A.; Schwartz, J.; Navarro-Blasco, I.; González-Peñas, E.; Sanmartín, C.; Irache, J.M.; Espuelas, S. Evaluation of Skin Permeation and Retention of Topical Dapsone in Murine Cutaneous Leishmaniasis Lesions. *Pharmaceutics* **2019**, *11*, 607. [CrossRef]
39. Patel, M.; Patel, D.H. Development and validation of RP-HPLC method for simultaneous estimation of Terbinafine hydrochloride and mometasone furoate in combined dosage form. *Int. J. Pharm. Pharm. Sci.* **2014**, *6*, 106–109.
40. Elkomy, M.H.; El Menshawe, S.F.; Eid, H.M.; Ali, A.M.A. Development of a nanogel formulation for transdermal delivery of tenoxicam: A pharmacokinetic-pharmacodynamic modeling approach for quantitative prediction of skin absorption. *Drug Dev. Ind. Pharm.* **2016**, *43*, 531–544. [CrossRef]

41. Iqbal, B.; Ali, J.; Baboota, S. Silymarin loaded nanostructured lipid carrier: From design and dermatokinetic study to mechanistic analysis of epidermal drug deposition enhancement. *J. Mol. Liq.* **2018**, *255*, 513–529. [CrossRef]
42. Karri, V.V.S.R.; Raman, S.K.; Kuppusamy, G.; Sanapalli, B.K.R.; Wadhwani, A.; Patel, V.; Malayandi, R. In vitro Antifungal Activity of a Novel Allylamine Antifungal Nanoemulsion Gel. *J. Nanosci. Curr. Res.* **2018**, *3*, 1–5. [CrossRef]

Publisher's Note: MDPI stays neutral with regard to jurisdictional claims in published maps and institutional affiliations.

© 2020 by the authors. Licensee MDPI, Basel, Switzerland. This article is an open access article distributed under the terms and conditions of the Creative Commons Attribution (CC BY) license (http://creativecommons.org/licenses/by/4.0/).

Article

Three-Step Synthesis of a Redox-Responsive Blend of PEG–*block*–PLA and PLA and Application to the Nanoencapsulation of Retinol

Louise Van Gheluwe [1], Eric Buchy [2], Igor Chourpa [1] and Emilie Munnier [1,*]

1 EA 6295 Nanomédicaments et Nanosondes, Faculté de Pharmacie, Université de Tours, 31 Avenue Monge, 37 200 Tours, France; louise.vangheluwe@univ-tours.fr (L.V.G.); igor.chourpa@univ-tours.fr (I.C.)
2 Laboratoires Eriger, 39 Rue des Granges Galand, 37550 Saint-Avertin, France; eric.buchy@laboratoires-eriger.com
* Correspondence: emilie.munnier@univ-tours.fr

Received: 14 September 2020; Accepted: 12 October 2020; Published: 14 October 2020

Abstract: Smart polymeric nanocarriers have been developed to deliver therapeutic agents directly to the intended site of action, with superior efficacy. Herein, a mixture of poly(lactide) (PLA) and redox-responsive poly(ethylene glycol)–*block*–poly(lactide) (PEG–*block*–PLA) containing a disulfide bond was synthesized in three steps. The nanoprecipitation method was used to prepare an aqueous suspension of polymeric nanocarriers with a hydrodynamic diameter close to 100 nm. Retinol, an anti-aging agent very common in cosmetics, was loaded into these smart nanocarriers as a model to measure their capacity to encapsulate and to protect a lipophilic active molecule. Retinol was encapsulated with a high efficiency with final loading close to 10% *w/w*. The stimuli-responsive behavior of these nanocarriers was demonstrated in vitro, in the presence of L-Glutathione, susceptible to break of disulfide bond. The toxicity was low on human keratinocytes in vitro and was mainly related to the active molecule. Those results show that it is not necessary to use 100% of smart copolymer in a nanosystem to obtain a triggered release of their content.

Keywords: redox responsive PEG-*block*-PLA; nanocarriers; disulfide bond; controlled release; retinol

1. Introduction

Stimuli-responsive nanocarriers (NCs) have been designed to control the release of active molecules into the intended site of action. The contents of the nanoparticle is released only if a biological or a physical stimulus occurs. The stimulus has to be adapted to the application. For example, light-responsive, ultrasound-responsive, temperature-responsive, pH-responsive or redox-responsive nanosystems can be found in the literature [1].

Lately, a lot of efforts have been focused on the development of redox-responsive nanosystems. They could be of major interest in the treatment of cancer, as the reduction potential of cancer cells is higher than the reduction potential of healthy cells [2]. The release is then triggered in the tumor, minimizing the side effects. It is the observed case in certain skin cancers, such as melanoma [3,4]. Moreover, when skin is exposed to UV light, cells react first by activating antioxidant mechanisms: the level of glutathione (GSH) and its accompanying enzymes is dramatically increased to counteract the appearance of reactive oxygen species [5]. The increased GSH activity could favor the nanocarriers' material cleavage and thus enhance the release of hydrophobic drugs [6,7].

Different NCs engineered with redox sensitive mechanisms for drug or gene delivery have been developed already like liposomes, polymer nanoparticles, micelles, or prodrug-based delivery systems [8,9]. The disulfide bond is the most used reduction-sensitive linker to prepare redox-responsive nanocarriers [10]. The disulfide bond can be localized in different places: it can make the link between two copolymers to form a stimulus-sensitive copolymer [11,12], it can make the link between the polymer and the drug to synthesize a redox-responsive conjugate [13], or it can be located in each repeating unit of a polymer [14,15]. In recent decades, significant efforts have been made in the development of stimuli-responsive polymers [16,17] to prepare efficient triggered release. Indeed, compared to conjugation, the synthesis of redox-responsive diblock or triblock copolymers is interesting as numerous active molecules can be encapsulated without developing a specific synthesis path for each one. Under the action of a stimulus, the copolymer disassembles into two or three parts, destabilizing the nanocarriers and leading to the release of the contents. Among the redox-responsive copolymers, the poly(ethylene glycol)–*block*–poly(lactide) (PEG–*block*–PLA) copolymer associated with a disulfide bond seems to still be the preferred iteration [6,7,18]. Indeed, both polymers PLA and PEG are approved by the pharmaceutical and cosmetic regulatory authorities. PLA is a synthetic linear aliphatic polyester that has proven its interest in the encapsulation of hydrophobic drugs [19,20]. Being biodegradable and biocompatible, it is used in various biomedical applications. PEG is a biocompatible polymer known for its hydrophilicity, flexibility, and non-toxicity. PEG confers to PEG–*block*–PLA-based nanocarriers a good stability in aqueous suspensions thanks to its hydrophilicity and sterical hindrance, and most of the time no adjunction of surfactant is necessary [21]. Moreover, PEG seems to play a positive role in the penetration of PLA-based nanosystems into the skin [22].

PEG–*block*–PLA-based nanocarriers have already proven their efficacy in the encapsulation of siRNA, active pharmaceutical or cosmetic ingredients [23–25], among which is retinol. Retinol is a low molecular weight anti-ageing agent very common in cosmetics [26]. It is highly sensitive to water, heat, oxygen and light [27–30]. Many authors work on its encapsulation to protect its integrity and increase its penetration into the skin [31–34]. NCs made of PEG–*block*–PLA copolymer showed very interesting performances for retinol delivery to the skin [34,35]. Retinol loaded PEG–*block*–PLA-based NCs displayed a significantly higher absorption into skin compared to PEG–*block*–PCL NCs, surfactant micelles, or oil solution [34]. As a powerful antioxidant molecule, it could benefit from an encapsulation in a redox-sensitive PEG–*block*–PLA to confer a triggered release in case of skin photooxidation.

Unfortunately, the syntheses of redox-responsive PEG–*block*–PLA described in the literature are multi-stage, with drastic anhydrous conditions, and time-consuming purification steps requiring specific equipment [7,18,36,37]. These constraints limit the potential interest of such delivery systems. Moreover, PEG–*block*–PLA nanocarriers are usually nanomicelles showing low physical stability in suspension and showing a low resistance to dilution, which is not compatible with the introduction into skin dedicated products like creams or gels. The stability can be increased by mixing it with another polymer that will reinforce the hydrophobic core of the nanocarriers, for example PEG–*block*–PLA mixed with PLA. In this case, both polymers are usually synthesized independently [22,38].

The objective of the present study was to prepare redox-responsive nanocarriers made of a blend of PLA and a PEG–*block*–PLA copolymer linked with a disulfide bond. Such nanocarriers should be stable in aqueous suspension but release their contents rapidly in skin undergoing oxidative stress. In order to attain this goal, the steps were (1) to prepare in as few steps as possible a redox-responsive blend of PEG–*block*–PLA and PLA; (2) to formulate retinol nanocarriers via a simple and efficient protocol. For the synthesis of the redox-responsive PEG–*block*–PLA copolymers blended with PLA, a new strategy in only three steps was developed. At the end, a redox-responsive blend of PEG–*block*–PLA and PLA was obtained and used directly, without further purification. It was chosen to work with short PLA chains in order to facilitate the release of retinol, which could be long in long-chain PLA nanocarriers despite the redox-responsive link, because of an increased affinity of retinol for the polymer [39]. To formulate the nanocarriers encapsulating retinol, the nanoprecipitation method was chosen because it is well known, fast and easy to scale up. This method is the simplest preparation method of nanocarriers as it takes place in only one step, with no need of surfactants or chlorinated solvents [40]. It leads to a narrowly dispersed particle size. The impact of GSH levels on redox-responsive release was investigated in vitro. Finally, cytotoxicity of the nanocarriers on human keratinocytes was evaluated to ensure that they could be safely used in dermatological or cosmetic applications.

2. Materials and Methods

2.1. Chemicals

2-mercaptoethanol, 2,2'-dithiobis(5-nitropyridine), anhydrous acetic acid, methanol, O-[2-(3-mercaptopropionylamino) ethyl]-O'-methylpolyethylene glycol (PEG-thiol, 5 kDa), 3,6-dimethyl-1,4-dioxane-2, 5-dione (D,L-lactide), stannous 2-ethylhexanoate, dichloromethane, anhydrous toluene, tetrahydrofuran (THF), Nile red, retinol and 3-(4,5-dimethylthiazol-2-yl)-2,5-diphenyltetrazolium bromide (MTT) were obtained from Sigma-Aldrich (St Quentin-Fallavier, France). Dialysis membrane (molecular weight cut-off (MWCO) 2 kDa, regenerated cellulose) was purchased from BioValley (Marne La Vallée, France). L-Glutathione reduced (GSH), Dulbecco's modified Eagle medium (DMEM), foetal bovine serum (SVF), dimethyl sulfoxide (DMSO) and penicillin/streptomycin solution were provided by Fisher Scientific (Illkirch, France). Ultrapure water was produced using a Milli-Q system, Millipore (Paris, France).

2.2. Synthesis and Characterization of Redox-Responsive (PEG–block–PLA)-Blend-(PLA)

2.2.1. Step 1: Synthesis of Ethanol, 2-[(5-Nitro-2-pyridinyl)dithio]-(Abbreviated as Compound 1)

In a three-necked flask equipped with a stir bar, under a nitrogen atmosphere and at room temperature, 1.2 g of 2,2'-dithiobis(5-nitropyridine) (310.31 g/mol, 3.87 mmol) were solubilized in 18 mL of dichloromethane. A solution of 2-mercaptoethanol was prepared in 1.5 mL of methanol (275 µL, 78.13 g/mol, 3.87 mmol) and added to the mixture. Finally, 25 µL of anhydrous acetic acid was added. The solution turned yellow and was stirred for 24 h. The solvent was then removed under vacuum and flash chromatography (silica: 60, 0.04–0.063 mm) was carried out using 30% ethyl acetate and 70% cyclohexane to give the compound 1 (232.28 g/mol, 468.5 mg, 2.02 mmol, 52.2% yield) as a yellow solid. ^{1}H NMR (300 MHz, CDCl$_3$, ppm): δ 9.34 (d, J = 2.1 Hz, 1H), 8.38 (dd, J = 8.8, 2.6 Hz, 1H), 7.69 (d, J = 8.8 Hz, 1H), 3.80 (t, 2H), 3.25 (s, 1H), 3.02 (t, 2H).

2.2.2. Step 2: Synthesis of Disulfide PEG (Abbreviated as P1)

In a three-necked flask equipped with a stirring bar, under a nitrogen atmosphere and at room temperature, 500 mg of PEG-thiol (5 kDa, 0.1 mmol), 71.3 mg of compound 1 (232.28 g/mol, 0.3 mmol) and 15 µL of anhydrous acetic acid were solubilized in 23 mL of methanol. The solution turned yellow and was stirred for at least 30 h. The progress of the reaction was followed with a UV-Visible spectrophotometer (Genesys 10S, Thermo Scientific, France) with the release of thionitropyridine in the reaction medium (385 nm). The solvent was then removed under vacuum and the residue was dissolved in distilled water and dialysis was performed against distilled water (regenerated cellulose, MWCO = 2 kDa). Finally, the product was freeze-dried for two days to obtain a white powder stocked at −20 °C. ^{1}H NMR (300 MHz, CDCl$_3$, ppm): δ 3.87 (t, 2H, J = 4.9 Hz), 3.64 (m, 4H), 3.46 (t, 2H), 3.40 (t, 2H, J = 4.9 Hz), 3.37 (s, 3H), 3.00 (t, 2H, J = 7.0 Hz), 2.87 (t, 2H, J = 5.7 Hz), 2.64 (t, 2H, J = 7.0 Hz).

2.2.3. Step 3: Synthesis of Redox-Responsive Copolymer (Abbreviated as P2)

Appropriate amounts of pure D,L-lactide and **P1** were placed in a two-neck flask with anhydrous toluene (4.5 mL) and the mixture was degassed by bubbling nitrogen for 20 min. Under a nitrogen atmosphere, the stannous 2-ethyl-hexanoate (10 mg) was added and the reaction mixture was refluxed in an oil bath (120 °C) for 1 h. During the reaction period, the condenser was cold enough to prevent the volatilization of toluene. The toluene was then removed under reduced pressure and the residue was solubilized in a minimum of tetrahydrofuran (THF) to perform purification by size exclusion chromatography (Sephadex LH-20, THF). The purification of the polymers **P2** produced from the residual lactide monomer was followed by infrared spectroscopy (Bruker Vector 22, ATR mode). After the collection and concentration of polymers under vacuum, it is solubilized in a minimum of THF and precipitated in cold water. After THF removal, freeze-drying was performed for 3 days to obtain a white powder stocked at −20 °C. ^{1}H NMR (300 MHz, CDCl$_3$, ppm): δ 5.18 (m, 1H), 4.38 (t, 2H, J = 6.6 Hz), 4.35 (q, 1H, J = 7.0 Hz), 3.87 (t, 2H, J = 4.9 Hz), 3.64 (m, 4H), 3.40 (t, 2H, J = 4.9 Hz), 3.37 (s, 3H), 2.97 (t, 2H, J = 7.1 Hz), 2.89 (t, 2H, J = 6.7 Hz), 2.58 (t, 2H, J = 7.2 Hz), 1.57 (m, 3H).

2.2.4. Characterizations

For each step, ^{1}H NMR spectra were recorded with a Bruker 300 MHz NMR spectrometer at 25 °C with CDCl$_3$ as solvent. MALDI-TOF mass spectra were acquired on a 4700 Proteomics Analyzer (Applied Biosystems). Dithranol (DT; 1,8-dihydroxy-9,10-dihydroanthracen-9-one) was used as the matrix for the ionization. Laser (YAG, 355 nm) power was adapted to obtain a significant signal-to-noise ratio and a good resolution of the mass peaks.

2.3. Preparation of Polymer Nanocarriers

Polymer nanocarriers were prepared by the nanoprecipitation method, in triplicate. Briefly, 0.6 mL of a solution of the redox-responsive blends of polymers in acetone (10 mg/mL) was added dropwise to 2 mL deionized water using a syringe pump (KR Analytical, UK) with a flow rate of 0.3 mL/min and under moderate stirring. The nanocarriers were formed instantly, and the organic solvent was evaporated overnight under stirring at room temperature. Blank nanoparticles labelled with Nile red as well as retinol-loaded nanocarriers were prepared by the same method after dissolving Nile red (13 µg/mL) or retinol (1.2 mg/mL), respectively, in the polymer acetone solution. The resulting suspensions were filtered through a 0.45 µm polyether sulfone membrane filter to remove unloaded Nile red or retinol. All the procedures were performed under dark conditions to preserve the integrity of the molecules.

2.4. Characterization of Copolymer-Based Nanocarriers

The morphological examination of nanocarriers was conducted by transmission electron microscopy (TEM). The nanocarriers' aqueous suspension was deposited onto a carbon coated grid and was negatively stained using uranyl acetate (3% w/v). Excess water was removed by filter paper and the sample was scanned in a JEOL 1011 TEM (Peabody, MA, USA).

The mean hydrodynamic diameter (D_H) and the polydispersity index (PdI) of the nanocarriers in aqueous suspensions were measured by the dynamic light scattering (DLS) technique, using a Nanosizer apparatus (Zetasizer®, Malvern Panalytical, UK) equipped with a He-Ne laser (633 nm, scatter angle 173°). Herein, the DLS method measures nanoparticle diameter by intensity. The same instrument was also used for zeta potential measurements, but the samples were put in dedicated micro-electrophoresis cells. Each sample was diluted 1:10 with deionized water and the measurement was performed three times at a temperature of 25 °C.

2.5. Redox-Responsivity of Copolymer-Based Nanocarriers

The redox-mediated response to reduced glutathione (GSH) was evaluated by in vitro tests, using blank nanocarriers labelled with environment-sensitive fluorescent dye Nile red (NR-NCs) or retinol-loaded nanocarriers (R-NCs). Typically, the nanocarriers' suspension was diluted 1:10 with deionized water. Then, 0.5 mL of diluted suspension was mixed with 0.5 mL of GSH solution (0.04 or 20 mM) or 0.5 mL of diluted nitric acid solution to ensure the same pH. Each tube was shaken at 37 °C and Nile red (NR) or retinol (R) fluorescence was measured at predetermined time intervals (0 h, 2 h, 4 h, 8 h, and 24 h) using a spectrofluorometer (A Hitachi F-4500). The fluorescence excitation wavelength was fixed to 535 nm (NR) or 327 nm (R), and the emission fluorescence to 555–700 nm (NR) or 350–650 (R); the range was monitored. Emission and excitation slits were adjusted as function of fluorescence intensity of the solution at t_0. To study the NR/R release as a function of time after GSH treatment, the following formula was used:

$$\% \text{ of content released at } t = \frac{\text{Fluorescence intensity } t_0 - \text{Fluorescence intensity } t}{\text{Fluorescence intensity } t_0} \times 100 \quad (1)$$

The encapsulation efficiency of retinol was measured using its absorbance (A) at 327 nm, after dissolution of the NCs in THF. Appropriate dilution of retinol-loaded nanocarriers' suspension was performed in THF and the solution was vortexed for 1 min to break down the particles and dissolve retinol. The absorbance of the resulting solution was measured in a 1 cm quartz cell placed in a Genesys 10 S (Thermo Scientific, France) UV-Vis spectrophotometer. From the absorbance values, retinol concentration was determined according to a standard calibration curve within the linear range. The encapsulation efficiency of retinol was calculated according to the following equation:

$$\text{Encapsulation Efficiency } (\%) = \frac{\text{Amount of retinol in the nanocarriers}}{\text{Feeding amount of retinol}} \times 100 \quad (2)$$

The theoretical loading was calculated by Equation (3):

$$\text{Molecule load } (\%) = \frac{\text{Determined weight of retinol}}{\text{Initial weight of copolymer}} \times 100 \quad (3)$$

Aqueous suspensions are divided into two samples and transferred into glass vials immediately after preparation and stored at 4 °C or room temperature. The chemical stability was monitored up to 1 month by drug content determination. Periodically, samples were taken and diluted with THF and analyzed with the UV-Vis spectrophotometer for residual retinol.

2.6. In Vitro Cytotoxicity Studies of Retinol-Loaded Nanocarriers

In vitro cytotoxicity of blank and retinol-loaded nanocarriers was evaluated by the 3-(4,5-dimethylthiazol-2-yl)-2,5-diphenyl-tetrazolium bromide (MTT) assay on the HaCaT cell line (human keratinocytes, a kind gift from Pr. G. Weber, U1069 N2C, Tours) [41]. Cells were cultured in DMEM completed with 10% of SVF and 1% of a penicillin–streptomycin mixture. For the cytotoxicity study, cells were seeded in 96-well plates (10,000 cells/well) and incubated 24 h at 37 °C in a humidified atmosphere with 5% CO_2. Cells were then treated with increasing concentrations of blank or retinol-loaded nanocarriers (1.64×10^{-7} to 1.64×10^{-2} mg/mL in retinol) during 24 h at 37 °C. MTT solution was added to each well (0.5 mg/mL) and the cells were incubated for further 4 h at 37 °C. Then, the supernatant was discarded and DMSO was added in each well to dissolve the formazan crystals. The absorbance A at 540 nm was measured with a microplate reader (Bio-tek EL800 plate reader). Results were expressed as viability percentages as function of the retinol concentration, and compared with Student's t-test.

3. Results and Discussion

3.1. Synthesis and Characterization of Redox-Responsive PEG–block–PLA

The redox-responsive PEG–block–PLA was synthesized in three steps, as described in Figure 1.

Unable to avoid the presence of traces of water after purification of **P1**, two compounds are expected at the end of step 3: redox-responsive PEG–block–PLA (**P2a**) and PLA (**P2b**).

The first step consists of producing an intermediate product, called compound 1, with a disulfide linkage. Its composition and structure were confirmed by ^{1}H NMR as shown in Figure 2.

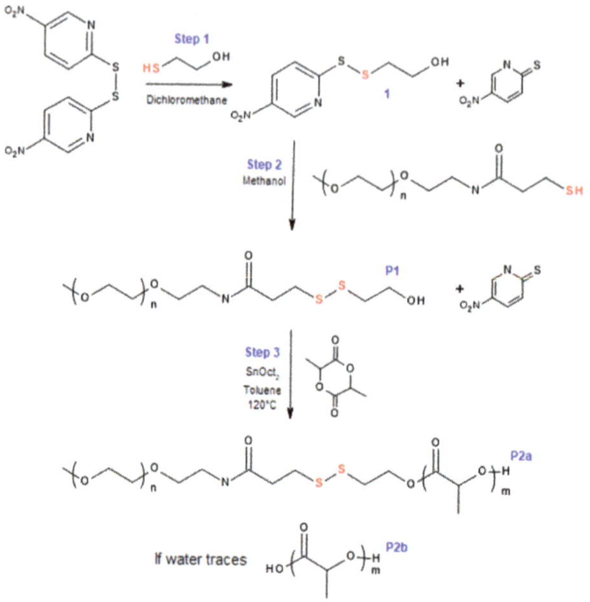

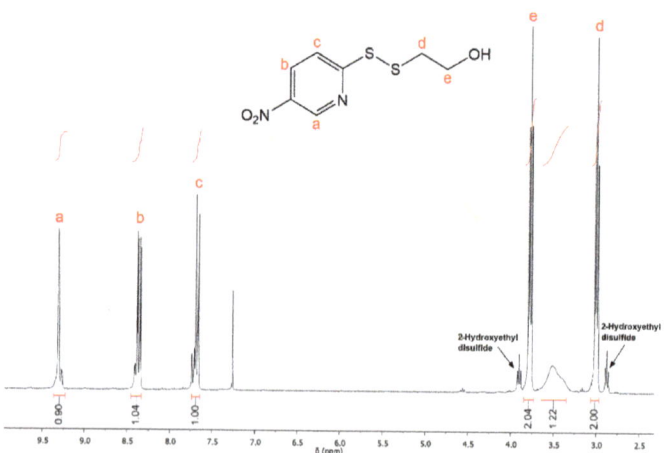

Figure 1. Synthesis of the amphiphilic redox-responsive poly(ethylene glycol)–*block*–poly(lactide) (PEG–*block*–PLA) in three steps.

Figure 2. ^{1}H NMR spectrum of compound 1 (step 1, CDCl$_3$, 25 °C).

The solvent peak of CDCl$_3$ was found at 7.26 ppm. Peaks at 2.87 and 3.90 ppm result from the oxidation of 2-mercaptoethanol to 2-Hydroethyldisulfide. Peaks at 3.8 and 3.02 ppm were assignable to the methylene (–CH$_2$–) protons of –CH$_2$–OH (e) and –S–CH$_2$– (d), respectively.

The second step consists of thiol–disulfide exchange between compound 1 and PEG-thiol in order to obtain the polymer **P1**. The reaction yield was established by following the release of thionitropyridine from compound 1 in the reaction medium with a UV-Vis spectrophotometer (λ_{max} = 385 nm). After approximately 30 h of reaction, the yield was close to 100%. At this stage, the elimination of the remaining quantity of compound 1 is crucial because the next step is the polymerization of PLA, possible from a hydroxyl group. To purify the hydrophilic polymer **P1**, a common laboratory technique was used, the dialysis method. This method is vulnerable to the emergence of traces of water in the reaction medium that would initiate the synthesis of PLA. The composition and the structure of the purified **P1** product were investigated with NMR and MALDI-TOF mass spectrometry. Figure 3 shows ^{1}H NMR spectrum of the modified PEG produced.

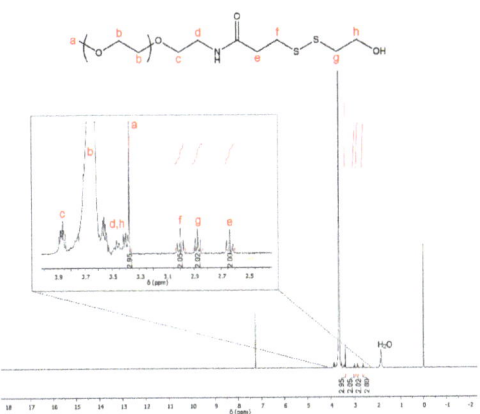

Figure 3. ^{1}H NMR spectrum of the modified polymer **P1** (step 2, CDCl$_3$, 25 °C).

In NMR investigation, the signal at 3.68 ppm was assigned to the methylene protons of PEG units (–O–CH$_2$–, b). The signal at 3.37 ppm can be assigned to the three chemically equivalent hydrogen atoms of the terminal methoxy group (–OCH$_3$, a) of the mPEG block. The triplet peaks at 3.87, 3.40, 3.00 and 2.64 ppm were attributed to the methylene protons (–CH$_2$–) of the PEG$_{5K}$-SH (c, d, f and e, respectively). Finally, peaks at 3.46 and 2.87 ppm were assignable to the insertion of the methylene (–CH$_2$–) protons of compound 1: –CH$_2$–OH (h) and –S–CH$_2$– (g), respectively. As expected, water traces are observed.

Mass spectrometry was used as a complementary method to conclude if the disulfide linkage was intact. Figure 4 shows the MALDI-TOF mass spectrum of the purified product of step 2.

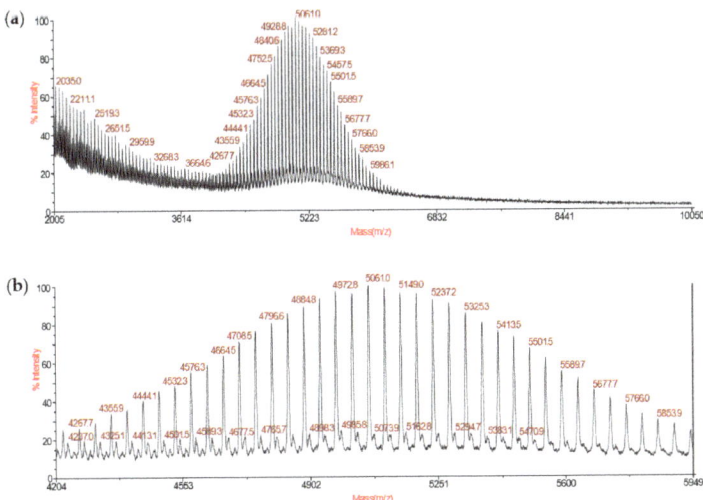

Figure 4. (a) Full MALDI-TOF mass spectrum from mass/charge (m/z) 2000 to 10,050 of **P1**. (b) Magnification of the mass range region from m/z 4204 to 5949.

The MALDI-TOF mass spectrum was well-resolved and the peaks were separated by 44 mass units, which corresponds to the molecular weight of the ethylene oxide (EO) unit of PEG. Two populations were observed in the mass range region from m/z 4204 to 5949 (Figure 4b). The major population corresponds to the modified PEG **P1** and the minor population corresponds to the unmodified mPEG-thiol, which has not reacted with compound 1. Thanks to mass spectrometry analysis, the integrity of the disulfide linkage of **P1** was confirmed.

The last step consists in producing the copolymer PEG–block–PLA (**P2a**) by ring-opening polymerization (ROP). The hydroxyl functional group at the end of the modified PEG **P1** allows the ROP of D,L-lactide using stannous octoate (SnOct$_2$) as the catalyst [6,7,42]. Even after freeze-drying of the modified PEG, traces of water persist and are susceptible to participate in the polymerization of PLA (**P2b**). The copolymer composition was confirmed by ^{1}H NMR as shown in Figure 5. The peaks at 1.57 and 5.18 ppm belong to methyl (–CH$_3$, j) and methine (–CH, i) protons of the lactate unit of PLA, respectively. The signal at 3.68 ppm (b) was assigned to the methylene protons of PEG units. The methylene protons signal (–CH$_2$–, h) of **P1**, linked to the hydroxyl group, shifts from 3.46 to 4.4 ppm after polymerization. This signal is superimposed with that corresponding to the PLA end-group methide proton (–CH, k). If the final product corresponded to only **P2a** copolymer, the integration at 4.4 ppm was then supposed to be three (=3H). Herein, the integration was 4.86 corresponding to 2H of methylene protons (–CH$_2$–, h) and 2.86H associated to PLA end-group methide protons. The assumption is as follows: among the 2.86 protons, 1H corresponds to the PLA end-group methide proton of **P2a** and 1.86H corresponds to the PLA end-group methide protons of PLA polymerized with water traces (**P2b**). The NMR investigation could be interpreted as follows: we have a blend of copolymer **P2a** and polymer **P2b** with an approximate ratio of 1:2. According to the initial amount of polymer **P1** involved in the reaction (~200 mg), the amount of water traces was evaluated as ~8.3 × 10^{-5} moles.

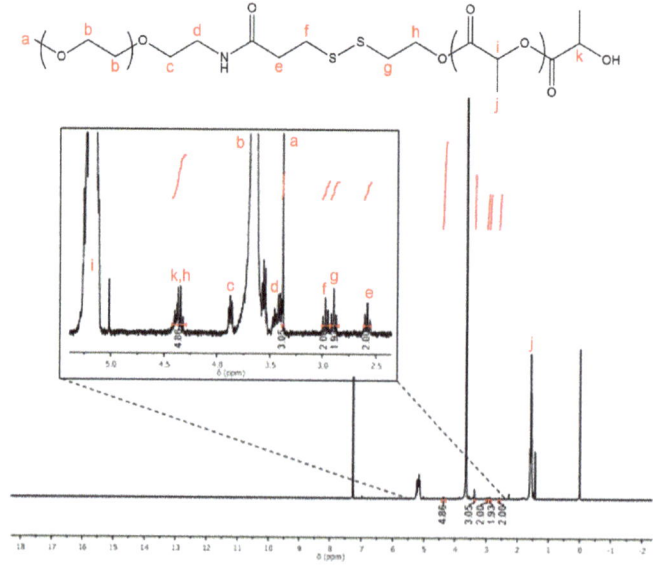

Figure 5. ^{1}H NMR spectrum of P2 copolymer (step 3, CDCl$_3$, 25 °C).

Usually, integrations of PEG and PLA protons are used to determine the molecular weight of the copolymer. Being in the presence of a blend, a MALDI-TOF mass spectrum was recorded as a complementary method to properly study the composition of the final product (Figure 6). The population below 5.5 kDa corresponds to the PLA polymer with an increment of 144 m/z corresponding to the lactate unit of PLA. This population is associated to the PLA polymerized from water traces with a molecular weight $Mn_{PLA} = 3637$ Da (PLA_{4K}). The population observed above 5.5 kDa corresponds to the copolymer PEG–block–PLA **P2a** with a molecular weight close to 8 kDa corresponding to a PEG of approximately 5 kDa and a PLA of approximately 3 kDa.

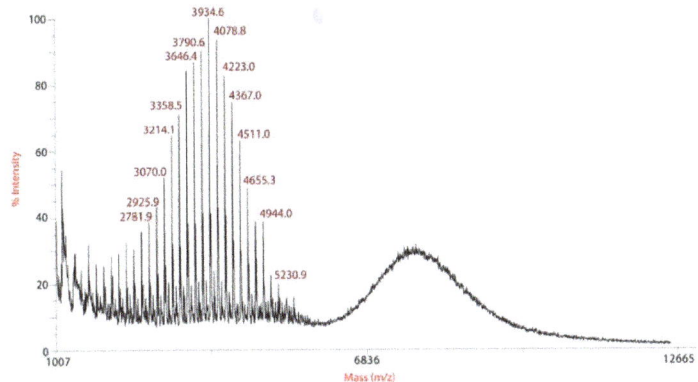

Figure 6. MALDI-TOF mass spectrum of purified product of step 3 (m/z 1007 to 12,665).

To conclude, thanks to NMR and mass investigations, the blend of polymers was characterized, and it is composed of the copolymer PEG–block–PLA (**P2a**) and the polymer PLA (**P2b**) with an approximate ratio 1:2. No further purification was performed to explore directly the interest of the mixture for nanocarrier development, minimizing steps for industrial manufacturing. This polymer will be referred to (PEG–block–PLA)-blend-(PLA) in the rest of the document.

3.2. Nanocarriers Characteristics

The nanoprecipitation method was used to produce (PEG–block–PLA)-blend-(PLA) nanocarriers [43]. This method allows a spontaneous particle formation with no surfactant and with low energy consumption [40,44]. The miscibility of PLA and PEG, added to the fact that both polymers possess a PLA chain, should ensure the coprecipitation of the polymer in mixed nanospheres [45]. The nanocarriers prepared from (PEG–block–PLA)-blend-(PLA) were stable in aqueous suspension without the addition of surfactant. This stability is in favor of a core-shell structure where the PLA blocks make up the inner core while the majority of the hydrophilic PEG blocks make up the outer hydrophilic shell [22,40]. Nevertheless, the nanoprecipitation method, compared to the emulsion/solvent evaporation or emulsion/solvent diffusion used to prepare nanosystems, may lead to less organized systems. Blank nanoparticles were labeled with Nile red (~0.1% w/w) in order to be able to follow their behavior in a reducing medium.

As can be seen in TEM images (Figure 7), Nile red labelled and retinol-loaded nanocarriers (NR-NCs and R-NCs, respectively) have similar spherical morphologies and sizes. The particle size distribution shows only one population of nanocarriers with narrow distribution, suggesting a monodispersed population of particles. This monodispersity is in line with the hypothesis that both polymers coprecipitate to form composite nanospheres and not two separate populations: one made of redox sensitive PEG-block-PLA and one made of PLA.

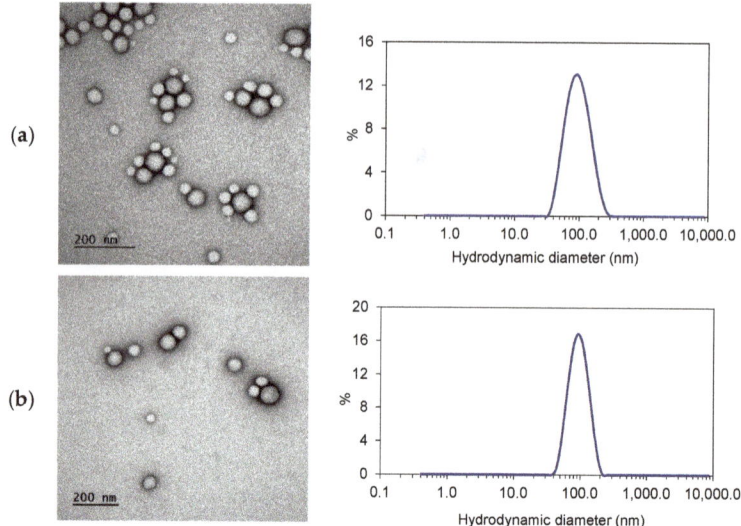

Figure 7. Typical TEM images and representative particle size distribution plots of (**a**) Nile red-labelled and (**b**) retinol-loaded (PEG–*block*–PLA)-blend-(PLA) nanocarriers.

The physicochemical characteristics of the labelled and retinol-loaded nanocarriers are compiled in Table 1. Depending on the copolymer composition, method and experimental conditions of formulation, the size of the PEG–*block*–PLA nanocarriers described in the literature may vary from 50 to 300 nm [46]. In this study, the hydrodynamic diameter of blank nanocarriers labelled with Nile red or loaded with retinol was close to 100 nm, which can be considered suitable for skin administration [47]. As expected, D_H values obtained from DLS measurements were higher than NP diameters observed in TEM images: the former includes hydrated layers of PEG shell, while the latter may not reveal PEG, because of its low electronic contrast.

Table 1. Characteristics of nanocarriers (mean values ± SD, n = 3).

	Characteristics	Nile Red-Labelled Nanocarriers	Retinol-Loaded Nanocarriers
Day 1	Hydrodynamic diameter (nm)	100.5 ± 2.7	98.7 ± 2.2
	Polydispersity Index	0.126 ± 0.013	0.101 ± 0.011
	Zeta potential (mV)	−25.7 ± 2.3	−21.6 ± 1.6
	Retinol (mg/mL)	-	0.28 ± 0.04
1 month 4 °C	Hydrodynamic diameter (nm)	98.0 ± 2.4	94.1 ± 1.6
	Polydispersity Index	0.115 ± 0.017	0.093 ± 0.010
	Zeta potential (mV)	−27.8 ± 3.1	−20.7 ± 3.6
	Retinol (mg/mL)	-	0.12 ± 0.01

NCs' aqueous suspensions exhibit low PdI values (<0.2) indicative of populations of rather narrow size distribution. They show high negative zeta potentials (−22 to −28 mV), similar to those of PEG–*block*–PLA-based nanosystems already described [48,49]. These surface charges suggest that electrostatic repulsion of NCs should significantly favor their colloidal stability and reinforce PEG steric hindrance.

As the particle size is linked to the size of the droplets generated during the nanoprecipitation, it is not surprising that blank and retinol-loaded NCs show similar sizes. Nevertheless, their internal structure is in all likelihood different. Indeed, R-NCs show a high loading of ~9.2% *w/w* of retinol. The introduction of nearly 10% of hydrophobic chains could lead to a different stability or release kinetics of blank and loaded NCs. The chosen length of the PLA chains, deliberately short, led to an effective retinol encapsulation. The encapsulation efficiency (EE) of retinol was of 76.9 ± 11.6% *w/w*, which is satisfactory compared to the literature data on retinol nanocarriers such as (PEG–*block*–PLA)-based NCs (PLA of ≈15 kDa, EE = 100% [34]), chitosan nanocarriers (EE = 76%, [32]), silicone and silica particles (EE = 85% [31] and 31% [33], respectively), and solid lipid nanocarriers (EE = 74% or up to 97% depending on formulation conditions [50,51]). The nanoprecipitation method and experimental conditions described in this study allow a high encapsulation efficiency of retinol into (PEG–*block*–PLA)-blend-(PLA) nanocarriers, with a final retinol concentration in water of 28% *w/v* (Table 1).

The size, PdI and zeta potential of the nanocarriers do not vary significantly over one month, showing the physical stability of such a suspension of nanocarriers. Since retinol is known to be chemically unstable, the stability of retinol is one of the most important factors during storage [27–29]. Retinol-loaded nanocarrier suspensions were stored in darkness at room temperature (RT) or 4 °C for 1 month. The chemical stability of retinol is reported in Figure 8.

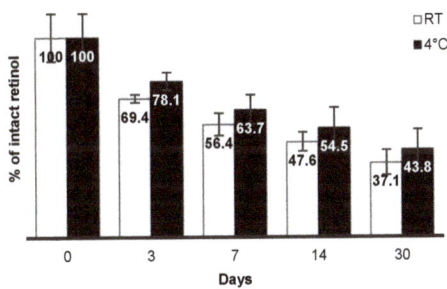

Figure 8. Chemical stability at room temperature (white) and 4 °C (black) of retinol incorporated into nanocarriers.

Stabilizers are usually required to improve the long-term stability of retinol, such as butyl hydroxy toluene (BHT), an antioxidant which protects retinol from chemical degradation [52]. Herein, no stabilizer was added and, after one month in darkness at room temperature or 4 °C, there remained ~40% of non degraded retinol. These results are encouraging compared to the literature. Indeed, several formulations of retinol have been developed with variable chemical stabilities specific to nanocarriers and development conditions. For example, retinol was encapsulated with BHT in silicone particles for topical delivery and after 14 days in darkness at 45 °C, there remained ~25% of non-degraded retinol [31]. Retinol was also encapsulated in silica particles where the amount of preserved retinol depended on the concentration of surfactants and PEG polymer. With 13 and 1.3% *w/w* of Span®80 and Tween®80, respectively, there remained ~85% of intact retinol after 6 days [33]. Solid lipid nanocarriers (SLNs) showed almost 100% of residual retinol after 4 weeks, at room temperature (shade) or 4 °C [50]. In another study, 43% of retinol remained intact in SLNs after 12 h at room temperature (shade). However, the instability of retinol could be overcome by co-loading of antioxidants such as BHT in SLNs [51]. Furthermore, the protection of retinol in R-NCs could be enhanced by the inclusion of stabilizers like BHT or retinol-palmitate in the organic phase before nanoprecipitation.

3.3. Nanocarriers Behaviour in a Reductive Medium

Glutathione (GSH) is a physiological detoxification molecule showing a highly reactive thiol function. It has been reported that, in the cytosol and nuclei, the concentration of GSH can reach 10 mM, while outside the cell the concentration is about 2–20 μM [53–55]. The insertion of the disulfide linkage into the copolymer-based nanocarriers is supposed to accelerate the release of the cargo in a GSH-rich environment [6]. Nile red-labelled blank NCs and retinol loaded NCs were used to investigate their redox response in vitro. Fluorescence spectra (λ_{ex} = 535 nm for NR and 327 nm for retinol) were recorded from diluted aqueous suspensions, exposed or not to GSH treatment, at different incubation times, at 37 °C. NR fluorescence is strongly influenced by its molecular environment. If NR is released from the hydrophobic environment of nanocarriers in the surrounding hydrophilic environment, its fluorescence yield dramatically decreases [56]. In addition, when retinol is released from the hydrophobic environment of nanocarriers, it is very quickly degraded and loses its fluorescence properties in the measurement conditions described in the Materials and Methods section. Indeed, the UV-visible spectrum of free retinol displays a band shift from 327 nm to 282 nm after a few minutes in water, revealing the disappearance of highly conjugated structures (See Figure S1 in Supplementary Materials). This phenomenon is not disrupted by the presence of GSH. It allowed us to track, by fluorescence, the relative amount of retinol still encapsulated. The NR or R fraction released from (PEG–*block*–PLA)-blend-(PLA) was estimated as described in the Experimental section and release kinetics curves were established (Figure 9). These curves show a spontaneous release of encapsulated molecules from the particles with a basal concentration of GSH, but this release is accelerated by the presence of high a concentration of GSH. During the first 2 h of incubation, the NR or R release was similar to the GSH basal level or with GSH activation level. The release seems thus independent of the response to a stimulus over the first 2 h. It is supposed that fluorescent molecules present near the surface of the nanocarriers might be released at early stages. Nevertheless, this release is progressive and cannot be qualified as burst release. From 4 h of incubation, the cumulative release of NR or retinol was higher in GSH-enriched environment compared to the control. This difference becomes even more significant after 24 h: 89 ± 3% vs. 36 ± 2% for Nile red and 91 ± 2% vs. 63 ± 0.1% for retinol. Retinol is released faster than Nile red, which could be explained by the difference in initial content and in NCs' internal structures. Indeed, retinol being more widely present in nanocarriers, their polymeric structure may be degraded more rapidly or the structure more easily destabilized than blank nanoparticles comprised of 99% polymer.

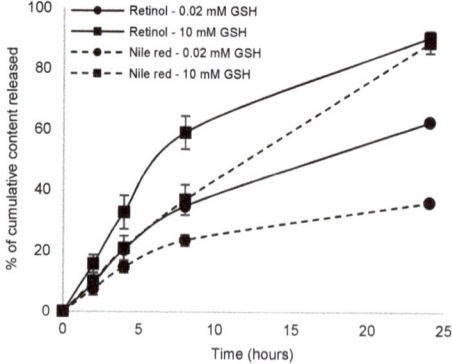

Figure 9. In vitro release of Nile red (dashes) and retinol (full line) from (PEG–*block*–PLA)-blend-(PLA) nanocarriers at 37 °C with glutathione (GSH basal level (0.02 mM) and GSH activation level (10 mM).

The release profile of retinol from (PEG-*block*-PLA)-based nanocarriers has not yet been reported in the literature, certainly because of retinol instability. However, (PEG–*block*–PLA)-based nanocarriers have been described for the protection and release of all-*trans* retinoic acid (atRA, log P ~ 5) [48,49]. The atRA cumulative release depends on hydrophobic and hydrophilic chain lengths. In Tiwari et al.'s study, atRA-loaded (PEG–*block*–PLA)-based nanocarriers produced by the nanoprecipitation method reached a size close to 150 nm. The atRA cumulative release reached ~7.5% after 24 h [49]. In fact, the strategy described in this paper could be applied to retinoids in general to achieve an accelerated release. Even if those studies are not directly comparable, the GSH-enhanced release of hydrophobic content is consistent with the supposed mechanism of the NC degradation via disulfide bonds cleavage [57,58]. The more plausible explanation according to the literature is that the PEG hydrophilic chains are then detached from the hydrophobic PLA core causing the destabilization of the NCs leading to the higher release of the contents [6]. This destabilization is even more efficient if the structure of the nanocarriers includes residual PEG chains in the core of the NCs. TEM images of R-NCs degraded after GSH treatment are shown as supplementary information (Figure S2 in Supplementary Materials) and are consistent with this hypothesis. Those results show that a ratio 1:2 between (PEG–block–PLA) and PLA permits to prepare stimuli-responsive nanosystems.

3.4. In Vitro Cytotoxicity Studies of Retinol-Loaded Nanocarriers

Cytotoxicity of (PEG–block–PLA)-blend-(PLA) nanocarriers towards skin cells is a key index for their topical use. The results on cell viability of human keratinocytes (HaCaT cells) exposed to increasing concentrations of blank and R-NCs (see Figure 10).

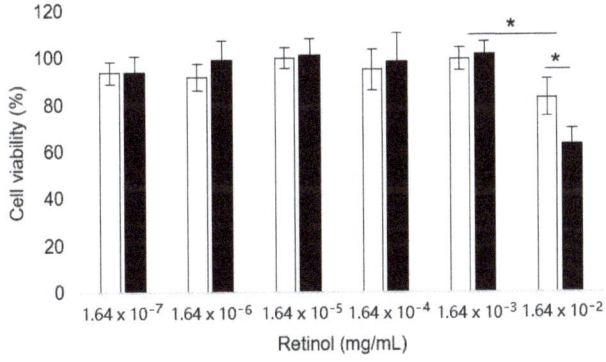

Figure 10. Cell viability of human keratinocytes (HaCaT) exposed to increasing concentrations of blank nanocarriers (open bars) and retinol-loaded nanocarriers (filled bars). Mean ± SD, $n = 6$. * $p < 0.05$.

According to the literature, blank and retinol-loaded nanocarriers suspensions can be considered non-toxic to human keratinocytes up to a corresponding retinol concentration of 1.64×10^{-3} mg/mL (viability > 70%) [59]. At higher retinol concentrations (1.64×10^{-2} mg/mL), human keratinocyte viability decreased to 83% for blank nanocarriers and 63% for retinol-loaded nanocarriers. It is not surprising because retinol is often described as toxic depending on the concentration, generating skin irritation [31].

To conclude, retinol-loaded (PEG–block–PLA)-blend-(PLA) nanocarriers show a satisfactory biocompatibility for their skin application. Nonetheless, the toxicity can be increased by the encapsulation of active molecules with intrinsic toxicities.

4. Conclusions

Thanks to the simplified synthesis described in this study, it was possible to prepare a redox-responsive blend of PEG–block–PLA and PLA with a 1:2 ratio. Smart nanocarriers were formulated by the nanoprecipitation method. Retinol, a hydrophobic and unstable anti-ageing agent, was used as a model active molecule. The strategy to deliberately choose short-length polymers to encapsulate retinol permitted the reaching of high loading and rapid release in a reducing medium in vitro. Nevertheless, the co-encapsulation of stabilizers could increase the chemical stability of the system. Blank NCs showed a good biocompatibility with respect to human keratinocytes, but (PEG–block–PLA)-blend-(PLA) NCs can become toxic as function of the concentration of encapsulated molecule, depending on the toxicity of the molecule itself. These nanocarriers could be used to deliver other retinoids for treatment of skin diseases like acne, photoageing, psoriasis vulgaris, melisma or skin cancers [60,61]. In a more general manner, such redox-responsive nanocarriers could have a high interest for biologically controlled delivery of hydrophobic active molecules in dermatological or cosmetic applications. This study shows that it is not necessary to work in drastic anhydrous conditions and to purify the redox-responsive polymer to obtain a triggered release of disulfide-linked (PEG–*block*–PLA)-blend-(PLA)-based nanocarriers. This simplified copolymer synthesis coupled to nanoprecipitation offers perspective to the industrialization of stimuli-responsive nanosystems.

Supplementary Materials: The following are available online at http://www.mdpi.com/2073-4360/12/10/2350/s1, Figure S1: Absorbance spectra of retinol-loaded copolymer-based nanoparticles (black); retinol in THF (violet); retinol in distilled water (blue, full line) and retinol in distilled water containing glutathione, 10 mM (blue, dashes); Figure S2. Typical TEM images of retinol-loaded nanocarriers incubated 24 h at 37 °C (a) without and (b) with GSH treatment (10 mM).

Author Contributions: Conceptualization and methodology, E.B. and E.M.; Supervision and Funding acquisition, E.M.; Investigation and formal analysis, L.V.G.; Writing—original draft, L.V.G.; Writing—review and editing, I.C., E.B. and E.M.; project administration, E.M. All authors have read and agreed to the published version of the manuscript.

Funding: This work is part of the project MISTIC 2017-00118114 (Cosmétosciences program). This research received financial support from Conseil Régional Centre Val-de-Loire and European Regional Development Fund (FEDER, N°EX003257).

Acknowledgments: Authors are grateful to Julien Burlaud-Gaillard (the IBiSA Electron Microscopy Facility, University of Tours) for TEM images and to Emmanuelle Sachon and Gilles Clodic (IBPS platform, UMR7203, Sorbonne University) for MALDI-TOF mass spectrometry measurement.

Conflicts of Interest: The authors declare no conflict of interest.

References

1. Joglekar, M.; Trewyn, B.G. Polymer-based stimuli-responsive nanosystems for biomedical applications. *Biotechnol. J.* **2013**, *8*, 931–945. [CrossRef] [PubMed]
2. Yang, M.; Ding, H.; Zhu, Y.; Ge, Y.; Li, L. Co-delivery of paclitaxel and doxorubicin using mixed micelles based on the redox sensitive prodrugs. *Colloids Surf. B Biointerfaces* **2019**, *175*, 126–135. [CrossRef] [PubMed]
3. Balendiran, G.K.; Dabur, R.; Fraser, D. The role of glutathione in cancer. *Cell Biochem. Funct.* **2004**, *22*, 343–352. [CrossRef] [PubMed]
4. Révész, L.; Edgren, M.R.; Wainson, A.A. Selective toxicity of buthionine sulfoximine (BSO) to melanoma cells in vitro and in vivo. *Int. J. Radiat. Oncol. Biol. Phys.* **1994**, *29*, 403–406. [CrossRef]
5. Traverso, N.; Ricciarelli, R.; Nitti, M.; Marengo, B.; Furfaro, A.L.; Pronzato, M.A.; Marinari, U.M.; Domenicotti, C. Role of Glutathione in Cancer Progression and Chemoresistance. *Oxidative Med. Cell. Longev.* **2013**, *2013*, 972913. [CrossRef]
6. Fuoco, T.; Pappalardo, D.; Finne-Wistrand, A. Redox-Responsive Disulfide Cross-Linked PLA–PEG Nanoparticles. *Macromolecules* **2017**, *50*, 7052–7061. [CrossRef]
7. Song, N.; Liu, W.; Tu, Q.; Liu, R.; Zhang, Y.; Wang, J. Preparation and in vitro properties of redox-responsive polymeric nanoparticles for paclitaxel delivery. *Colloids Surf. B Biointerfaces* **2011**, *87*, 454–463. [CrossRef]
8. Wen, H. Redox Sensitive Nanoparticles with Disulfide Bond Linked Sheddable Shell for Intracellular Drug Delivery. *Med. Chem.* **2014**, *4*, 748–755. [CrossRef]
9. Raza, A.; Hayat, U.; Rasheed, T.; Bilal, M.; Iqbal, H.M.N. Redox-responsive nano-carriers as tumor-targeted drug delivery systems. *Eur. J. Med. Chem.* **2018**, *157*, 705–715. [CrossRef]
10. Rao, N.V.; Ko, H.; Lee, J.; Park, J.H. Recent Progress and Advances in Stimuli-Responsive Polymers for Cancer Therapy. *Front. Bioeng. Biotechnol.* **2018**, *6*, 110. [CrossRef]
11. Tong, R.; Xia, H.; Lu, X. Fast release behavior of block copolymer micelles under high intensity focused ultrasound/redox combined stimulus. *J. Mater. Chem. B* **2013**, *1*, 886–894. [CrossRef] [PubMed]
12. Jia, L.; Li, Z.; Zhang, D.; Zhang, Q.; Shen, J.; Guo, H.; Tian, X.; Liu, G.; Zheng, D.; Qi, L. Redox-responsive catiomer based on PEG-ss-chitosan oligosaccharide-ss-polyethylenimine copolymer for effective gene delivery. *Polym. Chem.* **2013**, *4*, 156–165. [CrossRef]

13. Li, X.Q.; Wen, H.Y.; Dong, H.Q.; Xue, W.M.; Pauletti, G.M.; Cai, X.J.; Xia, W.J.; Shi, D.; Li, Y.Y. Self-assembling nanomicelles of a novel camptothecin prodrug engineered with a redox-responsive release mechanism. *Chem. Commun.* **2011**, *47*, 8647. [CrossRef] [PubMed]
14. Han, H.S.; Thambi, T.; Choi, K.Y.; Son, S.; Ko, H.; Lee, M.C.; Jo, D.-G.; Chae, Y.S.; Kang, Y.M.; Lee, J.Y.; et al. Bioreducible Shell-Cross-Linked Hyaluronic Acid Nanoparticles for Tumor-Targeted Drug Delivery. *Biomacromolecules* **2015**, *16*, 447–456. [CrossRef]
15. Xu, Z.; Liu, S.; Kang, Y.; Wang, M. Glutathione-Responsive Polymeric Micelles Formed by a Biodegradable Amphiphilic Triblock Copolymer for Anticancer Drug Delivery and Controlled Release. *ACS Biomater. Sci. Eng.* **2015**, *1*, 585–592. [CrossRef]
16. Motornov, M.; Roiter, Y.; Tokarev, I.; Minko, S. Stimuli-responsive nanoparticles, nanogels and capsules for integrated multifunctional intelligent systems. *Prog. Polym. Sci.* **2010**, *35*, 174–211. [CrossRef]
17. Liu, X.; Yang, Y.; Urban, M.W. Stimuli-Responsive Polymeric Nanoparticles. *Macromol. Rapid Commun.* **2017**, *38*, 1700030. [CrossRef]
18. Yang, Q.; Bai, L.; Zhang, Y.; Zhu, F.; Xu, Y.; Shao, Z.; Shen, Y.-M.; Gong, B. Dynamic Covalent Diblock Copolymers: Instructed Coupling, Micellation and Redox Responsiveness. *Macromolecules* **2014**, *47*, 7431–7441. [CrossRef]
19. Kumari, A.; Yadav, S.K.; Yadav, S.C. Biodegradable polymeric nanoparticles based drug delivery systems. *Colloids Surf. B Biointerfaces* **2010**, *75*, 1–18. [CrossRef]
20. Cao-Hoang, L.; Fougère, R.; Waché, Y. Increase in stability and change in supramolecular structure of β-carotene through encapsulation into polylactic acid nanoparticles. *Food Chem.* **2011**, *124*, 42–49. [CrossRef]
21. Ferrari, R.; Cingolani, A.; Moscatelli, D. Solvent Effect in PLA-PEG Based Nanoparticles Synthesis through Surfactant Free Polymerization. *Macromol. Symp.* **2013**, *324*, 107–113. [CrossRef]
22. Lalloz, A.; Bolzinger, M.-A.; Briançon, S.; Faivre, J.; Rabanel, J.-M.; Ac, A.G.; Hildgen, P.; Banquy, X. Subtle and unexpected role of PEG in tuning the penetration mechanisms of PLA-based nano-formulations into intact and impaired skin. *Int. J. Pharm.* **2019**, *563*, 79–90. [CrossRef] [PubMed]
23. Liu, Y.; Zhu, Y.-H.; Mao, C.-Q.; Dou, S.; Shen, S.; Tan, Z.-B.; Wang, J. Triple negative breast cancer therapy with CDK1 siRNA delivered by cationic lipid assisted PEG-PLA nanoparticles. *J. Control. Release* **2014**, *192*, 114–121. [CrossRef] [PubMed]
24. El-Naggar, M.E.; Al-Joufi, F.; Anwar, M.; Attia, M.F.; El-Bana, M.A. Curcumin-loaded PLA-PEG copolymer nanoparticles for treatment of liver inflammation in streptozotocin-induced diabetic rats. *Colloids Surf. B Biointerfaces* **2019**, *177*, 389–398. [CrossRef] [PubMed]
25. Giovino, C.; Ayensu, I.; Tetteh, J.; Boateng, J.S. An integrated buccal delivery system combining chitosan films impregnated with peptide loaded PEG-b-PLA nanoparticles. *Colloids Surf. B Biointerfaces* **2013**, *112*, 9–15. [CrossRef]
26. Varani, J.; Warner, R.L.; Gharaee-Kermani, M.; Phan, S.H.; Kang, S.; Chung, J.; Wang, Z.; Datta, S.C.; Fisher, G.J.; Voorhees, J.J. Vitamin A Antagonizes Decreased Cell Growth and Elevated Collagen-Degrading Matrix Metalloproteinases and Stimulates Collagen Accumulation in Naturally Aged Human Skin1. *J. Investig. Dermatol.* **2000**, *114*, 480–486. [CrossRef]
27. Allwood, M.C.; Plane, J.H. The wavelength-dependent degradation of vitamin A exposed to ultraviolet radiation. *Int. J. Pharm.* **1986**, *31*, 1–7. [CrossRef]
28. Tan, X.; Meltzer, N.; Lindenbaum, S. Determination of the kinetics of degradation of 13-cis-retinoic acid and all-trans-retinoic acid in solution. *J. Pharm. Biomed. Anal.* **1993**, *11*, 817–822. [CrossRef]
29. Oyler, A.R.; Motto, M.G.; Naldi, R.E.; Facchine, K.L.; Hamburg, P.F.; Burinsky, D.J.; Dunphy, R.; Cotter, M.L. Characterizationf of autoxidation products of retinoic acid. *Tetrahedron* **1989**, *45*, 7679–7694. [CrossRef]
30. Manan, F.; Baines, A.; Stone, J.; Ryley, J. The kinetics of the loss of all-trans retinol at low and intermediate water activity in air in the dark. *Food Chem.* **1995**, *52*, 267–273. [CrossRef]
31. Shields, C.W.; White, J.P.; Osta, E.G.; Patel, J.; Rajkumar, S.; Kirby, N.; Therrien, J.-P.; Zauscher, S. Encapsulation and controlled release of retinol from silicone particles for topical delivery. *J. Control. Release* **2018**, *278*, 37–48. [CrossRef] [PubMed]
32. Kim, D.; Jeong, Y.; Choi, C.; Roh, S.; Kang, S.; Jang, M.; Nah, J. Retinol-encapsulated low molecular water-soluble chitosan nanoparticles. *Int. J. Pharm.* **2006**, *319*, 130–138. [CrossRef] [PubMed]
33. Hwang, Y.-J.; Oh, C.; Oh, S.-G. Controlled release of retinol from silica particles prepared in O/W/O emulsion: The effects of surfactants and polymers. *J. Control. Release* **2005**, *106*, 339–349. [CrossRef] [PubMed]

34. Laredj-Bourezg, F.; Bolzinger, M.-A.; Pelletier, J.; Valour, J.-P.; Rovère, M.-R.; Smatti, B.; Chevalier, Y. Skin delivery by block copolymer nanoparticles (block copolymer micelles). *Int. J. Pharm.* **2015**, *496*, 1034–1046. [CrossRef] [PubMed]
35. Visentini, F.F.; Sponton, O.E.; Perez, A.A.; Santiago, L.G. Biopolymer nanoparticles for vehiculization and photochemical stability preservation of retinol. *Food Hydrocoll.* **2017**, *70*, 363–370. [CrossRef]
36. Bourissou, D.; Martin-Vaca, B.; Dumitrescu, A.; Graullier, M.; Lacombe, F. Controlled Cationic Polymerization of Lactide. *Macromolecules* **2005**, *38*, 9993–9998. [CrossRef]
37. Hu, Y.; Daoud, W.; Cheuk, K.; Lin, C. Newly Developed Techniques on Polycondensation, Ring-Opening Polymerization and Polymer Modification: Focus on Poly (Lactic Acid). *Materials* **2016**, *9*, 133. [CrossRef]
38. Sasatsu, M.; Onishi, H.; Machida, Y. In vitro and in vivo characterization of nanoparticles made of MeO-PEG amine/PLA block copolymer and PLA. *Int. J. Pharm.* **2006**, *317*, 167–174. [CrossRef]
39. Liu, Q.; Yang, X.; Xu, H.; Pan, K.; Yang, Y. Novel nanomicelles originating from hydroxyethyl starch-g-polylactide and their release behavior of docetaxel modulated by the PLA chain length. *Eur. Polym. J.* **2013**, *49*, 3522–3529. [CrossRef]
40. Almoustafa, H.A.; Alshawsh, M.A.; Chik, Z. Technical aspects of preparing PEG-PLGA nanoparticles as carrier for chemotherapeutic agents by nanoprecipitation method. *Int. J. Pharm.* **2017**, *533*, 275–284. [CrossRef]
41. Mosmann, T. Rapid colorimetric assay for cellular growth and survival: Application to proliferation and cytotoxicity assays. *J. Immunol. Methods* **1983**, *65*, 55–63. [CrossRef]
42. Nahire, R.; Haldar, M.K.; Paul, S.; Ambre, A.H.; Meghnani, V.; Layek, B.; Katti, K.S.; Gange, K.N.; Singh, J.; Sarkar, K.; et al. Multifunctional polymersomes for cytosolic delivery of gemcitabine and doxorubicin to cancer cells. *Biomaterials* **2014**, *35*, 6482–6497. [CrossRef] [PubMed]
43. Fessi, H.; Puisieux, F.; Devissaguet, J.P.; Ammoury, N.; Benita, S. Nanocapsule formation by interfacial polymer deposition following solvent displacement. *Int. J. Pharm.* **1989**, *55*, R1–R4. [CrossRef]
44. Hornig, S.; Heinze, T.; Becer, C.R.; Schubert, U.S. Synthetic polymeric nanoparticles by nanoprecipitation. *J. Mater. Chem.* **2009**, *19*, 3838. [CrossRef]
45. Rathi, S.R.; Coughlin, E.B.; Hsu, S.L.; Golub, C.S.; Ling, G.H.; Tzivanis, M.J. Effect of midblock on the morphology and properties of blends of ABA triblock copolymers of PDLA-mid-block-PDLA with PLLA. *Polymer* **2012**, *53*, 3008–3016. [CrossRef]
46. Avgoustakis, K. Pegylated Poly(Lactide) and Poly(Lactide-Co-Glycolide) Nanoparticles: Preparation, Properties and Possible Applications in Drug Delivery. *CDD* **2004**, *1*, 321–333. [CrossRef] [PubMed]
47. Gupta, S.; Gupta, S.; Jindal, N.; Jindal, A.; Bansal, R. Nanocarriers and nanoparticles for skin care and dermatological treatments. *Indian Dermatol. Online J.* **2013**, *4*, 267. [CrossRef] [PubMed]
48. Li, Y.; Rong Qi, X.; Maitani, Y.; Nagai, T. PEG–PLA diblock copolymer micelle-like nanoparticles as all-trans-retinoic acid carrier: In vitro and in vivo characterizations. *Nanotechnology* **2009**, *20*, 055106. [CrossRef]
49. Tiwari, M.D.; Mehra, S.; Jadhav, S.; Bellare, J.R. All-trans retinoic acid loaded block copolymer nanoparticles efficiently induce cellular differentiation in HL-60 cells. *Eur. J. Pharm. Sci.* **2011**, *44*, 643–652. [CrossRef]
50. Jung, Y.J.; Truong, N.K.V.; Shin, S.; Jeong, S.H. A robust experimental design method to optimize formulations of retinol solid lipid nanoparticles. *J. Microencapsul.* **2013**, *30*, 1–9. [CrossRef]
51. Jee, J.-P.; Lim, S.-J.; Park, J.-S.; Kim, C.-K. Stabilization of all-trans retinol by loading lipophilic antioxidants in solid lipid nanoparticles. *Eur. J. Pharm. Biopharm.* **2006**, *63*, 134–139. [CrossRef] [PubMed]
52. Eskandar, N.G.; Simovic, S.; Prestidge, C.A. Chemical stability and phase distribution of all-trans-retinol in nanoparticle-coated emulsions. *Int. J. Pharm.* **2009**, *376*, 186–194. [CrossRef] [PubMed]
53. Jones, D.P. Redox potential of GSH/GSSG couple: Assay and biological significance. In *Methods in Enzymology*; Elsevier: Amsterdam, The Netherlands, 2002; Volume 348, pp. 93–112. ISBN 978-0-12-182251-4.
54. Griffith, O.W. Biologic and pharmacologic regulation of mammalian glutathione synthesis. *Free Radic. Biol. Med.* **1999**, *27*, 922–935. [CrossRef]
55. Smith, C.V.; Jones, D.P.; Guenthner, T.M.; Lash, L.H.; Lauterburg, B.H. Compartmentation of Glutathione: Implications for the Study of Toxicity and Disease. *Toxicol. Appl. Pharmacol.* **1996**, *140*, 1–12. [CrossRef]
56. Greenspan, P.; Fowler, S. Spectrofluorometric studies of the lipid probe, nile red. *J. Lipid Res.* **1985**, *26*, 781–789.
57. Bej, R.; Dey, P.; Ghosh, S. Disulfide chemistry in responsive aggregation of amphiphilic systems. *Soft Matter* **2020**, *16*, 11–26. [CrossRef]
58. Meng, F.; Hennink, W.E.; Zhong, Z. Reduction-sensitive polymers and bioconjugates for biomedical applications. *Biomaterials* **2009**, *30*, 2180–2198. [CrossRef]

59. Doktorovova, S.; Souto, E.B.; Silva, A.M. Nanotoxicology applied to solid lipid nanoparticles and nanostructured lipid carriers–A systematic review of in vitro data. *Eur. J. Pharm. Biopharm.* **2014**, *87*, 1–18. [CrossRef]
60. Beckenbach, L.; Baron, J.M.; Merk, H.F.; Löffler, H.; Amann, P.M. Retinoid treatment of skin diseases. *Eur. J. Dermatol.* **2015**, *25*, 384–391. [CrossRef]
61. Cullen, J.K.; Simmons, J.L.; Parsons, P.G.; Boyle, G.M. Topical treatments for skin cancer. *Adv. Drug Deliv. Rev.* **2020**, *153*, 54–64. [CrossRef]

Publisher's Note: MDPI stays neutral with regard to jurisdictional claims in published maps and institutional affiliations.

© 2020 by the authors. Licensee MDPI, Basel, Switzerland. This article is an open access article distributed under the terms and conditions of the Creative Commons Attribution (CC BY) license (http://creativecommons.org/licenses/by/4.0/).

Article

Characterization and Therapeutic Effect of a pH Stimuli Responsive Polymeric Nanoformulation for Controlled Drug Release

Maria Victoria Cano-Cortes [1,2,3], Jose Antonio Laz-Ruiz [1,2,3], Juan Jose Diaz-Mochon [1,2,3,*] and Rosario Maria Sanchez-Martin [1,2,3,*]

1. GENYO, Centre for Genomics and Oncological Research, Pfizer/University of Granada/Andalusian Regional Government, PTS Granada, Avda. Ilustración 114, 18016 Granada, Spain; victoria.cano@genyo.es (M.V.C.-C.); josealazr@go.ugr.es (J.A.L.-R.)
2. Department of Medicinal & Organic Chemistry, Excellence Research Unit of "Chemistry Applied to Biomedicine and the Environment", Faculty of Pharmacy, University of Granada, Campus de Cartuja s/n, 18071 Granada, Spain
3. Biosanitary Research Institute of Granada (ibs. GRANADA), University Hospital, Av. del Conocimiento, s/n, 18016 Granada, Spain
* Correspondence: juanjose.diaz@genyo.es (J.J.D.-M.); rosario.sanchez@genyo.es or rmsanchez@go.ugr.es (R.M.S.-M.); Tel.: +34-958-715-500 (R.M.S.-M.)

Received: 15 April 2020; Accepted: 28 May 2020; Published: 1 June 2020

Abstract: Despite the large number of polymeric nanodelivery systems that have been recently developed, there is still room for improvement in terms of therapeutic efficiency. Most reported nanodevices for controlled release are based on drug encapsulation, which can lead to undesired drug leakage with a consequent reduction in efficacy and an increase in systemic toxicity. Herein, we present a strategy for covalent drug conjugation to the nanodevice to overcome this drawback. In particular, we characterize and evaluate an effective therapeutic polymeric PEGylated nanosystem for controlled pH-sensitive drug release on a breast cancer (MDA-MB-231) and two lung cancer (A549 and H520) cell lines. A significant reduction in the required drug dose to reach its half maximal inhibitory concentration (IC50 value) was achieved by conjugation of the drug to the nanoparticles, which leads to an improvement in the therapeutic index by increasing the efficiency. The genotoxic effect of this nanodevice in cancer cells was confirmed by nucleus histone H2AX specific immunostaining. In summary, we successfully characterized and validated a pH responsive therapeutic polymeric nanodevice in vitro for controlled anticancer drug release.

Keywords: covalent drug conjugation; therapeutic nanodevice; polymeric nanoparticles; cancer therapy; controlled drug delivery

---

1. Introduction

Nanomedicine for cancer therapy has become a promising therapeutic approach to overcome the various limitations of conventional small molecule chemotherapeutics by improving drug internalization and selective intracellular accumulation in cancer cells, easing the toxicity to normal tissues [1,2]. Polymeric nanoparticles possess remarkable properties when compared to other colloidal systems such as (i) higher stability, particularly in body fluids; (ii) a larger contact area between the nanoparticle and the biological target; and (iii) a rapid adsorption rate and accumulation in the tumor cellular interstices due to the enhanced permeability and retention (EPR) effect [3,4]. Moreover, polymeric nanoparticle–drug conjugates present advantages when compared to polymer–drug conjugates, such as tunability and high and predefined drug loading based on efficient conjugation of the active agents to polymeric nanocarriers [5].

One of the main advantages offered by nanoparticles (NPs) is their ability to release drugs in a controlled manner [6]. This controlled release can be achieved by implementing a stimulus-sensitive approach involving a two-step process: first, the nanosystem is preferentially accumulated at the target site through the EPR effect; then, the drug-loaded nanoparticles are directly activated by an external (light, temperature, etc.) or internal (pH, enzymatic, redox, etc.) stimulus to produce the local release of the drug [7,8]. In particular, pH has been used for a long time as a critical feature for the differentiation between healthy tissues and abnormal tissues. Although fluctuations may occur, the pH in most solid tumors is between 6 and 7 [9]. This pH difference opened a new pathway for the release of tumor-specific drugs in tumors and simultaneously reduces undesirable effects in healthy tissues. Several examples of pH-sensitive nanodevices such as amorphous calcium carbonate–silica nanoparticles (core/shell), N- (2-hydroxypropyl) ethacrylamide (HPMA), dendrimers, and gold nanoparticles have been reported [10–14].

The chemotherapeutic drug doxorubicin (DOX) has been widely used in clinic settings for the treatment of different types of cancer. However, its toxicity to healthy tissue with effects such as cardiotoxicity and the development of resistance to multiple drugs during prolonged treatment have limited its therapeutic use [15]. Doxil®, the first nanopharmaceutical approved by the U.S. Food and Drug Administration (FDA) in 1995, takes advantage of the EPR effect and moves passively to the tumors where the encapsulated doxorubicin is released [16]. Recently, many nanotechnology-based drug delivery systems have been reported for the selective release of doxorubicin [17–19].

However, there is still room for improvement in terms of the therapeutic efficiency, as compared with free doxorubicin. Most of these nanodevices are based on drug encapsulation, which can lead to undesired drug leakage, causing loss of efficiency and systemic toxicity. This drawback can be overcome by covalent conjugation of the drug to the nanoparticle.

We have previously reported the use of polystyrene-based nanoparticles for the efficient conjugation of bioactive molecules of different types, such as sensors, proteins, and nucleic acids. In addition, polystyrene nanoparticles have been implemented for imaging, biosensing, tracking cellular proliferation using fluorescent nanoparticles, metallofluorescent nanoparticles for multimodal applications, and in cellulo proteomics using drug-loaded fluorescent nanoparticles [20–22]. These polymeric particles are inherently attractive as a delivery system due to certain advantages, such as being easy to handle and robust, with a defined drug loading capacity, tunebility, and lack of toxicity. These nanosystems can achieve efficient delivery through a passive but rapid mechanism, without significant alterations involving cellular gene profiling or proteomics [23,24]. To overcome the limitations of current encapsulation-based polymeric nanodevices, we developed an efficient loading strategy based on the covalent conjugation of doxorubicin to cross-linked polystyrene nanoparticles for selective drug release. Herein, we characterize this PEGylated polystyrene nanodevice loaded covalently with doxorubicin in a pH labile linker controlled manner, and evaluate the therapeutic efficiency in several cancer cell lines in vitro.

2. Materials and Methods

2.1. Materials

All solvents and chemicals were purchased from Sigma-Aldrich (Haverhill, United Kingdom). Dulbecco's modified Eagle's medium (DMEM), Roswell Park Memorial Institute medium (RPMI), L-glutamine, 1% penicillin/streptomycin, trypsin-EDTA, and fetal bovine serum (FBS) were purchased from Gibco (Thermo Fisher Scientific, Allschwil, Switzerland).

2.2. Synthesis of Aminomethyl Polystyrene Nanoparticles (NPs)

Aminomethyl NPs (NAKED-NPs (**1**)) were prepared by dispersion polymerization as previously described [25]. Briefly, polyvinylpyrrolidone (PVP) (Mw 29000) was dissolved in ethanol/water (86:14) and deoxygenated via argon bubbling. Azobisisobutyronitrile (AIBN) was dissolved in styrene (freshly washed with 4-vinylbenzylamine hydrochloride (VBAH) and 2% divinylbenzene (DVB). The dispersion was deoxygenated via argon bubbling before addition to the PVP/ethanol solution. The mixture was stirred under argon for 30 min before heating to 65 °C for 15 h. Nanoparticles were obtained by centrifugation and washed with methanol (2×) and water (2×). Finally, the nanoparticles were stored in water at 4 °C. Particle size distribution: 460 nm (PDI = 0.042); amino quantification: 0.057 mmol g^{-1} of amino groups; N° of particles per gram: 1.96×10^{13}; solid content (SC): 2%.

2.3. PEGylation of NPs

A double PEGylation of aminomethyl NPs (**1**) was carried out. Briefly, aminomethyl NPs (**1**) were conditioned in N,N-Dimethylformamide (DMF). Separately, an Fmoc-4, 7, 10-trioxa-1, 13-tridecanediamine succinamic acid (Fmoc-PEG-COOH) spacer (MW = 542 g/mol) (15 eq) was dissolved in DMF, then Oxyma (15 eq) and N,N'-Diisopropylcarbodiimide (DIC) (15 eq) was added and mixed for 10 min at room temperature. Then, this mixture solution was added to NPs and mixed for 2 h at 60 °C. Subsequently, the NPs were washed by centrifugation (13,400 rpm, 3–10 min) with DMF, methanol, and water to obtain Fmoc-PEGylated NPs (**2**) (100% yield, 0.054 mmol g^{-1} of amino groups). Next, Fmoc deprotection was achieved by treating nanoparticles with 20% piperidine/DMF (3 × 20 min). PEGylated NPs (**3**) were obtained by centrifugation and subsequently washed with DMF, MeOH, and deionized water. Next, a second PEGylation and deprotection step was carried out to obtain the double-PEGylated NPs (**4**) (100% yield, 0.053 mmol g^{-1} of amino groups).

2.4. Preparation of pH Responsive Therapeutic Polymeric Nanodevice: DOX-NPs (7)

A solution of succinic anhydride (15 eq) and DIPEA (15 eq) in DMF was added to the double PEGylated NPs (**4**), sonicated, and mixed for 2 h at 60 °C. Then, carboxyl functionalized NPs (**5**) were activated with Oxyme (15 eq) and DIC (15 eq) for 4 h. Then, they were centrifuged and a solution of 55% v/v hydrazine (15 eq) in DMF was added and mixed at 25 °C for 15 h. Next, hydrazide-NPs (**6**) were washed and conditioned in PBS. Finally, doxorubicin (1 eq) was dissolved in pH 6 PBS and added to hydrazine-NPs (**6**) and mixed at 50 °C for 15 h to yield DOX NPs (**7**) (CE − 97%, 5.2×10^{-9} nmol DOX per NP).

2.5. Characterization of DOX-NPs (7)

2.5.1. Nanoparticle Size Distribution, Zeta Potential, and Morphology

Particle mean size, size distribution, and zeta potential of DOX-NPs (**7**) were determined by dynamic light scattering (DLS) and were measured on a Zetasizer Nano ZS ZEN 3500 (NanoMalvern Panalytical, UK) in biological grade water in a disposable cuvette for size measurements or in a transparent disposable cuvette for zeta potential measurements. The shape and morphology of the NPs were observed using a LIBRA 120 PLUS de Carl Zeiss SMT transmission electron microscopy (TEM, Oberkochen, Germany). The conjugation of DOX to NPs was checked by flow cytometry using FACSCanto II (Becton Dickinson & Co., New Yersey, USA) and Flowjo® 10 software for data analysis.

2.5.2. Determination of Nanoparticle Concentration by Spectrophotometry

The concentration of nanoparticles per microliter (NPs/µL) was determined by a standard spectrophotometric method described previously by our group [26]. Briefly, a measurement of the turbidity optical density at 600 nm was performed for each of the preparations based on nephelometric principals. In this way, a calibration standard curve was obtained for aminomethyl NPs (**1**) using a set of concentrations (Supplementary Figure S1, see Supplementary Material). Calibration curves fitted linear regression models by which the number of NPs/µL corresponding to one unit of OD600 for each size could be determined. Thus, these curves using known concentrations of aminomethyl NPs (1) allowed us to estimate the number of NPs in the final batches, even after multiple handling procedures, by OD600 measurements of 1 µL of each preparation (Supplementary Figure S2, see Supplementary Material). This spectrophotometric method was used to calculate number of NPs/µL in all the preparations used in this study.

2.5.3. Amino Quantification of Nanoparticles

To determine the capacity of conjugation of each batch of nanoparticles, the amount of reactive amino groups on the nanoparticle surface was calculated by the conjugation of a glycine with amino group protected with 9H-fluoren-9-yl-methoxycarbonyl (Fmoc) (Fmoc-Gly-OH) and quantification using the Fmoc test. The amount of released piperidine-dibenzofulvene adduct was quantified spectrophotometrically by measuring the absorbance of the solution at 302 nm using UV–vis light [27].

The amino quantification of the nanoparticles was calculated according to the equation

$$\text{Amino quantification } (\mu mol \times g^{-1}) = \frac{(A_{302} \times V)}{(\varepsilon_{302} \times m \times d)} \times 1 \times 10^6 \quad (1)$$

where A302 is the absorbance of the supernatant measured at 302 nm, V is the volume of the measured solution (mL), ε302 is the molar extinction coefficient at 302 nm (7800 M^{-1} cm^{-1}), d is the diameter of the cuvette (1 cm), and m is the mass of the NPs analyzed (mg).

2.5.4. Evaluation of Drug-Loading Efficiency

To determine the amount of DOX loaded on the nanoparticle surface (loading capacity, LC) and to evaluate the efficiency of the conjugation process (conjugation efficiency, CE), the concentration of free DOX in the supernatant obtained after the centrifugation of NPs was measured by UV spectroscopy at 480 nm. A calibration curve, with the lineal ratio between the optic density of DOX and its concentration, was generated (Supplementary Figure S3, see Supplementary Material). Then, the DOX loading capacity (LC) and DOX conjugation efficiency (CE %) were calculated according to the following formulas

$$LC = \frac{[\text{DOX conjugated on surface of nanoparticle}]}{\text{Number of NPs}} \times N_A \quad (2)$$

where N_A is Avogadro's number.

$$CE(\%) = \frac{[\text{DOX conjugated on surface of nanoparticle}]}{\text{Total concentration of DOX added}} \times 100 \quad (3)$$

2.5.5. Drug Release Profile

The drug release profile of the NPs was determined by analyzing the efficiency of the hydrolysis of the hydrazone bond of the DOX-NPs (7) samples at acidic pH. Briefly, 4.80 × 10^{10} DOX-NPs (7) were incubated in a phosphate solution at pH 6 and at pH 7.4 for 168 h (7 days) in an incubator at 37 °C. The supernatants were collected by centrifuging each sample at t = 1, 3, 6, 24, 48, 72, and 168 h, and they were analyzed using high performance liquid chromatography (HPLC) (Agilent 1200 series HPLC system). A calibration curve of Doxorubicin was generated using standard samples (Supplementary Figure S4, see Supplementary Material). Cumulative release was determined using the equation

$$(\%) = \frac{D_t}{D_T} \times 100 \quad (4)$$

where D_t is the concentration of DOX released from the DOX-NPs (7) at time t and D_T is the concentration of DOX-loaded onto the DOX-NPs (7).

2.6. Cell Cultures

Cell lines were provided by the cell bank of the CIC of the University of Granada. Three different cell lines were used in this study. The non-small cell lung cancer cell line H520 was cultured in RPMI supplemented with 10% (v/v) FBS, 1% L-Glutamine, and 1% penicillin/Streptomycin. The A549 (lung cancer) cell line and MDA MB 231 (human breast cancer) cell line were cultured in DMEM supplemented with 10% (v/v) FBS, 1% L-Glutamine, and 1% Penicillin/Streptomycin. All cell lines were grown in a humidified incubator at 5% CO_2 and 37 °C. All cell lines tested negative for mycoplasma infection.

2.7. Nanofection of Cancer Cell Lines

A549, H520, and MDA MB 231 cell lines were incubated with DOX-NPs (7) (NPs/cells) for the established incubation times in a humidified incubator at 5% CO_2 and 37 °C. Untreated cells and cells treated with NAKED-NPs (1) were used as control. After the incubation time, cells were detached and washed with PBS 1×. Then, samples were fixed in 2% paraformaldehyde and analyzed via flow cytometry using a FACSCanto II flow cytometer and confocal microscopy (see details of the nanofection protocol in **Supplementary Material**).

2.8. Cell Viability

The cellular cytotoxicity of the DOX-NPs (7) was determined using the resazurin cell viability assay (Sigma Aldrich). This quantitative fluorometric method is based on the ability of living cells to convert resazurin (a redox dye) into a fluorescent end product, which is measured at 570 nm directly from 96-well plates. The cells were seeded in a 96-well plate at pre-optimized concentrations, depending on the assay performed. The results were evaluated according to the manufacturer's protocol, and the amount of fluorescence obtained was proportional to the number of viable cells. Viability was expressed with respect to the percentage of untreated cells (100%). Control wells were included in each plate to measure the fluorescence of the culture medium with nanoparticles added in the absence of cells.

2.9. Determination of DNA Damage in Cancer Cells by Immunostaining of Phospho-H2A.X Foci

A549, H520, and MDA MB 231 cells were cultured in DMEM supplemented with 10% FBS, L-glutamine, and penicillin/streptomycin on coverslips in 24-well plates. Cells were incubated with 5000 DOX-NPs (7) per cell. After 24 h of incubation, the media was replaced with fresh full DMEM. One hour after the change of the medium, the cells were fixed with 4% PFA for 10 min at room temperature. After fixation, the cells were washed with PBS and incubated with blocking buffer containing 5% BSA and 0.3% Triton X-100 in PBS for 1 h at room temperature. Then, cells were incubated with a 1:500 solution of primary antiphospho-H2A.X antibodies in blocking buffer (Cell Signaling, 20E3, 1:500) at +4 °C overnight (300 µL/well). The next day, cells were washed with PBS and stained with a 1:1000 solution of secondary Alexa Fluor 488 conjugated antibodies (Invitrogen, A-11034) for 1 h at room temperature. After washing with PBS, the preparations were mounted with mounting medium including antifade and DAPI (Invitrogen) [28].

2.10. Statistical Analysis

The data are presented as the mean ± the standard deviation in the error bars. The sample size (n) indicates the experimental repeats of a single representative experiment (3, unless otherwise specified). The results of the experiments were validated by independent repetitions. Graphs and statistical difference data were made with GraphPad Prism 6.0 (GraphPad Software Inc.). Statistical significance was determined using Student's *t*-test in paired groups of samples with a known median and two-way analysis of variance (ANOVA,), and with a Bonferroni's post-hoc test when comparing more than two groups of samples. A *p*-value of ≤0.05 was considered significant.

3. Results and Discussion

3.1. Preparation of DOX-NPs (7)

Following a previously described protocol, a 500 nm monodisperse population of polystyrene amino-functionalized NPs (**1**) with 2% divinylbenzene (DVB) crosslinking were synthesized by dispersion polymerization (using 4-vinylbenzylamine hydrochloride– VBAH as the monomer to functionalize the nanoparticle with the amino groups) [25]. Following an Fmoc solid phase protocol, aminomethyl NPs (**1**) were double PEGylated (**4**) (100% yield). The density of PEGylation was calculated based on the amount of amino groups on the nanoparticle, this value being 3.57×10^{-9} nmoles PEG/NP. The PEGylation increases the biocompatibility of the NPs, thereby facilitating their transport across cell membranes. It also reduces unfavorable interactions between NPs and the bioactive cargoes. Then, drug loading was carried out. The drug of choice was DOX due to its broad clinical use and it being one of the most effective chemotherapeutics for treating cancers, such as lung, breast, gastric, sarcoma, and pediatric cancers [29–31]. The designed strategy for the conjugation of DOX was via hydrazone bond. For this purpose, carboxylated NPs (**5**) were prepared using succinic anhydride; then, hydrazine functionalized NPs (**6**) were prepared, and the selective conjugation to the keto group in position C-13 of doxorubicin was carried out to yield DOX-NPs (**7**) (Scheme 1).

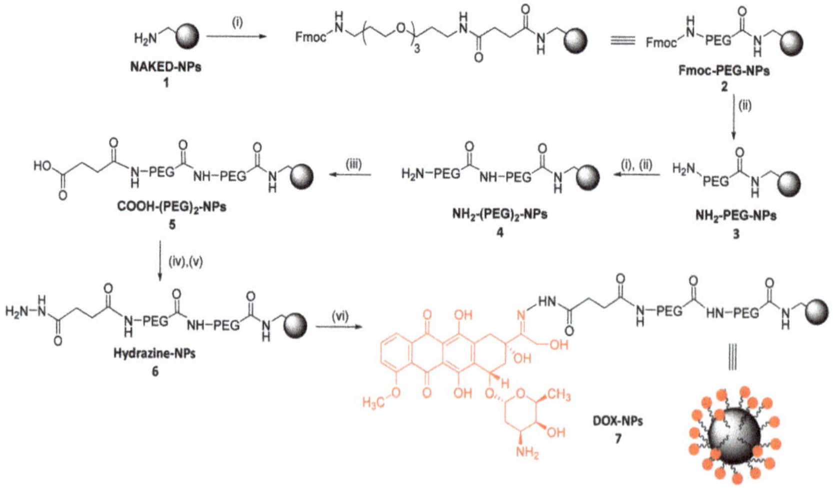

Scheme 1. Preparation of the pH responsive therapeutic polymeric nanodevice DOX-NPs (**7**). Reagents and conditions: (i) Fmoc-PEG-COOH (15 eq), Oxyma (15 eq), DIC (15 eq), DMF, 2 h, 60 °C; (ii) 20% piperidine/DMF, 3 × 20 min; (iii) Succinic anhydride (15 eq), DIPEA (15 eq), 2 h, 60 °C; (iv) Oxyma (15 eq), DIC (15 eq), 2 h, 25 °C; (v) Hydrated hydrazine 55% v/v (15 eq), 15 h, 25 °C; (vi) Doxorubicin (1 eq), PBS, 15 h, 50 °C.

Characterization of Drug-Loaded Nanoparticles (DOX-NPs) (7)

The size distribution and zeta potential of the nanoparticles loaded with doxorubicin, DOX-NPs (**7**), together with amino functionalized nanoparticles, NAKED-NPs (**1**) were determined quantitatively by DLS (Figure 1a,c). The results obtained show a hydrodynamic diameter of 464.2 ± 0.9 nm with a PDI of 0.047, demonstrating that nanoparticle population was monodisperse (Figure 1a). The size was corroborated by TEM analysis (Figure 1d). The zeta potential value of DOX-NPs (**7**) was slightly negative (−13.7 mV ± 0.9) in water, as compared to NAKED-NPs (**1**), which was +81.2 mV ± 0.8, and Hydrazine-NPs (**6**)) which was +31.4 mV ± 0.3. (Figure 1c)). Keeping in mind that it has been described that to maintain the stability through electrostatic repulsion, zeta potentials of a particle should be above −20 mV, it could be predicted that the stability of these particles could be compromised. However, we will like to highlight that these particles are stable. A stability assay has been runned (see Figure 1b) and DOX-NPs (**7**) were stable for six months at 4 °C and 25 °C. As the nature and number of molecules coated on the surfaces can affect the stability of polymeric particles, we could suggest that the balance between the loaded drug and the PEGylation on the surface has been a positive effect on the stability of these nanoparticles.

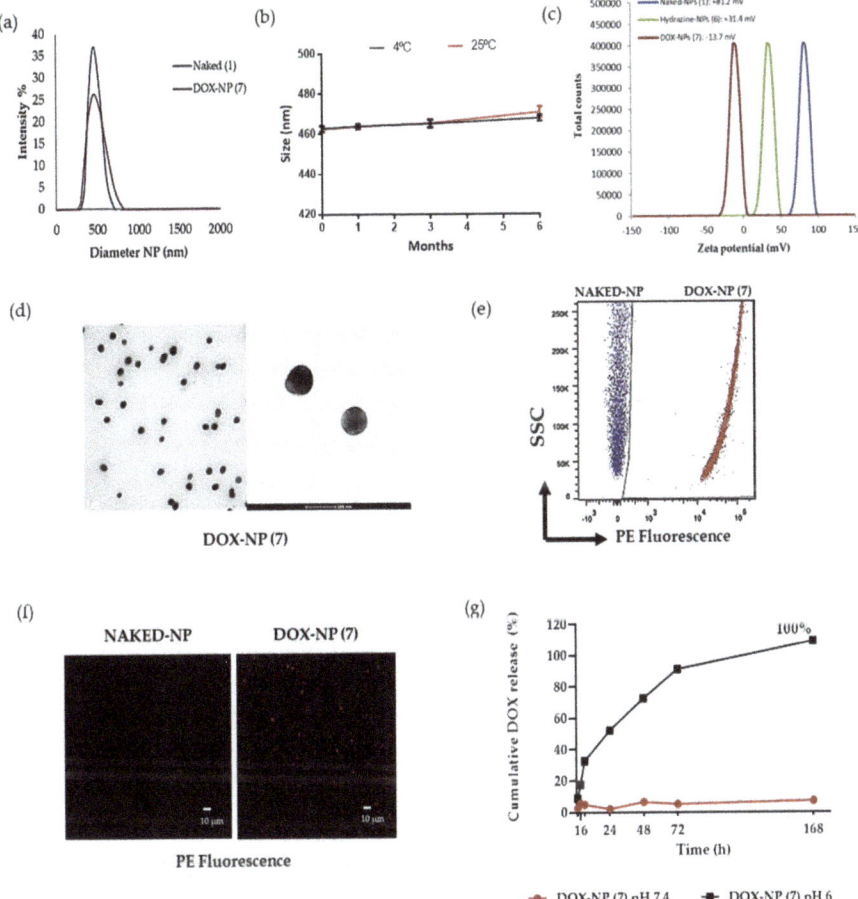

Figure 1. Characterization of the DOX-NPs (**7**). (**a**) Hydrodynamic diameter of DOX-NPs (**7**) versus Naked-NPs (**1**); (**b**) Stability of DOX-NPs (**7**) for six months at 4 °C and 25 °C; (**c**) Values of the zeta potential of DOX-NPs (**7**) (red) versus Hydrazine-NPs (**6**) (green) and NAKED-NPs (**1**) (blue); (**d**) Transmission electron microscopy (TEM) analysis; (**e**) Dot plots representative of flow cytometry of NAKED-NPs (**1**) (blue) and DOX-NPs (**7**) (red); (**f**) Images of confocal laser microscopy of naked nanoparticles and DOX-NPs (**7**). Images with an increase of 63x show the signal of red channel. Scale bar, 10 μm signal of red channel. Scale bar, 10 μm. (**g**) Cumulative DOX Release. DOX-NPs (**7**) were incubated in phosphate solution at pH 6 and PBS at pH 7.4 for 168 h at 37 °C. The results are expressed with the values of the mean ± SEM.

The effect of storage conditions on the long-term stability of the NPs was investigated. The stability of DOX-NPs (7) was evaluated by DLS (Figure 1b). The size of these nanodevices was measured after 1, 3, and 6 months at 4 °C and 25 °C, showing a constant size distribution (Figure 1a,b). DLS measurements revealed that the long-term stability of these nanoparticles was not influenced by storage temperature conditions. These results confirm that these particles are stable for a long time allowing them to be stored. This is a key property for further translation of this nanodevice.

On the basis of the fluorescence properties of DOX (λex = 470 nm, λem = 560 nm), the efficiency of drug conjugation onto the nanoparticles was monitored easily using fluorescence-based techniques such as flow cytometry and microscopy. Figure 1e shows a representative plot of the obtained results by flow cytometry analysis. An increase in fluorescence of the DOX-NPs (7) conjugates with respect to unloaded nanoparticles (NAKED-NPs (1)) can be observed. This result was corroborated by confocal microscopy (Figure 1f).

In order to evaluate the efficiency of the conjugation of doxorubicin to DOX-NPs (7), a spectrophotometric quantification of the remained unconjugated drug in the supernatant of the reaction was measured by UV spectroscopy (A480 nm). A calibration curve, with a lineal ratio between the optic density of doxorubicin and its concentration using a set of standard samples, was generated (Supplementary Figure S3, see Supplementary Material). To determine the value of loading capacity (LC), the amount of conjugated drug with respect to the number of nanoparticles was considered instead of the nanoparticle weight (as frequently reported) [31]. For this approach, an accurate spectrophotometric method to determine the number of nanoparticles per volume previously developed by our team was carried out [26]. The concentration of nanoparticles DOX-NPs (7) was estimated as 3.72×10^8 NPs/μL. The drug loading capacity is related to the number of nanoparticles, thereafter, the loading capacity per nanoparticle can be calculated. A loading capacity of 5.2×10^{-9} nmol DOX per NP was estimated. This parameter provides the value of drug dose with precision and accuracy; this fact being of extreme relevance for the clinical translation of nanomedicine. Taking into account the drug conjugated with respect to the total amount of the drug, the conjugation efficiency (CE) was 89%. These results shows the high efficiency of this nanodevice compared to previously reported therapeutic nanodevices based on drug encapsulation [18].

A major deficiency of the sustained release system is that the release is not specific. To ensure the release of the drug at the target site, to avoid a non-specific release, a pH-sensitive stimuli release strategy was implemented. The hydrazone is used in our approach as a cleavable bond that responds to stimuli depending on the pH; very useful in a tumor microenvironment that has a slightly acidic pH [32]. In order to release the drug in acidic conditions, doxorubicin was covalently conjugated to nanoparticles by a hydrazone bond sensitive to pH 6. Therefore, it is necessary to determine the pH-responsive drug release to evaluate this nanodevice. A high performance liquid chromatography (HPLC) assay was conducted to monitor the release process of doxorubicin in vitro. Release profiles were obtained by comparing the percentage of the released drug with respect to the amount of doxorubicin conjugated to the DOX-NPs (7) for seven days at pH 6 and pH 7.4 by HPLC analysis. As shown in Figure 1g (blackg line), in an acidic environment (pH 6 PBS), pH-sensitive cleavage of the hydrazone linker resulted in the exponential sustained release of the drug, and an accumulative release was obtained for up to one week (168 h). These results were expected as the hydrazone bond is sensitive to pH 6, and the drug is consequently released from the nanoparticles. A burst release was achieved within 24 h of incubation at pH 6 reaching a release rate of 52% ± 0.26. Additionally, a sustained release occurs for up to 168 h, achieving a maximum release value of 100%. Furthermore, in a simulated neutral physiological environment (pH 7.4 PBS), the amount of doxorubicin released from the nanodevice did not reach 10%, which indicated that the drug remained attached to the nanoparticles (Figure 1g, red line). This result indicated a remarkable stability and selectivity of the DOX-NPs (7). Therefore, this ability of DOX-NP (7) to release doxorubicin in a sustained manner in acidic tissues, such as tumoral tissues, could be a very beneficial feature to prolong and improve the therapeutic efficacy of NPs in the tumor. Consequently, the pH value of the medium has a clear effect on the release efficiency of doxorubicin, which validates the drug release strategy designed for this approach.

3.2. Evaluation of the Efficiency of Cellular Uptake of DOX-NP (7)

To investigate the efficiency of nanofection (cellular uptake and internalization) of DOX-NPs (7),) three cancer cell lines were chosen: two from lung cancer (A549 and H520) and one from triple negative breast cancer (MDA MB 231). The selection of these cell lines was based on the fact that doxorubicin is routinely used as a first-line chemotherapeutic agent in the treatment of these types of cancer [29–31]. For this purpose, the number of nanoparticles per cell needed to cause 50% of the cells to be nanofected (MNF50 index) was calculated in order to quantify the nanofection capacity of DOX-NPs (7)) following a protocol previously reported by us [26]. Nanofection efficiency of the DOX-NPs (7) in these three cell lines was quantitatively determined by flow cytometry based on the intrinsic red fluorescence of DOX. To this end, different concentrations of these nanoconjugates were incubated for 24 h, in an increasing gradient of concentration with between 50 and 10,000 NPs added per cell. After the incubation time, the cells were washed, trypsinized, and analyzed by flow cytometry. The results obtained on the cellular nanofection of DOX-NPs (7) revealed that they are internalized efficiently by the cancer cells studied. As shown in Figure 2a, a concentration-dependent effect on uptake efficiency was observed in the three cell lines. The analysis of the data showed that these three cell lines demonstrated a very similar MNF50 index, with values of 1720 ± 69.2 in the A549, 1328 ± 69 in the H520, and 1385 ± 68.4 for the MDA MB 231 cells (Figure 2b). However, the analysis of the increase in the median fluorescence (ΔMFI) indicated that, although the MNF50 was similar in the three cell lines and the internalization capacity gradually increased as the number of DOX-NPs (7) increased, starting at 2500 NPs/cell, the behavior of the lung cancer lines became different from that of the breast cancer cell line; the fluorescence intensity in the MDA MB 231 cells doubled by increasing the concentration range until reaching the saturation level. While the internalization of DOX-NPs (7) in the lung cells lines demonstrated a similar behavior, it was not the same, since the cellular entry was higher in the A549 without reaching the saturation point in both cases (Figure 2c). The cellular uptake of DOX-NPs (7) was also verified by confocal fluorescence microscopy. Figure 2d shows that cell internalization occurred efficiently in the three cell lines tested, confirming the cytoplasmic localization of DOX-NPs (7). Furthermore, to corroborate the intracellular location of DOX-NPs (7) and to discard any adsorption of these NPs onto the cell surface, a confocal microscopy analysis was carried out. Confocal microscopy image of A549 cells treated with DOX-NPs (7) showed that the location of these nanoparticles was intracellular (Figure 2e).

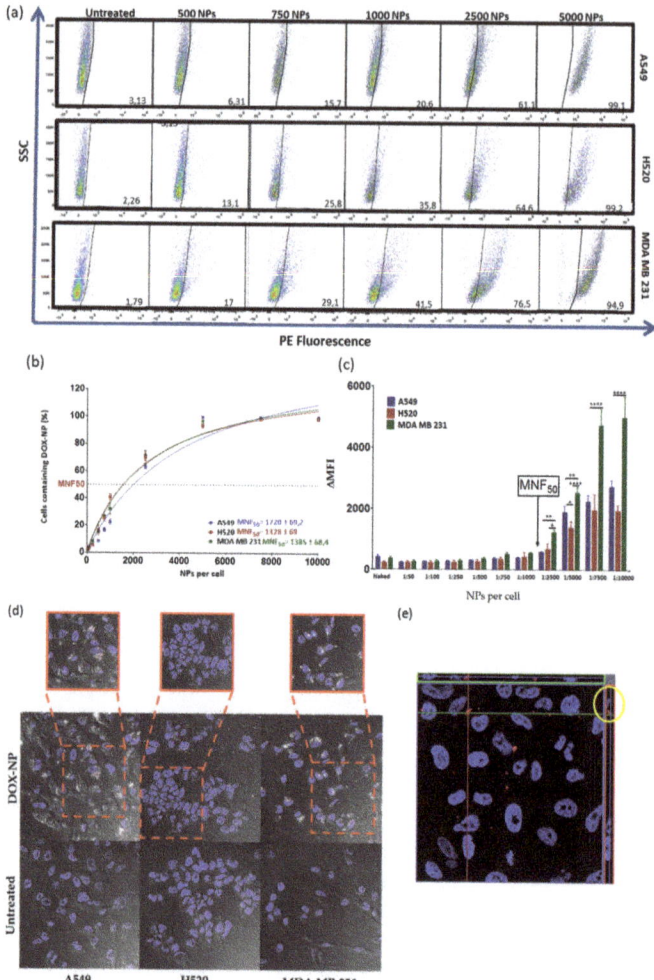

Figure 2. Evaluation of the cellular nanofection of DOX-NPs (**7**) in the cell lines A549, H520, and MDA MB 231 by flow cytometry. (**a**) Dot plots representative of flow cytometry obtained by incubating different concentrations of NPs for 24 h in the cell lines. (**b**) Percentage of cells containing the different concentrations of DOX-NPs (**7**). The data (mean ± SEM) are represented with a hyperbola equation model showing the different degrees of cellular internalization according to the number of DOX-NPs (**7**). (**c**) Representation of bars to compare ΔMFI between cell lines: A549 (blue), H520 (red), and MDA MB 231 (green). The arrow indicates the MNF50. The experiments were carried out in triplicate and the results are expressed with the values of the mean ± SEM. The statistical significance was determined by the analysis of the variance of a factor (ANOVA) using the Bonferroni multiple comparison (* p-value < 0.05, ** p-value < 0.001, **** p-value < 0.0001). MNF50: Multiplicity of nanofection 50. (**d**) Evaluation of uptake cellular of DOX-NPs (**7**) in the cell lines analyzed by confocal microscopy. The cells were incubated for 24 h with 2000 NPs/cell. The untreated cells were used as a negative control. Images with an increase of 63× show a composition of the three channels used: DIC; blue, DAPI for the nucleus; and red for DOX-NPs (**7**). (**e**) Orthogonal view (xy, xz, and yz) of the confocal microscope images showing the intersection planes at the position of the cross-line. Maximum intensity projection of the z-stack from blue (DAPI, nuclei) and red (DOX-NPs (**7**)) in the A549 cell line is displayed.

3.3. Evaluation of the Therapeutic Capacity of the DOX-NP (7)

In order to verify the enhanced anticancer effect of doxorubicin conjugated to DOX-NPs (7), the proliferation inhibition was tested by measuring the cell-mediated reduction of sodium resazurin, a standard colorimetric and quantitative method that determines the cell viability on the cell lines studied [33]. A549, H520, and MDA-MB-231 cells were treated with a range of different NPs concentrations for 96 h. Free DOX as well as nanoparticles without the drug loaded (NAKED-NPs (1)) were use as positive and negative controls, respectively. As shown in Figure 3a,b, the cell proliferation inhibition efficacy of DOX-NPs (7) exhibited a strongly dose-dependent pattern after culture for 96 h. It is important to remark that the naked NPs (1) without doxorubicin have no effect on the cell viability of the three cells lines (Figure 3b). Half the maximal inhibitory concentration (IC50) was determined (Figure 3a). The IC50 value of DOX-NPs (7) was calculated to be 42 nM in the A549 cells, 68 nM in the H520 cells, and 58 nM in the MDA MB 231 cells (Figure 3a), which correspond to 3061, 5012, and 4325 NPs added per cell, respectively (Supplementary Figure S5, see Supplementary Material). DOX-NPs (7) were found to be considerably more efficient than free DOX on an equimolar basis (IC50 = 100 nM in A549, 186 nM in H520, and 120 nM in MDA MB 231 cells) (Supplementary Figure S6, see Supplementary Material). These results are in agreement with previously reported studies with free doxorubicin [13,14]. These values indicated that the doxorubicin conjugated to the nanoparticles reduced the amount of drug required to achieve the IC50 value with respect to free doxorubicin treatment, suggesting that the nanoformulation had an enhancement effect. The higher cytotoxic activity of the DOX nanoformulation when compared with the free DOX, especially in the range of higher drug concentrations, is presumably due to selective pH stimuli release of the drug and the accumulation within tumor cells.

In order to evaluate the impact of the nanoformulation on the required dose of the drug to achieve the therapeutic effect, the effect on cell viability of the treatment with DOX-NPs (7) compared to a similar concentration of free doxorubicin was measured. In particular, cells were treated with a dose of 40,000 NPs/cell of DOX-NPs (7) corresponding to a concentration of 500 nM of doxorubicin. As seen in Figure 3c, DOX-NPs (7) showed a greater cytotoxicity than free doxorubicin under cell culture conditions, the results being statistically significant with a p-value of <0.0001. These results show that the release of the doxorubicin conjugated to the nanoparticles was sufficient to successfully inhibit the cell proliferation of the three cancer cell lines tested. DOX-NPs (7) showed a therapeutic efficiency twice that of free doxorubicin, with results comparable to other nanoformulations [11,13]. This fact suggests that this nanodevice is a promising tool for the selective release of drugs.

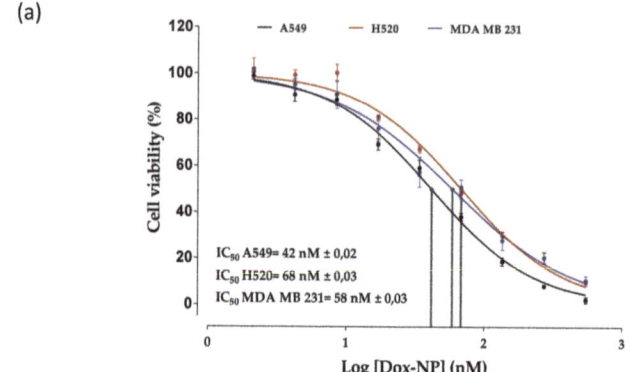

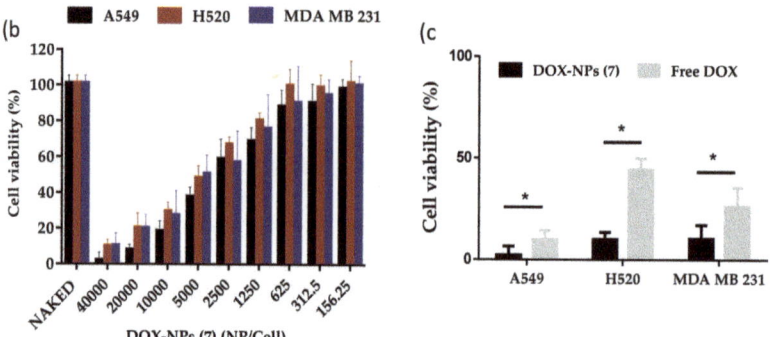

Figure 3. Effect of DOX-NPs (**7**) on cell viability. (**a**) Dose–response curves (percentage of cell viability versus concentration) of the treatment of A549, H520, and MDA-MB-231 cancer cells with DOX-NPs (**7**) represented in nM. The IC50 value was determined using the logarithm (inhibitor) versus normalized response: variable slope using the GraphPad software. (**b**) Bar graph showing cell viability of these three cell lines treated with DOX-NPs (**7**) and NAKED-NPs (**1**). (**c**) Comparison of therapeutic effect of DOX-NPs (**7**) compared to corresponding concentration of free doxorubicin (40,000 NPs/cell, 500 nM). Statistical significance was determined by Student's t-test (* p-value < 0.0001). The viability data represent the mean ± SEM of the results of three independent experiments with six points each.

3.4. Analysis of DOX-NPs (7)-Induced Genotoxic Effect in Cancer Cells

Two mechanisms of action have been proposed by which doxorubicin acts in cancer cells: (i) intercalation in DNA and disruption of DNA repair; and (ii) generation of free radicals with subsequent damage in cell membranes, DNA, and proteins [34]. However, the confirmation of a genotoxic effect of these DOX-NPs (**7**) nanoformulations loaded with DOX in cancer cells confirms the in situ cytoplasmic release of the drug from the nanoparticle, and confirms that it reaches the nucleus; thus, the efficacy of the pH release strategy is corroborated. The genotoxic effect of the DOX conjugated to NPs (DOX-NPs (**7**)) was evaluated in the cell models studied, focusing on the damage caused in the DNA through detection of the phosphorylated form of the variant histone H2AX (γ-H2AX). The cell responds to DNA damage through the phosphorylation of thousands of H2AX molecules flanking the damaged site. This highly amplified response can be visualized as a γ-H2AX focus in the chromatin that can be detected in situ with the appropriate antibody (antiphospho-H2A.X.) using confocal microscopy analysis [28]. For this purpose, DNA damage in MDA-M-231, H520, and A549 cancer cells was determined by immunostaining of phospho-H2A.X foci (Figure 4). Untreated cells

(UNT) and cells treated with NAKED-NPs (**1**) were used as negative controls for this experiment. As expected, DNA damage was not observed when cells were not treated or treated with nanoparticles without the drug being loaded (Figure 4a,b). Conversely, cells treated with DOX-NPs (**7**) but without primary staining with γ-H2Ax were analyzed to confirm the specificity of this assay. Noteworthy nanoparticles were detected (red dots) but no signal coming from unspecific immunostaining was observed (Figure 4c). In addition, the obtained results of specific immunostaining with γ-H2Ax following the treatment of the cells with DOX-NPs (**7**) show that DNA damage can be clearly observed following staining with a secondary antibody for the green channel (green foci, Figure 4d). This result corroborated the efficiency of the pH-sensitive release strategy of the drug from the nanoparticles, since the drug must be released cytoplasmically so that it is possible to enter the nucleus where DOX-NP (**7**)-induced DNA damage occurs.

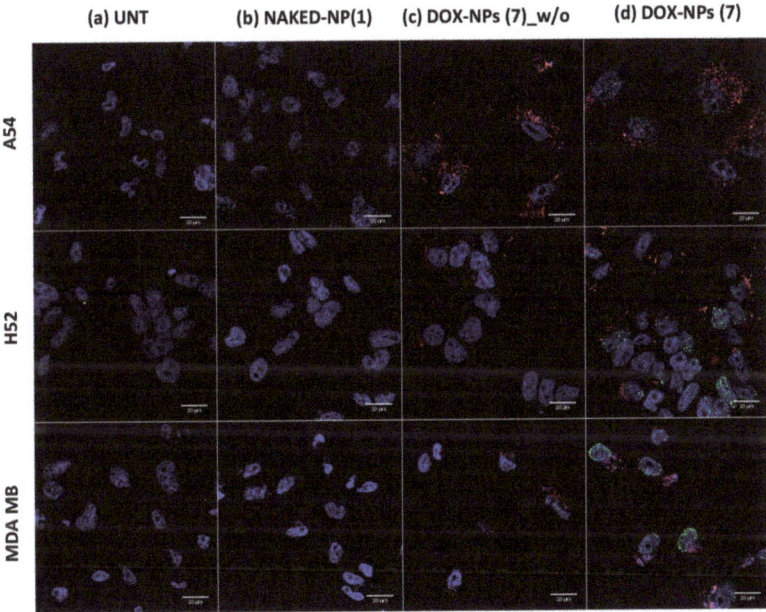

Figure 4. Evaluation of the DNA damage caused by the treatment with DOX-NPs (**7**) by immunostaining of γ-H2Ax in A549, H520, and MDA MB 231 cancer cell lines analyzed by confocal microcopy. The untreated cells (UNT) (**a**) and cells treated with NAKED-NPs (1) (**b**) were used as negative controls. Cells treated with DOX-NPs (**7**) were analyzed without (**c**) and with (**d**) a primary anti-γ-H2Ax antibody. Images with an increase of 63x show a composition of the four channels used: blue, DAPI for the nucleus staining; red, DOX-NPs (**7**); and green, for H2Ax staining with Alexa488-antiphospho-H2Ax.

4. Conclusions

In conclusion, a covalent strategy for the development of a PEGylated therapeutic nanosystem for pH-sensitive release was developed. Better anti-tumor activity was shown in cells treated with the nanoparticles than free DOX. The IC50 value was reduced by half by the conjugation of doxorubicin to the nanoparticle, compared to free doxorubicin, demonstrating the therapeutic capacity of DOX-NPs (**7**). This nanodevice has a proven capability of releasing the drug in a controlled manner at acidic pH and improving the drug therapeutic index by increasing efficacy compared to doxorubicin in solution. In addition, it was demonstrated to be stable for up to 6 months at 4 °C and 25 °C. The next step will involve conjugating other drugs in clinical use to this nanodevice using the same chemical strategy; thereafter, applying this therapeutic strategy to other types of cancer; and in the long term, to other pathologies. Future work will be focused on in vitro and in vivo preclinical characterization to move this nanodevice closer to clinical use.

Supplementary Materials: The following are available online at http://www.mdpi.com/2073-4360/12/6/1265/s1, supplementary figures: Figure S1. Calibration standard curve of concentration of nanoparticles (OD 600) by spectrophometry. Figure S2. Numbers of NPs per mL calculation. Figure S3. Calibration standard curve of doxorubicin solution (OD 480) by spectrophometry. Figure S4. Calibration standard curve of doxorubicin solution by HPLC. Figure S5. Dose–response curves (percentage of cell viability versus concentration) of treatment with DOX-NPs (7) in the cell models studied, represented in NPs/Cell. Figure S6. Dose–response curves (percentage of cell viability versus concentration) of treatment with free doxorobucin in the cell models studied. General protocol for cellular nanofection.

Author Contributions: Conceptualization, J.J.D.-M. and R.M.S.-M.; Methodology, M.V.C.-C. and R.M.S.-M.; Investigation, M.V.C.-C. and J.A.L.-R.; Formal analysis, M.V.C.-C. and J.A.L.-R.; Supervision, J.J.D.-M. and R.M.S.-M.; Writing—Original draft preparation, M.V.C.-C. and R.M.S.M.; Writing—Review and editing, J.J.D.-M. and R.M.S.-M.; Project administration, J.J.D.-M. and R.M.S.-M.; Funding acquisition, J.J.D.-M. and R.M.S.-M. All authors have read and agreed to the published version of the manuscript.

Funding: This research was funded by the Spanish Ministry of Economy and Competitiveness (MINECO), grant number BIO2016-80519 and the Health Institute Carlos III (ISCIII), grant number DTS18/00121 and the Andalusian Regional Government, grant number PAIDI-TC-PVT-PSETC-2.0.

Acknowledgments: The authors thank the Research Results Transfer Office (OTRI) of the University of Granada for their support for the technological development of this project. We thank the technical support of Microscopy and Molecular Image Unit and Flow Cytometry from GENYO. The authors thank the technical support of the Mass Spectrometry and Chromatography Unit of the Scientific Instrumentation Center of the University of Granada, in particular to Samuel Cantarero. The authors are members of the network NANOCARE (RED2018-102469-T) funded by the STATE INVESTIGATION AGENCY. J.A.L.R. thanks to the Fundación Benéfica Anticáncer San Francisco Javier y Santa Cándida for PhD funding.

Conflicts of Interest: The authors declare no conflict of interest.

References

1. Blau, R.; Krivitsky, A.; Epshtein, Y.; Satchi-Fainaro, R. Are nanotheranostics and nanodiagnostics-guided drug delivery stepping stones towards precision medicine? *Drug Resist. Updat.* **2016**, *27*, 39–58. [CrossRef] [PubMed]
2. Chen, G.; Roy, I.; Yang, C.; Prasad, P.N. Nanochemistry and Nanomedicine for Nanoparticle-based Diagnostics and Therapy. *Chem. Rev.* **2016**, *116*, 2826–2885. [CrossRef] [PubMed]
3. Cagel, M.; Grotz, E.; Bernabeu, E.; Moretton, M.A.; Chiappetta, D.A. Doxorubicin: Nanotechnological overviews from bench to bedside. *Drug Discov. Today* **2017**, *22*, 270–281. [CrossRef] [PubMed]
4. El-Say, K.M.; El-Sawy, H.S. Polymeric nanoparticles: Promising platform for drug delivery. *Int. J. Pharm.* **2017**, *528*, 675–691. [CrossRef]
5. Ekladious, I.; Colson, Y.L.; Grinstaff, M.W. Polymer–drug conjugate therapeutics: advances, insights and prospects. *Nat. Rev. Drug Discov.* **2019**, *18*, 273–294. [CrossRef]
6. Chenthamara, D.; Subramaniam, S.; Ramakrishnan, S.G.; Krishnaswamy, S.; Essa, M.M.; Lin, F.H.; Qoronfleh, M.W. Therapeutic efficacy of nanoparticles and routes of administration. *Biomater. Res.* **2019**, *23*, 1–29. [CrossRef]
7. Park, J.; Choi, Y.; Chang, H.; Um, W.; Ryu, J.H.; Kwon, I.C. Alliance with EPR effect: Combined strategies to improve the EPR effect in the tumor microenvironment. *Theranostics* **2019**, *9*, 8073–8090. [CrossRef]
8. Katz, J.S.; Burdick, J.A. Light-Responsive Biomaterials: Development and Applications. *Macromol. Biosci.* **2010**, *10*, 339–348. [CrossRef]
9. Karimi, M.; Sahandi, Z.P.; Ghasemi, A.; Amiri, M.; Bahrami, M.; Malekzad, H.; Ghahramanzadeh Asl, H.; Mahdieh, Z.; Bozorgomid, M.; Ghasemi, A.; et al. Temperature-Responsive Smart Nanocarriers for Delivery of Therapeutic Agents: Applications and Recent Advances. *ACS Appl. Mater. Interfaces* **2016**, *8*, 21107–21133. [CrossRef]
10. Zhao, Y.; Luo, Z.; Li, M.; Qu, Q.; Ma, X.; Yu, S.-H.; Zhao, Y. A Preloaded Amorphous Calcium Carbonate/Doxorubicin@Silica Nanoreactor for pH-Responsive Delivery of an Anticancer Drug. *Angew. Chem. Int. Ed.* **2015**, *54*, 919–922. [CrossRef]

11. Corbet, C.; Feron, O. Tumour acidosis: From the passenger to the driver's seat. *Nat. Rev. Cancer* **2017**, *17*, 577–593. [CrossRef] [PubMed]
12. Sun, T.; Zhang, Y.S.; Pang, B.; Hyun, D.C.; Yang, M.; Xia, Y. Engineered Nanoparticles for Drug Delivery in Cancer Therapy. *Angew. Chem. Int. Ed.* **2014**, 12320–12364. [CrossRef] [PubMed]
13. Zhang, Y.; Yang, C.; Wang, W.; Liu, J.; Liu, Q.; Huang, F.; Chu, L.; Gao, H.; Li, C.; Kong, D.; et al. Co-delivery of doxorubicin and curcumin by pH-sensitive prodrug nanoparticle for combination therapy of cancer. *Sci. Rep.* **2016**, *6*, 21225. [CrossRef] [PubMed]
14. Cui, T.; Liang, J.-J.; Chen, H.; Geng, D.-D.; Jiao, L.; Yang, J.-Y.; Qian, H.; Zhang, C.; Ding, Y. Performance of Doxorubicin-Conjugated Gold Nanoparticles: Regulation of Drug Location. *ACS Appl. Mater. Interfaces* **2017**, *9*, 8569–8580. [CrossRef]
15. Thorn, C.; Oshiro, C.; Marsh, S.; Hernandez-Boussard, T.; McLeod, H.; Klein, T.; Altman, R. Doxorubicin pathways:pharmacodynamics and adverse effects. *Pharmacogn. Genomics* **2012**, *21*, 440–446. [CrossRef]
16. Barenholz, Y.C. Doxil®—The first FDA-approved nano-drug: Lessons learned. *J. Control. Release* **2012**, *160*, 117–134. [CrossRef]
17. Huang, Y.; Yan, J.; Peng, S.; Tang, Z.; Tan, C.; Ling, J.; Lin, W.; Lin, X.; Zu, X.; Yi, G. pH/reduction dual-stimuli-responsive cross-linked micelles based on multi-functional amphiphilic star copolymer: Synthesis and controlled anti-cancer drug release. *Polymers* **2020**, *12*, 82. [CrossRef]
18. Raposo, C.D.; Costa, R.; Petrova, K.T.; Brito, C.; Scotti, M.T.; Cardoso, M.M. Development of novel galactosylated PLGA nanoparticles for hepatocyte targeting using molecular modelling. *Polymers* **2020**, *12*, 94. [CrossRef]
19. Gibbens-Bandala, B.; Morales-Avila, E.; Ferro-Flores, G.; Santos-Cuevas, C.; Luna-Gutiérrez, M.; Ramírez-Nava, G.; Ocampo-García, B. Synthesis and evaluation of 177Lu-DOTA-DN(PTX)-BN for selective and concomitant radio and drug-therapeutic effect on breast cancer cells. *Polymers* **2019**, *11*, 1572. [CrossRef]
20. Altea-Manzano, P.; Unciti-Broceta, J.D.; Cano-Cortes, V.; Ruiz-Blas, M.P.; Valero-Griñan, T.; Diaz-Mochon, J.J.; Sanchez-Martin, R. Tracking cell proliferation using a nanotechnology-based approach. *Nanomedicine* **2017**, *12*, 1591–1605. [CrossRef]
21. Valero, T.; Delgado-González, A.; Unciti-Broceta, J.D.; Cano-Cortés, V.; Pérez-López, A.M.; Unciti-Broceta, A.; Sánchez Martín, R.M. Drug "Clicking" on Cell-Penetrating Fluorescent Nanoparticles for *In Cellulo* Chemical Proteomics. *Bioconjug. Chem.* **2018**, *29*, 3154–3160. [CrossRef] [PubMed]
22. Delgado-Gonzalez, A.; Garcia-Fernandez, E.; Valero, T.; Cano-Cortes, M.V.; Ruedas-Rama, M.J.; Unciti-Broceta, A.; Sanchez-Martin, R.M.; Diaz-Mochon, J.J.; Orte, A. Metallofluorescent Nanoparticles for Multimodal Applications. *ACS Omega* **2018**, *3*, 144–153. [CrossRef] [PubMed]
23. Alexander, L.M.; Pernagallo, S.; Livigni, A.; Sánchez-Martín, R.M.; Brickman, J.M.; Bradley, M. Investigation of microsphere-mediated cellular delivery by chemical, microscopic and gene expression analysis. *Mol. Biosyst.* **2010**, *6*, 399–409. [CrossRef]
24. Pietrovito, L.; Cano-Cortés, V.; Gamberi, T.; Magherini, F.; Bianchi, L.; Bini, L.; Sánchez-Martín, R.M.; Fasano, M.; Modesti, A. Cellular response to empty and palladium-conjugated amino-polystyrene nanospheres uptake: A proteomic study. *Proteomics* **2015**, *15*, 34–43. [CrossRef]
25. Unciti-Broceta, A.; Johansson, E.M.V.; Yusop, M.R.; Sánchez-Martín, R.M.; Bradley, M. Synthesis of polystyrene microspheres and functionalization with Pd0 nanoparticles to perform bioorthogonal organometallic chemistry in living cells. *Nat. Protoc.* **2012**, *7*, 1207–1218. [CrossRef] [PubMed]
26. Unciti-Broceta, J.D.; Cano-Cortés, V.; Altea-Manzano, P.; Pernagallo, S.; Díaz-Mochón, J.J.; Sánchez-Martín, R.M. Number of nanoparticles per cell through a spectrophotometric method-A key parameter to assess nanoparticle-based cellular assays. *Sci. Rep.* **2015**, *5*, 1–10. [CrossRef] [PubMed]
27. Fields, G.B.; Noble, R.L. Solid phase peptide synthesis utilizing 9-fluorenylmethoxycarbonyl amino acids. *Int. J. Pept. Protein Res.* **1990**, *35*, 161–214. [CrossRef]
28. Francesco, C.; Larissa, L.; Antonio, G.; Joanna, C.; Alexey, P.; Alexandra, D.; Katarzyna, S.-K.; Alessandro, P.; Sergey, P.; Barlev, N.A. Specific Drug Delivery to Cancer Cells with Double-Imprinted Nanoparticles against Epidermal Growth Factor Receptor. *Nano Lett.* **2018**, *18*, 4641–4646.
29. Arcamone, F.; Cassinelli, G.; Fantini, G.; Grein, A.; Orezzi, P.; Pol, C.; Spalla, C. Adriamycin, 14-hydroxydaimomycin, a new antitumor antibiotic fromS. Peucetius var.caesius. *Biotechnol. Bioeng.* **1969**, *11*, 1101–1110. [CrossRef]

30. Cortés-Funes, H.; Coronado, C. Role of anthracyclines in the era of targeted therapy. *Cardiovasc. Toxicol.* **2007**, *7*, 56–60. [CrossRef]
31. Weiss, R.B. The anthracyclines: Will we ever find a better doxorubicin? *Semin. Oncol.* **1992**, *19*, 670–686. [PubMed]
32. No Manchun, S.; Dass, C.R.; Sriamornsak, P. Targeted Therapy for Cancer Using Ph-Responsive Nanocarrier Systems. *Life Sci.* **2012**, *90*, 381–387. [CrossRef] [PubMed]
33. O'Brien, J.; Wilson, I.; Orton, T.; Pognan, F. Investigation of the Alamar Blue (resazurin) fluorescent dye for the assessment of mammalian cell cytotoxicity. *Eur. J. Biochem.* **2000**, *267*, 5421–5426.
34. Gewirtz, D.A. A critical evaluation of the mechanisms of action proposed for the antitumor effects of the anthracycline antibiotics adriamycin and daunorubicin. *Biochem. Pharmacol.* **1999**, *57*, 727–741. [CrossRef]

© 2020 by the authors. Licensee MDPI, Basel, Switzerland. This article is an open access article distributed under the terms and conditions of the Creative Commons Attribution (CC BY) license (http://creativecommons.org/licenses/by/4.0/).

MDPI
St. Alban-Anlage 66
4052 Basel
Switzerland
Tel. +41 61 683 77 34
Fax +41 61 302 89 18
www.mdpi.com

Polymers Editorial Office
E-mail: polymers@mdpi.com
www.mdpi.com/journal/polymers

www.ingramcontent.com/pod-product-compliance
Lightning Source LLC
LaVergne TN
LVHW070659100526
838202LV00013B/1004